CAMBRIDGE STUDIES IN ADVANCED MATHEMATICS 123

Random Walk: A Modern Introduction

Random walks are stochastic processes formed by successive summation of independent, identically distributed random variables and are one of the most studied topics in probability theory. This contemporary introduction evolved from courses taught at Cornell University and the University of Chicago by the first author, who is one of the most highly regarded researchers in the field of stochastic processes. This text meets the need for a modern reference to the detailed properties of an important class of random walks on the integer lattice.

It is suitable for probabilists, mathematicians working in related fields, and for researchers in other disciplines who use random walks in modeling.

Gregory F. Lawler is Professor of Mathematics and Statistics at the University of Chicago. He received the George Pólya Prize in 2006 for his work with Oded Schramm and Wendelin Werner.

Vlada Limic works as a researcher for Centre National de la Recherche Scientifique (CNRS) at Université de Provence, Marseilles.

CAMBRIDGE STUDIES IN ADVANCED MATHEMATICS

All the titles listed below can be obtained from good booksellers or from Cambridge University Press. For a complete series listing visit: http://www.cambridge.org/series/sSeries.asp?code=CSAM

Already published
73 B. Bollobás *Random graphs (2nd Edition)*
74 R. M. Dudley *Real analysis and probability (2nd Edition)*
75 T. Sheil-Small *Complex polynomials*
76 C. Voisin *Hodge theory and complex algebraic geometry, I*
77 C. Voisin *Hodge theory and complex algebraic geometry, II*
78 V. Paulsen *Completely bounded maps and operator algebras*
79 F. Gesztesy & H. Holden *Soliton equations and their algebro-geometric solutions, I*
81 S. Mukai *An introduction to invariants and moduli*
82 G. Tourlakis *Lectures in logic and set theory, I*
83 G. Tourlakis *Lectures in logic and set theory, II*
84 R. A. Bailey *Association schemes*
85 J. Carlson, S. Müller-Stach & C. Peters *Period mappings and period domains*
86 J. J. Duistermaat & J. A. C. Kolk *Multidimensional real analysis, I*
87 J. J. Duistermaat & J. A. C. Kolk *Multidimensional real analysis, II*
89 M. C. Golumbic & A. N. Trenk *Tolerance graphs*
90 L. H. Harper *Global methods for combinatorial isoperimetric problems*
91 I. Moerdijk & J. Mrčun *Introduction to foliations and Lie groupoids*
92 J. Kollár, K. E. Smith & A. Corti *Rational and nearly rational varieties*
93 D. Applebaum *Lévy processes and stochastic calculus (1st Edition)*
94 B. Conrad *Modular forms and the Ramanujan conjecture*
95 M. Schechter *An introduction to nonlinear analysis*
96 R. Carter *Lie algebras of finite and affine type*
97 H. L. Montgomery & R. C. Vaughan *Multiplicative number theory, I*
98 I. Chavel *Riemannian geometry (2nd Edition)*
99 D. Goldfeld *Automorphic forms and L-functions for the group GL(n,R)*
100 M. B. Marcus & J. Rosen *Markov processes, Gaussian processes, and local times*
101 P. Gille & T. Szamuely *Central simple algebras and Galois cohomology*
102 J. Bertoin *Random fragmentation and coagulation processes*
103 E. Frenkel *Langlands correspondence for loop groups*
104 A. Ambrosetti & A. Malchiodi *Nonlinear analysis and semilinear elliptic problems*
105 T. Tao & V. H. Vu *Additive combinatorics*
106 E. B. Davies *Linear operators and their spectra*
107 K. Kodaira *Complex analysis*
108 T. Ceccherini-Silberstein, F. Scarabotti & F. Tolli *Harmonic analysis on finite groups*
109 H. Geiges *An introduction to contact topology*
110 J. Faraut *Analysis on Lie groups: An Introduction*
111 E. Park *Complex topological K-theory*
112 D. W. Stroock *Partial differential equations for probabilists*
113 A. Kirillov, Jr *An introduction to Lie groups and Lie algebras*
114 F. Gesztesy *et al. Soliton equations and their algebro-geometric solutions, II*
115 E. de Faria & W. de Melo *Mathematical tools for one-dimensional dynamics*
116 D. Applebaum *Lévy processes and stochastic calculus (2nd Edition)*
117 T. Szamuely *Galois groups and fundamental groups*
118 G. W. Anderson, A. Guionnet & O. Zeitouni *An introduction to random matrices*
119 C. Perez-Garcia & W. H. Schikhof *Locally convex spaces over non-Archimedean valued fields*
120 P. K. Friz & N. B. Victoir *Multidimensional stochastic processes as rough paths*
121 T. Ceccherini-Silberstein, F. Scarabotti & F. Tolli *Representation theory of the symmetric groups*
122 S. Kalikow & R. McCutcheon *An outline of ergodic theory*

Random Walk:
A Modern Introduction

GREGORY F. LAWLER
University of Chicago

VLADA LIMIC
Université de Provence

CAMBRIDGE
UNIVERSITY PRESS

CAMBRIDGE
UNIVERSITY PRESS

University Printing House, Cambridge CB2 8BS, United Kingdom

One Liberty Plaza, 20th Floor, New York, NY 10006, USA

477 Williamstown Road, Port Melbourne, VIC 3207, Australia

4843/24, 2nd Floor, Ansari Road, Daryaganj, Delhi - 110002, India

79 Anson Road, #06-04/06, Singapore 079906

Cambridge University Press is part of the University of Cambridge.

It furthers the University's mission by disseminating knowledge in the pursuit of education, learning and research at the highest international levels of excellence.

www.cambridge.org
Information on this title: www.cambridge.org/9780521519182

First published 2010

A catalogue record for this publication is available from the British Library

ISBN 978-0-521-51918-2 Hardback

Contents

Preface

Random walk – the stochastic process formed by successive summation of independent, identically distributed random variables – is one of the most basic and well-studied topics in probability theory. For random walks on the integer lattice $\mathbb{Z}^d$, the main reference is the classic book by Spitzer (1976). This text considers only a subset of such walks, namely those corresponding to increment distributions with zero mean and finite variance. In this case, one can summarize the main result very quickly: the central limit theorem implies that under appropriate rescaling the limiting distribution is normal, and the functional central limit theorem implies that the distribution of the corresponding path-valued process (after standard rescaling of time and space) approaches that of Brownian motion.

Researchers who work with perturbations of random walks, or with particle systems and other models that use random walks as a basic ingredient, often need more precise information on random walk behavior than that provided by the central limit theorems. In particular, it is important to understand the size of the error resulting from the approximation of random walk by Brownian motion. For this reason, there is need for more detailed analysis. This book is an introduction to the random walk theory with an emphasis on the error estimates. Although "mean zero, finite variance" assumption is both necessary and sufficient for normal convergence, one typically needs to make stronger assumptions on the increments of the walk in order to obtain good bounds on the error terms.

This project was embarked upon with an idea of writing a book on the simple, nearest neighbor random walk. Symmetric, finite range random walks gradually became the central model of the text. This class of walks, while being rich enough to require analysis by general techniques, can be studied without much additional difficulty. In addition, for some of the results, in particular, the local central limit theorem and the Green's function estimates, we have extended the discussion to include other mean zero, finite variance walks, while indicating the way in which moment conditions influence the form of the error.

The first chapter is introductory and sets up the notation. In particular, there are three main classes of irreducible walk in the integer lattice $\mathbb{Z}^d$ — $\mathcal{P}_d$ (symmetric, finite range), $\mathcal{P}'_d$ (aperiodic, mean zero, finite second moment), and $\mathcal{P}^*_d$ (aperiodic with no other assumptions). Symmetric random walks on other integer lattices such as the triangular lattice can also be considered by taking a linear transformation of the lattice onto $\mathbb{Z}^d$.

The local central limit theorem (LCLT) is the topic for Chapter 2. Its proof, like the proof of the usual central limit theorem, is done by using Fourier analysis to express the probability of interest in terms of an integral, and then estimating the integral. The error estimates depend strongly on the number of finite moments of the corresponding increment distribution. Some important corollaries are proved in Section 2.4; in particular, the fact that aperiodic random walks starting at different points can be coupled so that with probability $1 - O(n^{-1/2})$ they agree for all times greater than n is true for any aperiodic walk, without any finite moment assumptions. The chapter ends by a more classical, combinatorial derivation of LCLT for simple random walk using Stirling's formula, while again keeping track of error terms.

Brownian motion is introduced in Chapter 3. Although we would expect a typical reader to be familiar already with Brownian motion, we give the construction via the dyadic splitting method. The estimates for the modulus of continuity are also given. We then describe the Skorokhod method of coupling a random walk and a Brownian motion on the same probability space, and give error estimates. The dyadic construction of Brownian motion is also important for the dyadic coupling algorithm of Chapter 7.

Green's function and its analog in the recurrent setting, the potential kernel, are studied in Chapter 4. One of the main tools in the potential theory of random walk is the analysis of martingales derived from these functions. Sharp asymptotics at infinity for Green's function are needed to take full advantage of the martingale technique. We use the sharp LCLT estimates of Chapter 2 to obtain the Green's function estimates. We also discuss the number of finite moments needed for various error asymptotics.

Chapter 5 may seem somewhat out of place. It concerns a well-known estimate for one-dimensional walks called the gambler's ruin estimate. Our motivation for providing a complete self-contained argument is twofold. First, in order to apply this result to all one-dimensional projections of a higher dimensional walk simultaneously, it is important to show that this estimate holds for non-lattice walks uniformly in few parameters of the distribution (variance, probability of making an order 1 positive step). In addition, the argument introduces the reader to a fairly general technique for obtaining the overshoot estimates. The final two sections of this chapter concern variations of one-dimensional walk that arise naturally in the arguments for estimating probabilities of hitting (or avoiding) some special sets, for example, the half-line.

In Chapter 6, the classical potential theory of the random walk is covered in the spirit of Spitzer (1976) and Lawler (1996) (and a number of other sources). The difference equations of our discrete space setting (that in turn become matrix equations on finite sets) are analogous to the standard linear partial differential equations of (continuous) potential theory. The closed form of the solutions is important, but we emphasize here the estimates on hitting probabilities that one can obtain using them. The martingales derived from Green's function are very important in this analysis, and again special care is given to error terms. For notational ease, the discussion is restricted here to symmetric walks. In fact, most of the results of this chapter hold for nonsymmetric walks, but in this case one must distinguish between the "original" walk and the "reversed" walk, i.e. between an operator and its adjoint. An implicit exercise for a dedicated student would be to redo this entire chapter for nonsymmetric walks, changing the statements of the propositions as necessary. It would be more work to relax the finite range assumption, and the moment conditions would become a crucial component of the analysis in this general setting. Perhaps this will be a topic of some future book.

Chapter 7 discusses a tight coupling of a random walk (that has a finite exponential moment) and a Brownian motion, called the dyadic coupling or KMT or Hungarian coupling, originated in Kómlos *et al.* (1975a, b). The idea of the coupling is very natural (once explained), but hard work is needed to prove the strong error estimate. The sharp LCLT estimates from Chapter 2 are one of the key points for this analysis.

In bounded rectangles with sides parallel to the coordinate directions, the rate of convergence of simple random walk to Brownian motion is very fast. Moreover, in this case, exact expressions are available in terms of finite Fourier sums. Several of these calculations are done in Chapter 8.

Chapter 9 is different from the rest of this book. It covers an area that includes both classical combinatorial ideas and topics of current research. As has been gradually discovered by a number of researchers in various disciplines (combinatorics, probability, statistical physics) several objects inherent to a graph or network are closely related: the number of spanning trees, the determinant of the Laplacian, various measures on loops on the trees, Gaussian free field, and loop-erased walks. We give an introduction to this theory, using an approach that is focused on the (unrooted) random walk loop measure, and that uses Wilson's algorithm (1996) for generating spanning trees.

The original outline of this book put much more emphasis on the path-intersection probabilities and the loop-erased walks. The final version offers only a general introduction to some of the main ideas, in the last two chapters. On the one hand, these topics were already discussed in more detail in Lawler (1996), and on the other, discussing the more recent developments in the area would require familiarity with Schramm–Loewner evolution, and explaining this would take us too far from the main topic.

Most of the content of this text (the first eight chapters in particular) are well-known classical results. It would be very difficult, if not impossible, to give a detailed and complete list of references. In many cases, the results were obtained in several places at different occasions, as auxiliary (technical) lemmas needed for understanding some other model of interest, and were therefore not particularly noticed by the community. Attempting to give even a reasonably fair account of the development of this subject would have inhibited the conclusion of this project. The bibliography is therefore restricted to a few references that were used in the writing of this book. We refer the reader to Spitzer (1976) for an extensive bibliography on random walk, and to Lawler (1996) for some additional references.

This book is intended for researchers and graduate students alike, and a considerable number of exercises is included for their benefit. The appendix consists of various results from probability theory that are used in the first eleven chapters but are, however, not really linked to random walk behavior. It is assumed that the reader is familiar with the basics of measure-theoretic probability theory.

♣ The book contains quite a few remarks that are separated from the rest of the text by this typeface. They are intended to be helpful heuristics for the reader, but are not used in the actual arguments.

A number of people have made useful comments on various drafts of this book including students at Cornell University and the University of Chicago. We thank Christian Beneš, Juliana Freire, Michael Kozdron, José Truillijo Ferreras, Robert Masson, Robin Pemantle, Mohammad Abbas Rezaei, Nicolas de Saxcé, Joel Spencer, Rongfeng Sun, John Thacker, Brigitta Vermesi, and Xinghua Zheng. The research of Greg Lawler is supported by the National Science Foundation.

1
Introduction

1.1 Basic definitions

We will define the random walks that we consider in this book. We focus our attention on random walks in $\mathbb{Z}^d$ that have bounded symmetric increment distributions, although we occasionally discuss results for wider classes of walk. We also impose an irreducibility criterion to guarantee that all points in the lattice $\mathbb{Z}^d$ can be reached.

We start by setting some basic notation. We use x, y, z to denote points in the integer lattice $\mathbb{Z}^d = \{(x^1, \ldots, x^d) : x^j \in \mathbb{Z}\}$. We use superscripts to denote components and we use subscripts to enumerate elements. For example, $x_1, x_2, \ldots$ represents a sequence of points in $\mathbb{Z}^d$, and the point x_j can be written in component form $x_j = (x_j^1, \ldots, x_j^d)$. We write $\mathbf{e}_1 = (1, 0, \ldots, 0), \ldots, \mathbf{e}_d = (0, \ldots, 0, 1)$ for the standard basis of unit vectors in $\mathbb{Z}^d$. The prototypical example is (discrete time) simple random walk starting at $x \in \mathbb{Z}^d$. This process can be considered either as a sum of a sequence of independent, identically distributed random variables

$$S_n = x + X_1 + \cdots + X_n$$

where $\mathbb{P}\{X_j = \mathbf{e}_k\} = \mathbb{P}\{X_j = -\mathbf{e}_k\} = 1/(2d), k = 1, \ldots, d$, or it can be considered as a Markov chain with state space $\mathbb{Z}^d$ and transition probabilities

$$\mathbb{P}\{S_{n+1} = z \mid S_n = y\} = \frac{1}{2d}, \quad z - y \in \{\pm \mathbf{e}_1, \ldots \pm \mathbf{e}_d\}.$$

We call $V = \{x_1, \ldots, x_l\} \subset \mathbb{Z}^d \setminus \{0\}$ a *(finite) generating set* if each $y \in \mathbb{Z}^d$ can be written as $k_1 x_1 + \cdots + k_l x_l$ for some $k_1, \ldots, k_l \in \mathbb{Z}$. We let $\mathcal{G}$ denote the collection of generating sets V with the property that if $x = (x^1, \ldots, x^d) \in V$, then the first nonzero component of x is positive. An example of such a set is

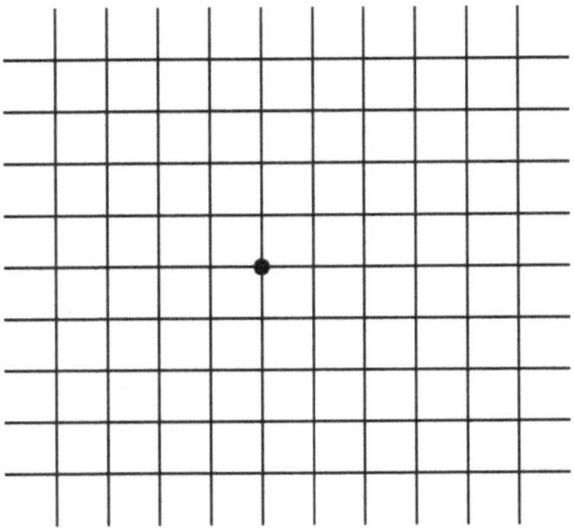

Figure 1.1 The square lattice $\mathbb{Z}^2$

$\{\mathbf{e}_1, \ldots, \mathbf{e}_d\}$. A *(finite range, symmetric, irreducible) random walk* is given by specifying a $V = \{x_1, \ldots, x_l\} \in \mathcal{G}$ and a function $\kappa : V \to (0, 1]$ with $\kappa(x_1) + \cdots + \kappa(x_l) \leq 1$. Associated to this is the symmetric probability distribution on $\mathbb{Z}^d$

$$p(x_k) = p(-x_k) = \frac{1}{2}\,\kappa(x_k), \qquad p(0) = 1 - \sum_{x \in V} \kappa(x).$$

We let $\mathcal{P}_d$ denote the set of such distributions p on $\mathbb{Z}^d$ and $\mathcal{P} = \cup_{d \geq 1} \mathcal{P}_d$. Given p, the corresponding random walk S_n can be considered as the time-homogeneous Markov chain with state space $\mathbb{Z}^d$ and transition probabilities

$$p(y, z) := \mathbb{P}\{S_{n+1} = z \mid S_n = y\} = p(z - y).$$

We can also write

$$S_n = S_0 + X_1 + \cdots + X_n$$

where $X_1, X_2, \ldots$ are independent random variables, independent of S_0, with distribution p. (Most of the time, we will choose S_0 to have a trivial distribution.) We will use the phrase $\mathcal{P}$-walk or $\mathcal{P}_d$-walk for such a random walk. We will use the term *simple random walk* for the particular p with

$$p(\mathbf{e}_j) = p(-\mathbf{e}_j) = \frac{1}{2d}, \quad j = 1, \ldots, d.$$

We call p the *increment distribution* for the walk. Given that $p \in \mathcal{P}$, we write p_n for the n-step distribution

$$p_n(x,y) = \mathbb{P}\{S_n = y \mid S_0 = x\}$$

and $p_n(x) = p_n(0,x)$. Note that $p_n(\cdot)$ is the distribution of $X_1 + \cdots + X_n$ where $X_1, \ldots, X_n$ are independent with increment distribution p.

♣ In many ways the main focus of this book is simple random walk, and a first-time reader might find it useful to consider this example throughout. We have chosen to generalize this slightly, because it does not complicate the arguments much and allows the results to be extended to other examples. One particular example is simple random walk on other regular lattices such as the planar triangular lattice. In Section 1.3, we show that walks on other d-dimensional lattices are isomorphic to p-walks on $\mathbb{Z}^d$.

If $S_n = (S_n^1, \ldots, S_n^d)$ is a $\mathcal{P}$-walk with $S_0 = 0$, then $\mathbb{P}\{S_{2n} = 0\} > 0$ for every even integer n; this follows from the easy estimate $\mathbb{P}\{S_{2n} = 0\} \geq [\mathbb{P}\{S_2 = 0\}]^n \geq p(x)^{2n}$ for every $x \in \mathbb{Z}^d$. We will call the walk *bipartite* if $p_n(0,0) = 0$ for every odd n, and we will call it *aperiodic* otherwise. In the latter case, $p_n(0,0) > 0$ for all n sufficiently large (in fact, for all $n \geq k$ where k is the first odd integer with $p_k(0,0) > 0$). Simple random walk is an example of a bipartite walk since $S_n^1 + \cdots + S_n^d$ is odd for odd n and even for even n. If p is bipartite, then we can partition $\mathbb{Z}^d = (\mathbb{Z}^d)_e \cup (\mathbb{Z}^d)_o$ where $(\mathbb{Z}^d)_e$ denotes the points that can be reached from the origin in an even number of steps and $(\mathbb{Z}^d)_o$ denotes the set of points that can be reached in an odd number of steps. In algebraic language, $(\mathbb{Z}^d)_e$ is an additive subgroup of $\mathbb{Z}^d$ of index 2 and $(\mathbb{Z}^d)_o$ is the nontrivial coset. Note that if $x \in (\mathbb{Z}^d)_o$, then $(\mathbb{Z}^d)_o = x + (\mathbb{Z}^d)_e$.

♣ It would suffice and would perhaps be more convenient to restrict our attention to aperiodic walks. Results about bipartite walks can easily be deduced from them. However, since our main example, simple random walk, is bipartite, we have chosen to allow such p.

If $p \in \mathcal{P}_d$ and $j_1, \ldots, j_d$ are nonnegative integers, the $(j_1, \ldots, j_d)$ moment is given by

$$\mathbb{E}[(X_1^1)^{j_1} \cdots (X_1^d)^{j_d}] = \sum_{x \in \mathbb{Z}^d} (x^1)^{j_1} \cdots (x^d)^{j_d}\, p(x).$$

We let Γ denote the *covariance matrix*

$$\Gamma = \left[\mathbb{E}[X_1^j X_1^k] \right]_{1 \le j,k \le d}.$$

The covariance matrix is symmetric and positive definite. Since the random walk is truly d-dimensional, it is easy to verify (see Proposition 1.1.1 (a)) that the matrix Γ is invertible. There exists a symmetric positive definite matrix Λ such that $\Gamma = \Lambda\,\Lambda^T$ (see Section A.3). There is a (not unique) orthonormal basis $u_1, \ldots, u_d$ of $\mathbb{R}^d$ such that we can write

$$\Gamma x = \sum_{j=1}^{d} \sigma_j^2\,(x \cdot u_j)\,u_j, \qquad \Lambda x = \sum_{j=1}^{d} \sigma_j\,(x \cdot u_j)\,u_j.$$

If X_1 has covariance matrix $\Gamma = \Lambda\,\Lambda^T$, then the random vector $\Lambda^{-1} X_1$ has covariance matrix I.

For future use, we define norms $\mathcal{J}^*, \mathcal{J}$ by

$$\mathcal{J}^*(x)^2 = |x \cdot \Gamma^{-1} x| = |\Lambda^{-1} x|^2 = \sum_{j=1}^{d} \sigma_j^{-2}\,(x \cdot u_j)^2, \qquad \mathcal{J}(x) = d^{-1/2}\,\mathcal{J}^*(x). \tag{1.1}$$

If $p \in \mathcal{P}_d$,

$$\mathbb{E}[\mathcal{J}(X_1)^2] = \frac{1}{d}\,\mathbb{E}[\mathcal{J}^*(X_1)^2] = \frac{1}{d}\,\mathbb{E}\left[|\Lambda^{-1} X_1|^2\right] = 1.$$

For simple random walk in $\mathbb{Z}^d$,

$$\Gamma = d^{-1} I, \qquad \mathcal{J}^*(x) = d^{1/2}\,|x|, \qquad \mathcal{J}(x) = |x|.$$

We will use $\mathcal{B}_n$ to denote the discrete ball of radius n,

$$\mathcal{B}_n = \{x \in \mathbb{Z}^d : |x| < n\},$$

and $\mathcal{C}_n$ to denote the discrete ball under the norm $\mathcal{J}$,

$$\mathcal{C}_n = \{x \in \mathbb{Z}^d : \mathcal{J}(x) < n\} = \{x \in \mathbb{Z}^d : \mathcal{J}^*(x) < d^{1/2}\,n\}.$$

We choose to use $\mathcal{J}$ in the definition of $\mathcal{C}_n$ so that for simple random walk, $\mathcal{C}_n = \mathcal{B}_n$. We will write $R = R_p = \max\{|x| : p(x) > 0\}$ and we will call R the *range* of p. The following is very easy, but it is important enough to state as a proposition.

Proposition 1.1.1 *Suppose that* $p \in \mathcal{P}_d$.

(a) There exists an $\epsilon > 0$ *such that for every unit vector* $u \in \mathbb{R}^d$,

$$\mathbb{E}[(X_1 \cdot u)^2] \geq \epsilon.$$

(b) If $j_1, \ldots, j_d$ *are nonnegative integers with* $j_1 + \cdots + j_d$ *odd, then*

$$\mathbb{E}[(X_1^1)^{j_1} \cdots (X_1^d)^{j_d}] = 0.$$

(c) There exists a $\delta > 0$ *such that for all* x,

$$\delta\, \mathcal{J}(x) \leq |x| \leq \delta^{-1}\, \mathcal{J}(x).$$

In particular,

$$\mathcal{C}_{\delta n} \subset \mathcal{B}_n \subset \mathcal{C}_{n/\delta}.$$

We note for later use that we can construct a random walk with increment distribution $p \in \mathcal{P}$ from a collection of independent one-dimensional simple random walks and an independent multinomial process. To be more precise, let $V = \{x_1, \ldots, x_l\} \in \mathcal{G}$ and let $\kappa : V \to (0, 1]^l$ be as in the definition of $\mathcal{P}$. Suppose that on the same probability space we have defined l independent one-dimensional simple random walks $S_{n,1}, S_{n,2}, \ldots, S_{n,l}$ and an independent multinomial process $L_n = (L_n^1, \ldots, L_n^l)$ with probabilities $\kappa(x_1), \ldots, \kappa(x_l)$. In other words,

$$L_n = \sum_{j=1}^{n} Y_j,$$

where $Y_1, Y_2, \ldots$ are independent $\mathbb{Z}^l$-valued random variables with

$$\mathbb{P}\{Y_k = (1, 0, \ldots, 0)\} = \kappa(x_1), \ldots, \mathbb{P}\{Y_k = (0, 0, \ldots, 1)\} = \kappa(x_l),$$

and $\mathbb{P}\{Y_k = (0, 0, \ldots, 0)\} = 1 - [\kappa(x_1) + \cdots + \kappa(x_l)]$. It is easy to verify that the process

$$S_n := x_1\, S_{L_n^1,1} + x_2\, S_{L_n^2,2} + \cdots + x_l\, S_{L_n^l,l} \tag{1.2}$$

has the distribution of the random walk with increment distribution p. Essentially, what we have done is to split the decision as to how to jump at time n into two decisions: first, to choose an element $x_j \in \{x_1, \ldots, x_l\}$ and then to decide whether to move by $+x_j$ or $-x_j$.

1.2 Continuous-time random walk

It is often more convenient to consider random walks in $\mathbb{Z}^d$ indexed by positive real times. Given that V, κ, p as in the previous section, the *continuous-time random walk with increment distribution p* is the continuous-time Markov chain $\tilde{S}_t$ with rates p. In other words, for each $x, y \in \mathbb{Z}^d$,

$$\mathbb{P}\{\tilde{S}_{t+\Delta t} = y \mid \tilde{S}_t = x\} = p(y-x)\,\Delta t + o(\Delta t), \quad y \neq x,$$

$$\mathbb{P}\{\tilde{S}_{t+\Delta t} = x \mid \tilde{S}_t = x\} = 1 - \left[\sum_{y \neq x} p(y-x)\right] \Delta t + o(\Delta t).$$

Let $\tilde{p}_t(x,y) = \mathbb{P}\{\tilde{S}_t = y \mid \tilde{S}_0 = x\}$, and $\tilde{p}_t(y) = \tilde{p}_t(0,y) = \tilde{p}_t(x, x+y)$. Then the expressions above imply that

$$\frac{d}{dt}\tilde{p}_t(x) = \sum_{y \in \mathbb{Z}^d} p(y)\,[\tilde{p}_t(x-y) - \tilde{p}_t(x)].$$

There is a very close relationship between the discrete time and continuous time random walks with the same increment distribution. We state this as a proposition which we leave to the reader to verify.

Proposition 1.2.1 *Suppose that S_n is a (discrete-time) random walk with increment distribution p and N_t is an independent Poisson process with parameter 1. Then $\tilde{S}_t := S_{N_t}$ has the distribution of a continuous-time random walk with increment distribution p.*

There are various technical reasons why continuous-time random walks are sometimes easier to handle than discrete-time walks. One reason is that in the continuous setting there is no periodicity. If $p \in \mathcal{P}_d$, then $\tilde{p}_t(x) > 0$ for every $t > 0$ and $x \in \mathbb{Z}^d$. Another advantage can be found in the following proposition which gives an analogous, but nicer, version of (1.2). We leave the proof to the reader.

Proposition 1.2.2 *Suppose that $p \in \mathcal{P}_d$ with generating set $V = \{x_1, \ldots, x_l\}$ and suppose that $\tilde{S}_{t,1}, \ldots, \tilde{S}_{t,l}$ are independent one-dimensional* *continuous-time random walks with increment distribution $q_1, \ldots, q_l$ where $q_j(\pm 1) = p(x_j)$. Then*

$$\tilde{S}_t := x_1 \tilde{S}_{t,1} + x_2 \tilde{S}_{t,2} + \cdots + x_l \tilde{S}_{t,l} \tag{1.3}$$

has the distribution of a continuous-time random walk with increment distribution p.

If p is the increment distribution for simple random walk, we call the corresponding walk $\tilde{S}_t$ the *continuous-time simple random walk in $\mathbb{Z}^d$*. From the previous proposition, we see that the coordinates of the continuous-time simple random walk are independent — this is clearly not true for the discrete-time simple random walk. In fact, we get the following. Suppose that $\tilde{S}_{t,1}, \ldots, \tilde{S}_{t,d}$ are independent one-dimensional continuous-time simple random walks. Then,

$$\tilde{S}_t := (\tilde{S}_{t/d,1}, \ldots, \tilde{S}_{t/d,d})$$

is a continuous time simple random walk in $\mathbb{Z}^d$. In particular, if $\tilde{S}_0 = 0$, then

$$\mathbb{P}\{\tilde{S}_t = (y^1, \ldots, y^d)\} = \mathbb{P}\{\tilde{S}_{t/d,1} = y^1\} \cdots \mathbb{P}\{\tilde{S}_{t/d,l} = y^l\}.$$

Remark To verify that a discrete-time process S_n is a random walk with distribution $p \in \mathcal{P}_d$ starting at the origin, it suffices to show for all positive integers $j_1 < j_2 < \cdots < j_k$ and $x_1, \ldots, x_k \in \mathbb{Z}^d$,

$$\mathbb{P}\{S_{j_1} = x_1, \ldots, S_{j_k} = x_k\} = p_{j_1}(x_1)\, p_{j_2 - j_1}(x_2 - x_1) \cdots p_{j_k - j_{k-1}}(x_k - x_{k-1}).$$

To verify that a continuous-time process $\tilde{S}_t$ is a continuous-time random walk with distribution p starting at the origin, it suffices to show that the paths are right-continuous with probability one, and that for all real $t_1 < t_2 < \cdots < t_k$ and $x_1, \ldots, x_k \in \mathbb{Z}^d$,

$$\mathbb{P}\{\tilde{S}_{t_1} = x_1, \ldots, \tilde{S}_{t_k} = x_k\} = \tilde{p}_{t_1}(x_1)\, \tilde{p}_{t_2 - t_1}(x_2 - x_1) \cdots \tilde{p}_{t_k - t_{k-1}}(x_k - x_{k-1}).$$

1.3 Other lattices

A *lattice* $\mathbb{L}$ is a discrete additive subgroup of $\mathbb{R}^d$. The term discrete means that there is a real neighborhood of the origin whose intersection with $\mathbb{L}$ is just the origin. While this book will focus on the lattice $\mathbb{Z}^d$, we will show in this section that this also implies results for symmetric, bounded random walks on other lattices. We start by giving a proposition that classifies all lattices.

Proposition 1.3.1 *If $\mathbb{L}$ is a lattice in $\mathbb{R}^d$, then there exists an integer $k \leq d$ and elements $x_1, \ldots, x_k \in \mathbb{L}$ that are linearly independent as vectors in $\mathbb{R}^d$ such that*

$$\mathbb{L} = \{j_1 x_1 + \cdots + j_k x_k, \quad j_1, \ldots, j_k \in \mathbb{Z}\}.$$

In this case we call $\mathbb{L}$ a k-dimensional lattice.

Proof Suppose first that $\mathbb{L}$ is contained in a one-dimensional subspace of $\mathbb{R}^d$. Choose $x_1 \in \mathbb{L} \setminus \{0\}$ with minimal distance from the origin. Clearly $\{jx_1 : j \in \mathbb{Z}\} \subset \mathbb{L}$. Also, if $x \in \mathbb{L}$, then $jx_1 \leq x < (j+1)x_1$ for some $j \in \mathbb{Z}$, but if $x > jx_1$, then $x - jx_1$ would be closer to the origin than x_1. Hence, $\mathbb{L} = \{jx_1 : j \in \mathbb{Z}\}$.

More generally, suppose that we have chosen linearly independent $x_1, \ldots, x_j$ such that the following holds: if $\mathbb{L}_j$ is the subgroup generated by $x_1, \ldots, x_j$, and V_j is the real subspace of $\mathbb{R}^d$ generated by the vectors $x_1, \ldots, x_j$, then $\mathbb{L} \cap V_j = \mathbb{L}_j$. If $\mathbb{L} = \mathbb{L}_j$, we stop. Otherwise, let $w_0 \in \mathbb{L} \setminus \mathbb{L}_j$ and let

$$\begin{aligned} U &= \{tw_0 : t \in \mathbb{R}, tw_0 + y_0 \in \mathbb{L} \text{ for some } y_0 \in V_j\} \\ &= \{tw_0 : t \in \mathbb{R}, tw_0 + t_1 x_1 + \cdots + t_j x_j \in \mathbb{L} \text{ for some } t_1, \ldots, t_j \in [0, 1]\}. \end{aligned}$$

The second equality uses the fact that $\mathbb{L}$ is a subgroup. Using the first description, we can see that U is a subgroup of $\mathbb{R}^d$ (although not necessarily contained in $\mathbb{L}$). We claim that the second description shows that there is a neighborhood of the origin whose intersection with U is exactly the origin. Indeed, the intersection of $\mathbb{L}$ with every bounded subset of $\mathbb{R}^d$ is finite (why?), and hence there are only a finite number of lattice points of the form

$$tw_0 + t_1 x_1 + \cdots + t_j x_j$$

with $0 < t \leq 1$; and $0 \leq t_1, \ldots, t_j \leq 1$. Hence, there is an $\epsilon > 0$ such that there are no such lattice points with $0 < |t| \leq \epsilon$. Therefore, U is a one-dimensional lattice, and hence there is a $w \in U$ such that $U = \{kw : k \in \mathbb{Z}\}$. By definition, there exists a $y_1 \in V_j$ (not unique, but we just choose one) such that $x_{j+1} := w + y_1 \in \mathbb{L}$. Let $\mathbb{L}_{j+1}, V_{j+1}$ be as above using $x_1, \ldots, x_j, x_{j+1}$. Note that V_{j+1} is also the real subspace generated by $\{x_1, \ldots, x_j, w_0\}$. We claim that $\mathbb{L} \cap V_{j+1} = \mathbb{L}_{j+1}$. Indeed, suppose that $z \in \mathbb{L} \cap V_{j+1}$, and write $z = s_0 w_0 + y_2$ where $y_2 \in V_j$. Then $s_0 w_0 \in U$, and hence, $s_0 w_0 = lw$ for some integer l. Hence, we can write $z = lx_{j+1} + y_3$ with $y_3 = y_2 - ly_1 \in V_j$. But, $z - lx_{j+1} \in V_j \cap \mathbb{L} = \mathbb{L}_j$. Hence, $z \in \mathbb{L}_{j+1}$. □

♣ The proof above seems a little complicated. At first glance it seems that one might be able to simplify the argument as follows. Using the notation in the proof, we start by choosing x_1 to be a nonzero point in $\mathbb{L}$ at minimal distance from the origin, and then inductively to choose x_{j+1} to be a nonzero point in $\mathbb{L} \setminus \mathbb{L}_j$ at minimal distance from the origin. This selection method produces linearly independent $x_1, \ldots, x_k$; however, it is not always the case that

$$\mathbb{L} = \{j_1 x_1 + \cdots + j_k x_k : j_1, \ldots, j_k \in \mathbb{Z}\}.$$

As an example, suppose that $\mathbb{L}$ is the five-dimensional lattice generated by

$$2\mathbf{e}_1, \ 2\mathbf{e}_2, \ 2\mathbf{e}_3, \ 2\mathbf{e}_4, \ \mathbf{e}_1 + \mathbf{e}_2 + \cdots + \mathbf{e}_5.$$

Note that $2\mathbf{e}_5 \in \mathbb{L}$ and the only nonzero points in $\mathbb{L}$ that are within distance two of the origin are $\pm 2\mathbf{e}_j, j = 1, \ldots, 5$. Therefore, this selection method would choose (in some order) $\pm 2\mathbf{e}_1, \ldots, \pm 2\mathbf{e}_5$. But, $\mathbf{e}_1 + \cdots + \mathbf{e}_5$ is not in the subgroup generated by these points.

It follows from the proposition that if $k \leq d$ and $\mathbb{L}$ is a k-dimensional lattice in $\mathbb{R}^d$, then we can find a linear transformation $A : \mathbb{R}^d \to \mathbb{R}^k$ that is an isomorphism of $\mathbb{L}$ onto $\mathbb{Z}^k$. Indeed, we define A by $A(x_j) = \mathbf{e}_j$ where $x_1, \ldots, x_k$ is a basis for $\mathbb{L}$ as in the proposition. If S_n is a bounded, symmetric, irreducible random walk taking values in $\mathbb{L}$, then $S_n^* := AS_n$ is a random walk with increment distribution $p \in \mathcal{P}_k$. Hence, results about walks on $\mathbb{Z}^k$ immediately translate to results about walks on $\mathbb{L}$. If $\mathbb{L}$ is a k-dimensional lattice in $\mathbb{R}^d$ and A is the corresponding transformation, we will call $|\det A|$ the *density* of the lattice. The term comes from the fact that as $r \to \infty$, the cardinality of the intersection of the lattice and ball of radius r in $\mathbb{R}^d$ is asymptotically equal to $|\det A|\, r^k$ times the volume of the unit ball in $\mathbb{R}^k$. In particular, if $j_1, \ldots, j_k$ are positive integers, then $(j_1\mathbb{Z}) \times \cdots \times (j_k\mathbb{Z})$ has density $(j_1, \ldots, j_k)^{-1}$.

Examples

- The *triangular lattice*, considered as a subset of $\mathbb{C} = \mathbb{R}^2$ is the lattice generated by 1 and $e^{i\pi/3}$,

$$\mathbb{L}_{\mathrm{T}} = \{k_1 + k_2\, e^{i\pi/3} : k_1, k_2 \in \mathbb{Z}\}$$

  Note that $e^{2i\pi/3} = e^{i\pi/3} - 1 \in \mathbb{L}_{\mathrm{T}}$. The triangular lattice is also considered as a graph with the above vertices and with edges connecting points that are Euclidean distance one apart. In this case, the origin has six nearest neighbors, the six sixth roots of unity. Simple random walk on the triangular lattice is

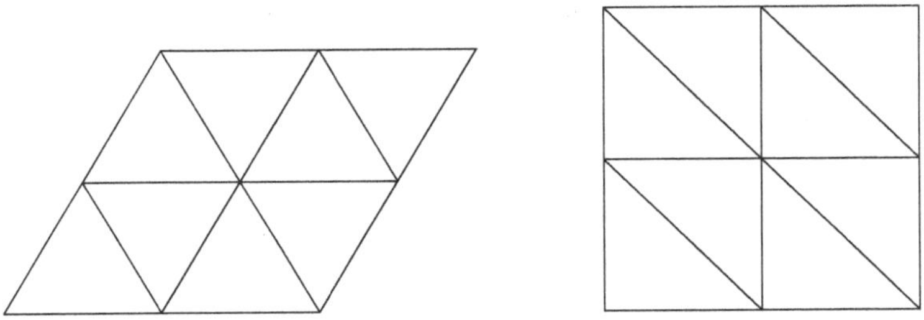

Figure 1.2 The triangular lattice $\mathbb{L}_T$ and its transformation $A\mathbb{L}_T$

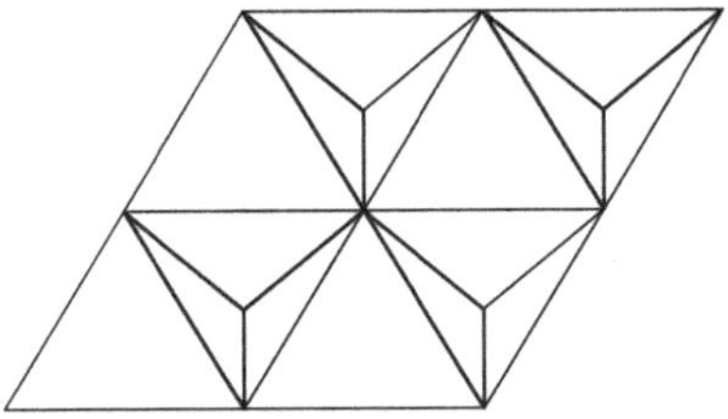

Figure 1.3 The hexagons within $\mathbb{L}_T$

the process that chooses among these six nearest neighbors equally likely. Note that this is a symmetric walk with bounded increments. The matrix

$$A = \begin{bmatrix} 1 & -1/\sqrt{3} \\ 0 & 2/\sqrt{3} \end{bmatrix}$$

maps $\mathbb{L}_T$ to $\mathbb{Z}^2$ sending $\{1, e^{i\pi/3}, e^{2i\pi/3}\}$ to $\{\mathbf{e}_1, \mathbf{e}_2, \mathbf{e}_2 - \mathbf{e}_1\}$. The transformed random walk gives probability $1/6$ to the following vectors: $\pm\mathbf{e}_1, \pm\mathbf{e}_2, \pm(\mathbf{e}_2 - \mathbf{e}_1)$. Note that our transformed walk has lost some of the symmetry of the original walk.

- The *hexagonal or honeycomb lattice* is not a lattice in our sense but rather, a dual graph to the triangular lattice. It can be constructed in a number of ways. One way is to start with the triangular lattice $\mathbb{L}_T$. The lattice partitions the plane into triangular regions, of which some point up and some point down. We add a vertex in the center of each triangle pointing down. The edges of this graph are the line segments from the center points to the vertices of these triangles (see Fig. 1.3).

  Simple random walk on this graph is the process that at each time step moves to one of the three nearest neighbors. This is not a random walk in our strict sense because the increment distribution depends on whether the current position is a "center" point or a "vertex" point. However, if we start at a vertex in $\mathbb{L}_T$, the two-step distribution of this walk is the same as the walk on the triangular lattice with step distribution $p(\pm 1) = p(\pm e^{i\pi/3}) = p(\pm e^{2i\pi/3}) = 1/9; p(0) = 1/3$.

When studying random walks on other lattices $\mathbb{L}$, we can map the walk to another walk on $\mathbb{Z}^d$. However, since this might lose useful symmetries of the walk, it is sometimes better to work on the original lattice.

1.4 Other walks

Although we will focus primarily on $p \in \mathcal{P}$, there are times when we will want to look at more general walks. There are two classes of distribution we will be considering.

Definition

- $\mathcal{P}_d^*$ denotes the set of p that generate aperiodic, irreducible walks supported on $\mathbb{Z}^d$, i.e. the set of p such that for all $x, y \in \mathbb{Z}^d$ there exists an N such that $p_n(x, y) > 0$ for $n \geq N$.
- $\mathcal{P}_d'$ denotes the set of $p \in \mathcal{P}_d^*$ with mean zero and finite second moment.

We write $\mathcal{P}^* = \cup_d \mathcal{P}_d^*, \mathcal{P}' = \cup \mathcal{P}_d'$.

Note that under our definition $\mathcal{P}$ is not a subset of $\mathcal{P}'$ since $\mathcal{P}$ contains bipartite walks. However, if $p \in \mathcal{P}$ is aperiodic, then $p \in \mathcal{P}'$.

1.5 Generator

If $f : \mathbb{Z}^d \to \mathbb{R}$ is a function and $x \in \mathbb{Z}^d$, we define the first and second difference operators in x by

$$\nabla_x f(y) = f(y + x) - f(y),$$
$$\nabla_x^2 f(y) = \frac{1}{2} f(y + x) + \frac{1}{2} f(y - x) - f(y).$$

Note that $\nabla_x^2 = \nabla_{-x}^2$. We will sometimes write just ∇_j, ∇_j^2 for $\nabla_{\mathbf{e}_j}, \nabla_{\mathbf{e}_j}^2$. If $p \in \mathcal{P}_d$ with generator set V, then the generator $\mathcal{L} = \mathcal{L}_p$ is defined by

$$\mathcal{L} f(y) = \sum_{x \in \mathbb{Z}^d} p(x)\, \nabla_x f(y) = \sum_{x \in V} \kappa(x)\, \nabla_x^2 f(y) = -f(y) + \sum_{x \in \mathbb{Z}^d} p(x) f(x + y).$$

In the case of simple random walk, the generator is often called the *discrete Laplacian* and we will represent it by Δ_D,

$$\Delta_D f(y) = \frac{1}{d} \sum_{j=1}^{d} \nabla_j^2 f(y).$$

Remark We have defined the discrete Laplacian in the standard way for probability. In graph theory, the discrete Laplacian of f is often defined to be

$$2d\, \Delta_D f(y) = \sum_{|x-y|=1} [f(x) - f(y)].$$

♣ We can define

$$\mathcal{L} f(y) = \sum_{x \in \mathbb{Z}^d} p(x)\, [f(x+y) - f(y)]$$

for any $p \in \mathcal{P}_d^*$. If p is not symmetric, one often needs to consider

$$\mathcal{L}^R f(y) = \sum_{x \in \mathbb{Z}^d} p(-x)\, [f(x+y) - f(y)].$$

The R stands for "reversed"; this is the generator for the random walk obtained by looking at the walk with time reversed.

The generator of a random walk is very closely related to the walk. We will write $\mathbb{E}^x, \mathbb{P}^x$ to denote expectations and probabilities for random walk (both discrete and continuous time) assuming that $S_0 = x$ or $\tilde{S}_0 = x$. Then, it is easy to check that

$$\mathcal{L} f(y) = \mathbb{E}^y[f(S_1)] - f(y) = \left. \frac{d}{dt} \mathbb{E}^y[f(\tilde{S}_t)] \right|_{t=0}.$$

(In the continuous-time case, some restrictions on the growth of f at infinity are needed.) Also, the transition probabilities $p_n(x), \tilde{p}_t(x)$ satisfy the following "heat equations":

$$p_{n+1}(x) - p_n(x) = \mathcal{L} p_n(x), \qquad \frac{d}{dt} \tilde{p}_t(x) = \mathcal{L} \tilde{p}_t(x).$$

The derivation of these equations uses the symmetry of p. For example, to derive the first, we write

$$\begin{aligned} p_{n+1}(x) &= \sum_{y\in\mathbb{Z}^d} \mathbb{P}\{S_1 = y; S_{n+1} - S_1 = x - y\} \\ &= \sum_{y\in\mathbb{Z}^d} p(y)\, p_n(x-y) \\ &= \sum_{y\in\mathbb{Z}^d} p(-y)\, p_n(x-y) = p_n(x) + \mathcal{L}p_n(x). \end{aligned}$$

The generator $\mathcal{L}$ is also closely related to a second-order differential operator. If $u \in \mathbb{R}^d$ is a unit vector, we write ∂_u^2 for the second partial derivative in the direction u. Let $\hat{\mathcal{L}}$ be the operator

$$\hat{\mathcal{L}}f(y) = \frac{1}{2}\sum_{x\in V} \kappa(x)\, |x|^2\, \partial^2_{x/|x|} f(y).$$

In the case of simple random walk, $\hat{\mathcal{L}} = (2d)^{-1}\,\Delta$, where Δ denotes the usual Laplacian,

$$\Delta f(x) = \sum_{j=1}^{d} \partial_{x_j x_j} f(y);$$

Taylor's theorem shows that there is a c such that if $f : \mathbb{R}^d \to \mathbb{R}$ is C^4 and $y \in \mathbb{Z}^d$,

$$|\mathcal{L}f(y) - \hat{\mathcal{L}}f(y)| \le c\, R^4\, M_4, \tag{1.4}$$

where R is the range of the walk and $M_4 = M_4(f, y)$ is the maximal absolute value of a fourth derivative of f for $|x - y| \le R$. If the covariance matrix Γ is diagonalized,

$$\Gamma x = \sum_{j=1}^{d} \sigma_j^2\, (x\cdot u_j)\, u_j,$$

where $u_1, \ldots, u_d$ is an orthonormal basis, then

$$\hat{\mathcal{L}}f(y) = \frac{1}{2}\sum_{j=1}^{d} \sigma_j^2\, \partial^2_{u_j} f(y).$$

For future reference, we note that if $y \neq 0$,

$$\hat{\mathcal{L}}[\log \mathcal{J}^*(y)^2] = \hat{\mathcal{L}}[\log \mathcal{J}(y)^2]$$
$$= \hat{\mathcal{L}}\left[\log \sum_{j=1}^{d} \sigma_j^{-2}\,(y \cdot u_j)^2\right] = \frac{d-2}{\mathcal{J}^*(y)^2} = \frac{d-2}{d\,\mathcal{J}(y)^2}. \quad (1.5)$$

♣ The estimate (1.4) uses the symmetry of p. If p is mean zero and finite range, but not necessarily symmetric, we can relate its generator to a (purely) second-order differential operator, but the error involves the third derivatives of f. This only requires f to be C^3 and hence can be useful in the symmetric case as well.

1.6 Filtrations and strong Markov property

The basic property of a random walk is that the increments are independent and identically distributed. It is useful to set up a framework that allows more "information" at a particular time than just the value of the random walk. This will not affect the distribution of the random walk, provided that this extra information is independent of the future increments of the walk.

A *(discrete-time) filtration* $\mathcal{F}_0 \subset \mathcal{F}_1 \subset \cdots$ is an increasing sequence of σ-algebras. If $p \in \mathcal{P}_d$, then we say that S_n is a random walk with increment distribution p with respect to $\{\mathcal{F}_n\}$ if:

- for each n, S_n is $\mathcal{F}_n$-measurable;
- for each $n > 0$, $S_n - S_{n-1}$ is independent of $\mathcal{F}_{n-1}$ and $\mathbb{P}\{S_n - S_{n-1} = x\} = p(x)$.

Similarly, we define a *(right continuous, continuous-time) filtration* to be an increasing collection of σ-algebras $\mathcal{F}_t$ satisfying $\mathcal{F}_t = \cap_{\epsilon>0}\mathcal{F}_{t+\epsilon}$. If $p \in \mathcal{P}_d$, then we say that $\tilde{S}_t$ is a continuous-time random walk with increment distribution p with respect to $\{\mathcal{F}_t\}$ if:

- for each t, $\tilde{S}_t$ is $\mathcal{F}_t$-measurable;
- for each $s < t$, $\tilde{S}_t - \tilde{S}_s$ is independent of $\mathcal{F}_s$ and $\mathbb{P}\{\tilde{S}_t - \tilde{S}_s = x\} = \tilde{p}_{t-s}(x)$.

We let $\mathcal{F}_\infty$ denote the σ-algebra generated by the union of the $\mathcal{F}_t$ for $t > 0$.

If S_n is a random walk with respect to $\mathcal{F}_n$, and T is a random variable independent of $\mathcal{F}_\infty$, then we can add information about T to the filtration and still retain the properties of the random walk. We will describe one example of this in detail here; later on, we will similarly add information without being explicit. Suppose that T has an exponential distribution with parameter λ, i.e. $\mathbb{P}\{T > \lambda\} = e^{-\lambda}$. Let $\mathcal{F}'_n$ denote the σ-algebra generated by $\mathcal{F}_n$ and the events $\{T \leq t\}$ for $t \leq n$. Then $\{\mathcal{F}'_n\}$ is a filtration, and S_n is a random walk with respect to $\mathcal{F}'_n$. Also, given that $\mathcal{F}'_n$, then on the event that $\{T > n\}$, the random variable $T - n$ has an exponential distribution with parameter λ. We can do similarly for the continuous-time walk $\tilde{S}_t$.

We will discuss stopping times and the strong Markov property. We will only do the slightly more difficult continuous-time case leaving the discrete-time analogue to the reader. If $\{\mathcal{F}_t\}$ is a filtration, then a *stopping time* with respect to $\{\mathcal{F}_t\}$ is a $[0, \infty]$-valued random variable τ such that for each t, $\{\tau \leq t\} \in \mathcal{F}_t$. Associated to the stopping time τ is a σ-algebra $\mathcal{F}_\tau$ consisting of all events A such that for each t, $A \cap \{\tau \leq t\} \in \mathcal{F}_t$. (It is straightforward to check that the set of such A is a σ-algebra.)

Theorem 1.6.1 (strong Markov property) *Suppose that $\tilde{S}_t$ is a continuous-time random walk with increment distribution p with respect to the filtration $\{\mathcal{F}_t\}$. Suppose that τ is a stopping time with respect to the process. Then on the event that $\{\tau < \infty\}$, the process*

$$Y_t = \tilde{S}_{t+\tau} - \tilde{S}_\tau,$$

is a continuous-time random walk with increment distribution p independent of $\mathcal{F}_\tau$.

Proof **(sketch)** We will assume for ease that $\mathbb{P}\{\tau < \infty\} = 1$. Note that with probability one, Y_t has right-continuous paths. We first suppose that there exists $0 = t_0 < t_1 < t_2 < \ldots$ such that with probability one, $\tau \in \{t_0, t_1, \ldots\}$. Then, the result can be derived immediately, by considering the countable collection of events $\{\tau = t_j\}$. For more general τ, let τ_n be the smallest dyadic rational $l/2^n$ that is greater than τ. Then, τ_n is a stopping time and the result holds for τ_n. But,

$$Y_t = \lim_{n \to \infty} \tilde{S}_{t+\tau_n} - \tilde{S}_{\tau_n}.$$

□

We will use the strong Markov property throughout this book often without being explicit about its use.

Proposition 1.6.2 (reflection principle) *Suppose that S_n (resp., $\tilde{S}_t$) is a random walk (resp., continuous-time random walk) with increment distribution $p \in \mathcal{P}_d$ starting at the origin.*

(a) *If* $u \in \mathbb{R}^d$ *is a unit vector and* $b > 0$,

$$\mathbb{P}\{\max_{0\le j\le n} S_j \cdot u \ge b\} \le 2\,\mathbb{P}\{S_n \cdot u \ge b\},$$

$$\mathbb{P}\{\sup_{s\le t} \tilde{S}_s \cdot u \ge b\} \le 2\,\mathbb{P}\{\tilde{S}_t \cdot u \ge b\}.$$

(b) *If* $b > 0$,

$$\mathbb{P}\{\max_{0\le j\le n} |S_j| \ge b\} \le 2\,\mathbb{P}\{|S_n| \ge b\},$$

$$\mathbb{P}\{\sup_{0\le s\le t} |\tilde{S}_t| \ge b\} \le 2\,\mathbb{P}\{|\tilde{S}_t| \ge b\}.$$

Proof We will do the continuous-time case. To prove (a), fix $t > 0$ and a unit vector u and let $A_n = A_{n,t,b}$ be the event

$$A_n = \left\{\max_{j=1,\ldots,2^n} \tilde{S}_{jt2^{-n}} \cdot u \ge b\right\}.$$

The events A_n are increasing in n and right continuity implies that w.p.1,

$$\lim_{n\to\infty} A_n = \left\{\sup_{s\le t} \tilde{S}_s \cdot u \ge b\right\}.$$

Hence, it suffices to show that for each n, $\mathbb{P}(A_n) \le 2\,\mathbb{P}\{\tilde{S}_t \cdot u \ge b\}$. Let $\tau = \tau_{n,t,b}$ be the smallest j such that $\tilde{S}_{jt2^{-n}} \cdot u \ge b$. Note that

$$\bigcup_{j=1}^{2^n} \left\{\tau = j; (\tilde{S}_t - \tilde{S}_{jt2^{-n}}) \cdot u \ge 0\right\} \subset \{\tilde{S}_t \cdot u \ge b\}.$$

Since $p \in \mathcal{P}$, symmetry implies that for all t, $\mathbb{P}\{\tilde{S}_t \cdot u \ge 0\} \ge 1/2$. Therefore, using independence, $\mathbb{P}\{\tau = j; (\tilde{S}_t - \tilde{S}_{jt2^{-n}}) \cdot u \ge 0\} \ge (1/2)\,\mathbb{P}\{\tau = j\}$, and hence

$$\mathbb{P}\{\tilde{S}_t \cdot u \ge b\} \ge \sum_{j=1}^{2^n} \mathbb{P}\left\{\tau = j; (\tilde{S}_t - \tilde{S}_{jt2^{-n}}) \cdot u \ge 0\right\}$$

$$\ge \frac{1}{2} \sum_{j=1}^{2^n} \mathbb{P}\{\tau = j\} = \frac{1}{2}\,\mathbb{P}(A_n).$$

Part (b) is done similarly, by letting τ be the smallest j with $\{|\tilde{S}_{jt2^{-n}}| \geq b\}$ and writing

$$\bigcup_{j=1}^{2^n} \left\{\tau = j; (\tilde{S}_t - \tilde{S}_{jt2^{-n}}) \cdot \tilde{S}_{jt2^{-n}} \geq 0\right\} \subset \{|\tilde{S}_t| \geq b\}.$$

$\square$

Remark The only fact about the distribution p that is used in the proof is that it is symmetric about the origin.

1.7 A word about constants

Throughout this book c will denote a positive constant that depends on the dimension d and the increment distribution p but does not depend on any other constants. We write

$$f(n,x) = g(n,x) + O(h(n)),$$

to mean that there exists a constant c such that for all n,

$$|f(n,x) - g(n,x)| \leq c\,|h(n)|.$$

Similarly, we write

$$f(n,x) = g(n,x) + o(h(n)),$$

if for every $\epsilon > 0$ there is an N such that

$$|f(n,x) - g(n,x)| \leq \epsilon\,|h(n)|, \quad n \geq N.$$

Note that implicit in the definition is the fact that c, N can be chosen uniformly for all x. If f, g are positive functions, we will write

$$f(n,x) \asymp g(n,x), \quad n \to \infty,$$

if there exists a c (again, independent of x) such that for all n, x,

$$c^{-1}\,g(n,x) \leq f(n,x) \leq c\,g(n,x).$$

We will write similarly for asymptotics of $f(t,x)$ as $t \to 0$.

As an example, let $f(z) = \log(1-z)$, $|z| < 1$, where log denotes the branch of the complex logarithm function with $\log 1 = 0$. Then f is analytic in the unit disk with Taylor series expansion

$$\log(1-z) = -\sum_{j=1}^{\infty} \frac{z^j}{j}.$$

By the remainder estimate, for every $\epsilon > 0$,

$$\left| \log(1-z) + \sum_{j=1}^{k} \frac{z^j}{j} \right| \leq \frac{|z|^{k+1}}{\epsilon^k\,(k+1)}, \qquad |z| \leq 1-\epsilon.$$

For a fixed value of k we can write this as

$$\log(1-z) = -\left[\sum_{j=1}^{k} \frac{z^j}{j}\right] + O(|z|^{k+1}), \qquad |z| \leq 1/2, \tag{1.6}$$

or

$$\log(1-z) = -\left[\sum_{j=1}^{k} \frac{z^j}{j}\right] + O_\epsilon(|z|^{k+1}), \qquad |z| \leq 1-\epsilon, \tag{1.7}$$

where we write O_ϵ to indicate that the constant in the error term depends on ϵ.

Exercises

Exercise 1.1 Show that there are exactly $2^d - 1$ additive subgroups of $\mathbb{Z}^d$ of index 2. Describe them and show that they can all arise from some $p \in \mathcal{P}$. (A subgroup G of $\mathbb{Z}^d$ has index two if $G \neq \mathbb{Z}^d$ but $G \cup (x+G) = \mathbb{Z}^d$ for some $x \in \mathbb{Z}^d$.)

Exercise 1.2 Show that if $p \in \mathcal{P}_d$, n is a positive integer, and $x \in \mathbb{Z}^d$, then $p_{2n}(0) \geq p_{2n}(x)$.

Exercise 1.3 Show that if $p \in \mathcal{P}_d^*$, then there exists a finite set $\{x_1, \ldots, x_k\}$ such that:

- $p(x_j) > 0, \quad j = 1, \ldots, k,$

- for every $y \in \mathbb{Z}^d$, there exist (strictly) positive integers $n_1, \ldots, n_k$ with

$$n_1 x_1 + \cdots + n_k x_k = y. \tag{1.8}$$

(Hint: first write each unit vector $\pm \mathbf{e}_j$ in the above form with perhaps different sets $\{x_1, \ldots, x_k\}$. Then add the equations together.)

Use this to show that there exist $\epsilon > 0, q \in \mathcal{P}'_d, q' \in \mathcal{P}^*_d$ such that q has finite support and

$$p = \epsilon\, q + (1 - \epsilon)\, q'.$$

Note that (1.8) is used with $y = 0$ to guarantee that q has zero mean.

Exercise 1.4 Suppose that $S_n = X_1 + \cdots + X_n$ where $X_1, X_2, \ldots$ are independent $\mathbb{R}^d$-valued random variables with mean zero and covariance matrix Γ. Show that

$$M_n := |S_n|^2 - (\mathrm{tr}\Gamma)\, n$$

is a martingale.

Exercise 1.5 Suppose that $p \in \mathcal{P}'_d \cup \mathcal{P}_d$ with covariance matrix $\Gamma = \Lambda \Lambda^T$ and S_n is the corresponding random walk. Show that

$$M_n := \mathcal{J}(S_n)^2 - n$$

is a martingale.

Exercise 1.6 Let $\mathbb{L}$ be a two-dimensional lattice contained in $\mathbb{R}^d$ and suppose that $x_1, x_2 \in \mathbb{L}$ are points such that

$$|x_1| = \min\{|x| : x \in \mathbb{L} \setminus \{0\}\},$$
$$|x_2| = \min\{|x| : x \in \mathbb{L} \setminus \{jx_1 : j \in \mathbb{Z}\}\}.$$

Show that

$$\mathbb{L} = \{j_1 x_1 + j_2 x_2 : j_1, j_2 \in \mathbb{Z}\}.$$

You may wish to compare this to the remark after Proposition 1.3.1.

Exercise 1.7 Let S_n^1, S_n^2 be independent simple random walks in $\mathbb{Z}$ and let

$$Y_n = \left(\frac{S_n^1 + S_n^2}{2}, \quad \frac{S_n^1 - S_n^2}{2} \right).$$

Show that Y_n is a simple random walk in $\mathbb{Z}^2$.

Exercise 1.8 Suppose that S_n is a random walk with increment distribution $p \in \mathcal{P}^* \cup \mathcal{P}$. Show that there exists an $\epsilon > 0$ such that for every unit vector $\theta \in \mathbb{R}^d$, $\mathbb{P}\{S_1 \cdot \theta \geq \epsilon\} \geq \epsilon$.

2
Local central limit theorem

2.1 Introduction

If $X_1, X_2, \ldots$ are independent, identically distributed random variables in $\mathbb{R}$ with mean zero and variance σ^2, then the central limit theorem (CLT) states that the distribution of

$$\frac{X_1 + \cdots + X_n}{\sqrt{n}} \tag{2.1}$$

approaches that of a normal distribution with mean zero and variance σ^2. In other words, for $-\infty < r < s < \infty$,

$$\lim_{n\to\infty} \mathbb{P}\left\{ r \le \frac{X_1 + \cdots + X_n}{\sqrt{n}} \le s \right\} = \int_r^s \frac{1}{\sqrt{2\pi\sigma^2}} e^{-\frac{y^2}{2\sigma^2}}\, dy.$$

If $p \in \mathcal{P}_1$ is aperiodic with variance σ^2, we can use this to motivate the following approximation:

$$\begin{aligned} p_n(k) &= \mathbb{P}\{S_n = k\} = \mathbb{P}\left\{ \frac{k}{\sqrt{n}} \le \frac{S_n}{\sqrt{n}} < \frac{k+1}{\sqrt{n}} \right\} \\ &\approx \int_{k/\sqrt{n}}^{(k+1)/\sqrt{n}} \frac{1}{\sqrt{2\pi\sigma^2}} e^{-\frac{y^2}{2\sigma^2}}\, dy \approx \frac{1}{\sqrt{2\pi\sigma^2 n}} \exp\left\{ -\frac{k^2}{2\sigma^2 n} \right\}. \end{aligned}$$

Similarly, if $p \in \mathcal{P}_1$ is bipartite, we can conjecture that

$$\begin{aligned} &p_n(k) + p_n(k+1) \\ &\approx \int_{k/\sqrt{n}}^{(k+2)/\sqrt{n}} \frac{1}{\sqrt{2\pi\sigma^2}} e^{-\frac{y^2}{2\sigma^2}}\, dy \approx \frac{2}{\sqrt{2\pi\sigma^2 n}} \exp\left\{ -\frac{k^2}{2\sigma^2 n} \right\}. \end{aligned}$$

The local central limit theorem (LCLT) justifies this approximation.

♣ One gets a better approximation by writing

$$\mathbb{P}\{S_n = k\} = \mathbb{P}\left\{\frac{k-\frac{1}{2}}{\sqrt{n}} \leq \frac{S_n}{\sqrt{n}} < \frac{k+\frac{1}{2}}{\sqrt{n}}\right\} \approx \int_{(k-\frac{1}{2})/\sqrt{n}}^{(k+\frac{1}{2})/\sqrt{n}} \frac{1}{\sqrt{2\pi\sigma^2 n}}\, e^{-\frac{y^2}{2\sigma^2}}\, dy.$$

If $p \in \mathcal{P}_d$ with covariance matrix $\Gamma = \Lambda\,\Lambda^T = \Lambda^2$, then the normalized sums (2.1) approach a joint normal random variable with covariance matrix Γ, i.e. a random variable with density

$$f(x) = \frac{1}{(2\pi)^{d/2}\,(\det\Lambda)}\, e^{-|\Lambda^{-1}x|^2/2} = \frac{1}{(2\pi)^{d/2}\sqrt{\det\Gamma}}\, e^{-(x\cdot\Gamma^{-1}x)/2}.$$

(See Section A.3 for a review of the joint normal distribution.) A similar heuristic argument can be given for $p_n(x)$. Recall from (1.1) that $\mathcal{J}^*(x)^2 = x\cdot\Gamma^{-1}x$. Let $\overline{p}_n(x)$ denote the estimate of $p_n(x)$ that one obtains by the CLT argument,

$$\overline{p}_n(x) = \frac{1}{(2\pi n)^{d/2}\sqrt{\det\Gamma}}\, e^{-\frac{\mathcal{J}^*(x)^2}{2n}} = \frac{1}{(2\pi)^d\, n^{d/2}} \int_{\mathbb{R}^d} e^{i\frac{s\cdot x}{\sqrt{n}}}\, e^{-\frac{s\cdot\Gamma s}{2}}\, d^d s. \quad (2.2)$$

The second equality is a straightforward computation; see (A.14). We define $\overline{p}_t(x)$ for real $t > 0$ in the same way. The LCLT states that for large n, $p_n(x)$ is approximately $\overline{p}_n(x)$. To be more precise, we will say that an aperiodic p satisfies the LCLT if

$$\lim_{n\to\infty} n^{d/2} \sup_{x\in\mathbb{Z}^d} |p_n(x) - \overline{p}_n(x)| = 0.$$

A bipartite p satisfies the LCLT if

$$\lim_{n\to\infty} n^{d/2} \sup_{x\in\mathbb{Z}^d} |p_n(x) + p_{n+1}(x) - 2\overline{p}_n(x)| = 0.$$

In this weak form of the LCLT we have not made any estimate of the error term $|p_n(x) - \overline{p}_n(x)|$ other than that it goes to zero faster than $n^{-d/2}$ uniformly in x. Note that $\overline{p}_n(x)$ is bounded by $c\,n^{-d/2}$ uniformly in x. This is the correct order of magnitude for $|x|$ of order $\sqrt{n}$ but $\overline{p}_n(x)$ is much smaller for larger $|x|$. We will prove a LCLT for any mean zero distribution with finite second moment. However, the LCLT we state now for $p \in \mathcal{P}_d$ includes error estimates that do not hold for all $p \in \mathcal{P}'_d$.

Theorem 2.1.1 (local central limit theorem) *If $p \in \mathcal{P}_d$ is aperiodic, and $\overline{p}_n(x)$ is as defined in (2.2), then there is a c and for every integer $k \geq 4$ there is a $c(k) < \infty$ such that for all integers $n > 0$ and $x \in \mathbb{Z}^d$ the following hold where $z = x/\sqrt{n}$:*

$$|p_n(x) - \overline{p}_n(x)| \leq \frac{c(k)}{n^{(d+2)/2}} \left[(|z|^k + 1)\, e^{-\frac{\mathcal{J}^*(z)^2}{2}} + \frac{1}{n^{(k-3)/2}} \right], \tag{2.3}$$

$$|p_n(x) - \overline{p}_n(x)| \leq \frac{c}{n^{(d+2)/2}\, |z|^2}. \tag{2.4}$$

We will prove this result in a number of steps in Section 2.3. Before doing so, let us consider what the theorem states. Plugging $k = 4$ into (2.3) implies that

$$|p_n(x) - \overline{p}_n(x)| \leq \frac{c}{n^{(d+2)/2}}. \tag{2.5}$$

For "typical" x with $|x| \leq \sqrt{n}$, $\overline{p}_n(x) \asymp n^{-d/2}$. Hence, (2.5) implies that

$$p_n(x) = \overline{p}_n(x) \left[1 + O\left(\frac{1}{n}\right) \right], \qquad |x| \leq \sqrt{n}.$$

The error term in (2.5) is uniform over x, but as $|x|$ grows, the ratio between the error term and $p_n(x)$ grows. The inequalities (2.3) and (2.4) are improvements on the error term for $|x| \geq \sqrt{n}$. Since $\overline{p}_n(x) \asymp n^{-d/2}\, e^{-\mathcal{J}^*(x)^2/2n}$, (2.3) implies that

$$p_n(x) = \overline{p}_n(x) \left[1 + \frac{O_k(|x/\sqrt{n}|^k)}{n} \right] + O_k\left(\frac{1}{n^{(d+k-1)/2}} \right), \qquad |x| \geq \sqrt{n},$$

where we write O_k to emphasize that the constant in the error term depends on k.

An even better improvement is established in Section 2.3.1 where it is shown that

$$p_n(x) = \overline{p}_n(x)\, \exp\left\{ O\left(\frac{1}{n} + \frac{|x|^4}{n^3} \right) \right\}, \qquad |x| < \epsilon\, n.$$

Although Theorem 2.1.1 is not as useful for atypical x, simple large deviation results as given in the next propositions often suffice to estimate probabilities.

Proposition 2.1.2 *(a) Suppose that $p \in \mathcal{P}'_d$ and S_n is a p-walk starting at the origin. Suppose that k is a positive integer such that $\mathbb{E}[|X_1|^{2k}] < \infty$. There exists $c < \infty$ such that for all $s > 0$*

$$\mathbb{P}\left\{ \max_{0 \leq j \leq n} |S_j| \geq s\sqrt{n} \right\} \leq c\, s^{-2k}. \tag{2.6}$$

(b) *Suppose that* $p \in \mathcal{P}_d$ *and* S_n *is a p-walk starting at the origin. There exist* $\beta > 0$ *and* $c < \infty$ *such that for all n and all* $s > 0$,

$$\mathbb{P}\left\{\max_{0 \le j \le n} |S_j| \ge s\sqrt{n}\right\} \le c\, e^{-\beta s^2}. \tag{2.7}$$

Proof It suffices to prove the results for one-dimensional walks. See Corollaries A.2.6 and A.2.7. □

♣ The statement of the LCLT given here is stronger than is needed for many applications. For example, to determine whether the random walk is recurrent or transient, we only need the following corollary. If $p \in \mathcal{P}_d$ is aperiodic, then there exist $0 < c_1 < c_2 < \infty$ such that for all x, $p_n(x) \le c_2\, n^{-d/2}$, and for $|x| \le \sqrt{n}$, $p_n(x) \ge c_1\, n^{-d/2}$. The exponent $d/2$ is important to remember and can be understood easily. In n steps, the random walk tends to go distance $\sqrt{n}$. In $\mathbb{Z}^d$, there are of order $n^{d/2}$ points within distance $\sqrt{n}$ of the origin. Therefore, the probability of being at a particular point should be of order $n^{-d/2}$.

The proof of Theorem 2.1.1 (see Section 2.2) will use the characteristic function. We discuss LCLTs for $p \in \mathcal{P}'_d$, where, as before, $\mathcal{P}'_d$ denotes the set of aperiodic, irreducible increment distributions p in $\mathbb{Z}^d$ with mean zero and finite second moment. In the proof of Theorem 2.1.1, we will see that we do not need to assume that the increments are bounded. For fixed $k \ge 4$, (2.3) holds for $p \in \mathcal{P}'_d$ provided that $\mathbb{E}[|X|^{k+1}] < \infty$ and the third moments of p vanish. The inequalities (2.5) and (2.4) need only finite fourth moments and vanishing third moments. If $p \in \mathcal{P}'_d$ has finite third moments that are nonzero, we can prove a weaker version of (2.3). Suppose that $k \ge 3$, and $\mathbb{E}[|X_1|^{k+1}] < \infty$. There exists $c(k) < \infty$ such that

$$|p_n(x) - \overline{p}_n(x)| \le \frac{c(k)}{n^{(d+1)/2}} \left[(|z|^k + 1)\, e^{-\frac{\mathcal{J}^*(z)^2}{2}} + \frac{1}{n^{(k-2)/2}} \right].$$

Also, for any $p \in \mathcal{P}'_d$ with $\mathbb{E}[|X_1|^3] < \infty$,

$$|p_n(x) - \overline{p}_n(x)| \le \frac{c}{n^{(d+1)/2}}, \qquad |p_n(x) - \overline{p}_n(x)| \le \frac{c}{n^{(d-1)/2}\, |x|^2}.$$

We focus our discussion in Section 2.2 on aperiodic, discrete-time walks, but the next theorem shows that we can deduce the results for bipartite and continuous-time walks from LCLT for aperiodic, discrete-time walks. We state the analogue of (2.3); the analogue of (2.4) can be proved similarly.

Theorem 2.1.3 *If $p \in \mathcal{P}_d$ and $\overline{p}_n(x)$ is as defined in (2.2), then for every $k \geq 4$ there is a $c = c(k) < \infty$ such that the follwing holds for all $x \in \mathbb{Z}^d$.*

- *If n is a positive integer and $z = x/\sqrt{n}$, then*

$$|p_n(x) + p_{n+1}(x) - 2\,\overline{p}_n(x)| \leq \frac{c}{n^{(d+2)/2}} \left[(|z|^k + 1)\, e^{-\mathcal{J}^*(z)^2/2} + \frac{1}{n^{(k-3)/2}} \right]. \tag{2.8}$$

- *If $f\ t > 0$ and $z = x/\sqrt{t}$,*

$$|\tilde{p}_t(x) - \overline{p}_t(x)| \leq \frac{c}{t^{(d+2)/2}} \left[(|z|^k + 1)\, e^{-\mathcal{J}^*(y)^2/2} + \frac{1}{t^{(k-3)/2}} \right]. \tag{2.9}$$

Proof (assuming Theorem 2.1.1) We only sketch the proof. If $p \in \mathcal{P}_d$ is bipartite, then $S_n^* := S_{2n}$ is an aperiodic walk on the lattice $\mathbb{Z}_e^d$. We can establish the result for S_n^* by mapping $\mathbb{Z}_e^d$ to $\mathbb{Z}^d$ as described in Section 1.3. This gives the asymptotics for $p_{2n}(x), x \in \mathbb{Z}_e^d$ and for $x \in \mathbb{Z}_o^d$, we know that

$$p_{2n+1}(x) = \sum_{y \in \mathbb{Z}^d} p_{2n}(x-y)\, p(y).$$

The continuous-time walk viewed at integer times is the discrete-time walk with increment distribution $\tilde{p} = \tilde{p}_1$. Since $\tilde{p}$ satisfies all the moment conditions, (2.3) holds for $\tilde{p}_n(x), n = 0, 1, 2, \ldots$ If $0 < t < 1$, we can write

$$\tilde{p}_{n+t}(x) = \sum_{y \in \mathbb{Z}^d} \tilde{p}_n(x-y)\, \tilde{p}_t(y),$$

and deduce the result for all t. □

2.2 Characteristic functions and LCLT

2.2.1 Characteristic functions of random variables in $\mathbb{R}^d$

One of the most useful tools for studying the distribution of the sums of independent random variables is the characteristic function. If $X = (X^1, \ldots, X^d)$ is a random variable in $\mathbb{R}^d$, then its *characteristic function* $\phi = \phi_X$ is the function from $\mathbb{R}^d$ into $\mathbb{C}$ given by

$$\phi(\theta) = \mathbb{E}[\exp\{i\theta \cdot X\}].$$

Proposition 2.2.1 *Suppose that $X = (X^1, \ldots, X^d)$ is a random variable in $\mathbb{R}^d$ with characteristic function ϕ.*

(a) ϕ is a uniformly continuous function with $\phi(0) = 1$ and $|\phi(\theta)| \leq 1$ for all $\theta \in \mathbb{R}^d$.
(b) If $\theta \in \mathbb{R}^d$ then $\phi_{X,\theta}(s) := \phi(s\theta)$ is the characteristic function of the one-dimensional random variable $X \cdot \theta$.
(c) Suppose that $d = 1$ and m is a positive integer with $\mathbb{E}[|X|^m] < \infty$. Then $\phi(s)$ is a C^m function of s; in fact,

$$\phi^{(m)}(s) = i^m\, \mathbb{E}[X^m e^{isX}].$$

(d) If m is a positive integer, $\mathbb{E}[|X|^m] < \infty$, and $|u| = 1$, then

$$\left| \phi(su) - \sum_{j=0}^{m-1} \frac{i^j\, \mathbb{E}[(X \cdot u)^j]}{j!}\, s^j \right| \leq \frac{\mathbb{E}[|X \cdot u|^m]}{m!}\, |s|^m.$$

(e) If $X_1, X_2, \ldots, X_n$ are independent random variables in $\mathbb{R}^d$, with characteristic functions $\phi_{X_1}, \ldots, \phi_{X_n}$, then

$$\phi_{X_1+\cdots+X_n}(\theta) = \phi_{X_1}(\theta) \cdots \phi_{X_n}(\theta).$$

In particular, if $X_1, X_2, \ldots$ are independent, identically distributed with the same distribution as X, then the characteristic function of $S_n = X_1 + \cdots + X_n$ is given by

$$\phi_{S_n}(\theta) = [\phi(\theta)]^n.$$

Proof To prove uniform continuity, note that

$$|\phi(\theta_1 + \theta) - \phi(\theta)| = |\mathbb{E}[e^{iX(\theta_1+\theta)} - e^{iX\theta}]| \leq \mathbb{E}[|e^{iX\theta_1} - 1|],$$

and the dominated convergence theorem implies that

$$\lim_{\theta_1 \to 0} \mathbb{E}[|e^{iX\theta_1} - 1|] = 0.$$

The other statements in (a) and (b) are immediate. Part (c) is derived by differentiating; the condition $\mathbb{E}[|X|^m] < \infty$ is needed to justify the differentiation using the dominated convergence theorem (details omitted). Part (d) follows from (b), (c), and Taylor's theorem with remainder. Part (e) is immediate from the product rule for expectations of independent random variables. □

We will write $P_m(\theta)$ for the mth order Taylor series approximation of ϕ about the origin. Then the last proposition implies that if $\mathbb{E}[|X|^m] < \infty$, then

$$\phi(\theta) = P_m(\theta) + o(|\theta|^m), \qquad \theta \to 0. \tag{2.10}$$

Note that if $\mathbb{E}[X] = 0$ and $\mathbb{E}[|X|^2] < \infty$, then

$$P_2(\theta) = 1 - \frac{1}{2}\sum_{j=1}^{d}\sum_{k=1}^{d} \mathbb{E}[X^j X^k]\,\theta^j\,\theta^k = 1 - \frac{\theta \cdot \Gamma\theta}{2} = 1 - \frac{\mathbb{E}[(X \cdot \theta)^2]}{2}.$$

Here, Γ denotes the covariance matrix for X. If $\mathbb{E}[|X|^m] < \infty$, we write

$$P_m(\theta) = 1 - \frac{\theta \cdot \Gamma\theta}{2} + \sum_{j=3}^{m} q_j(\theta), \tag{2.11}$$

where q_j are homogeneous polynomials of degree j determined by the moments of X. If all the third moments of X exist and equal zero, $q_3 \equiv 0$. If X has a symmetric distribution, then $q_j \equiv 0$ for all odd j for which $\mathbb{E}[|X|^j] < \infty$.

2.2.2 Characteristic functions of random variables in $\mathbb{Z}^d$

If $X = (X^1, \ldots, X^d)$ is a $\mathbb{Z}^d$-valued random variable, then its characteristic function has period 2π in each variable, i.e. if $k_1, \ldots, k_d$ are integers,

$$\phi(\theta^1, \ldots, \theta^d) = \phi(\theta^1 + 2k_1\pi, \ldots, \theta^d + 2k_d\pi).$$

The characteristic function determines the distribution of X; in fact, the next proposition gives a simple inversion formula. Here, and for the remainder of this section, we will write $d\theta$ for $d\theta^1, \ldots, d\theta^d$.

Proposition 2.2.2 *If $X = (X^1, \ldots, X^d)$ is a $\mathbb{Z}^d$-valued random variable with characteristic function ϕ, then for every $x \in \mathbb{Z}^d$,*

$$\mathbb{P}\{X = x\} = \frac{1}{(2\pi)^d} \int_{[-\pi,\pi]^d} \phi(\theta)\, e^{-ix\cdot\theta}\, d\theta.$$

Proof Since

$$\phi(\theta) = \mathbb{E}[e^{iX\cdot\theta}] = \sum_{y \in \mathbb{Z}^d} e^{iy\cdot\theta}\, \mathbb{P}\{X = y\},$$

we get

$$\int_{[-\pi,\pi]^d} \phi(\theta)\, e^{-ix\cdot\theta}\, d\theta = \sum_{y\in\mathbb{Z}^d} \mathbb{P}\{X = y\} \int_{[-\pi,\pi]^d} e^{i(y-x)\cdot\theta}\, d\theta.$$

(The dominated convergence theorem justifies the interchange of the sum and the integral.) But, if $x, y \in \mathbb{Z}^d$,

$$\int_{[-\pi,\pi]^d} e^{i(y-x)\cdot\theta}\, d\theta = \begin{cases} (2\pi)^d, & y = x \\ 0, & y \neq x. \end{cases}$$

□

Corollary 2.2.3 *Suppose that $X_1, X_2, \ldots$ are independent, identically distributed random variables in $\mathbb{Z}^d$ with characteristic function ϕ. Let $S_n = X_1 + \cdots + X_n$. Then, for all $x \in \mathbb{Z}^d$,*

$$\mathbb{P}\{S_n = x\} = \frac{1}{(2\pi)^d} \int_{[-\pi,\pi]^d} \phi^n(\theta)\, e^{-ix\cdot\theta}\, d\theta.$$

2.3 LCLT – characteristic function approach

In some sense, Corollary 2.2.3 completely solves the problem of determining the distribution of a random walk at a particular time n. However, the integral is generally hard to evaluate and estimation of oscillatory integrals is tricky. Fortunately, we can use this corollary as a starting point for deriving the LCLT. We will consider $p \in \mathcal{P}'$ in this section. Here, as before, we write $p_n(x)$ for the distribution of $S_n = X_1 + \cdots + X_n$ where $X_1, \ldots, X_n$ are independent with distribution p. We also write $\tilde{S}_t$ for a continuous time walk with rates p. We let ϕ denote the characteristic function of p,

$$\phi(\theta) = \sum_{x\in\mathbb{Z}^d} e^{i\theta\cdot x}\, p(x).$$

We have noted that the characteristic function of S_n is ϕ^n.

Lemma 2.3.1 *The characteristic function of $\tilde{S}_t$ is*

$$\phi_{\tilde{S}_t}(\theta) = \exp\{t[\phi(\theta) - 1]\}.$$

Proof Since $\tilde{S}_t$ has the same distribution as S_{N_t} where N_t is an independent Poisson process with parameter 1, we get

$$\phi_{\tilde{S}_t}(\theta) = \mathbb{E}[e^{i\theta\cdot\tilde{S}_t}]$$
$$= \sum_{j=0}^{\infty} e^{-t}\,\frac{t^j}{j!}\,\mathbb{E}[e^{i\theta\cdot S_j}] = \sum_{j=0}^{\infty} e^{-t}\,\frac{t^j}{j!}\,\phi(\theta)^j = \exp\{t[\phi(\theta)-1]\}.$$

□

Corollary 2.2.3 gives the formulas

$$p_n(x) = \frac{1}{(2\pi)^d}\int_{[-\pi,\pi]^d} \phi^n(\theta)\, e^{-i\theta\cdot x}\, d\theta, \tag{2.12}$$
$$\tilde{p}_t(x) = \frac{1}{(2\pi)^d}\int_{[-\pi,\pi]^d} e^{t[\phi(\theta)-1]}\, e^{-i\theta\cdot x}\, d\theta.$$

Lemma 2.3.2 *Suppose that* $p \in \mathcal{P}'_d$.

(a) For every $\epsilon > 0$,

$$\sup\left\{|\phi(\theta)| : \theta \in [-\pi,\pi]^d, |\theta| \geq \epsilon\right\} < 1.$$

(b) There is a $b > 0$ *such that for all* $\theta \in [-\pi,\pi]^d$,

$$|\phi(\theta)| \leq 1 - b|\theta|^2. \tag{2.13}$$

In particular, for all $\theta \in [-\pi,\pi]^d$, *and* $r > 0$,

$$|\phi(\theta)|^r \leq \left[1 - b|\theta|^2\right]^r \leq \exp\left\{-br|\theta|^2\right\}. \tag{2.14}$$

Proof By continuity and compactness, to prove (a) it suffices to prove that $|\phi(\theta)| < 1$ for all $\theta \in [-\pi,\pi]^d \setminus \{0\}$. To see this, suppose that $|\phi(\theta)| = 1$. Then $|\phi(\theta)^n| = 1$ for all positive integers n. Since

$$\phi(\theta)^n = \sum_{z\in\mathbb{Z}^d} p_n(z)\, e^{iz\cdot\theta},$$

and for each fixed $z, p_n(z) > 0$ for all sufficiently large n, we see that $e^{iz\cdot\theta} = 1$ for all $z \in \mathbb{Z}^d$. (Here we use the fact that if $w_1, w_2, \ldots \in \mathbb{C}$ with $|w_1+w_2+\cdots| = 1$ and $|w_1|+|w_2|+\cdots = 1$, then there is a ψ such that $w_j = r_j e^{i\psi}$ with $r_j \geq 0$.)

The only $\theta \in [-\pi, \pi]^d$ that satisfies this is $\theta = 0$. Using (a), it suffices to prove (2.13) in a neighborhood of the origin, and this follows from the second-order Taylor series expansion (2.10). $\square$

♣ The last lemma does not hold for bipartite p. For example, for simple random walk $\phi(\pi i, \pi i, \ldots, \pi i) = -1$.

In order to illustrate the proof of the LCLT using the characteristic function, we will consider the one-dimensional case with $p(1) = p(-1) = 1/4$ and $p(0) = 1/2$. Note that this increment distribution is the same as the two-step distribution of (1/2 times) the usual simple random walk. The characteristic function for p is

$$\phi(\theta) = \frac{1}{2} + \frac{1}{4} e^{i\theta} + \frac{1}{4} e^{-i\theta} = \frac{1}{2} + \frac{1}{2} \cos\theta = 1 - \frac{\theta^2}{4} + O(\theta^4).$$

The inversion formula (2.12) tells us that

$$p_n(x) = \frac{1}{2\pi} \int_{-\pi}^{\pi} e^{-ix\theta} \, \phi(\theta)^n \, d\theta = \frac{1}{2\pi\sqrt{n}} \int_{-\pi\sqrt{n}}^{\pi\sqrt{n}} e^{-i(x/\sqrt{n})s} \, \phi(s/\sqrt{n})^n \, ds.$$

The second equality follows from the substitution $s = \theta\sqrt{n}$. For $|s| \leq \pi\sqrt{n}$, we can write

$$\phi\left(\frac{s}{\sqrt{n}}\right) = 1 - \frac{s^2}{4n} + O\left(\frac{s^4}{n^2}\right) = 1 - \frac{(s^2/4) + O(s^4/n)}{n}.$$

We can find $\delta > 0$ such that if $|s| \leq \delta\sqrt{n}$,

$$\left|\frac{s^2}{4} + O\left(\frac{s^4}{n}\right)\right| \leq \frac{n}{2}.$$

Therefore, using (A.3), if $|s| \leq \delta\sqrt{n}$,

$$\phi\left(\frac{s}{\sqrt{n}}\right)^n = \left[1 - \frac{s^2}{4n} + O\left(\frac{s^4}{n^2}\right)\right]^n = e^{-s^2/4} \, e^{g(s,n)},$$

where

$$|g(s,n)| \leq c \, \frac{s^4}{n}.$$

If $\epsilon = \min\{\delta, 1/\sqrt{8c}\}$ we also have

$$|g(s,n)| \leq \frac{s^2}{8}, \qquad |s| \leq \epsilon\sqrt{n}.$$

For $\epsilon\sqrt{n} < |s| \leq \pi\sqrt{n}$, (2.13) shows that $|e^{-i(x/\sqrt{n})s}\,\phi(s/\sqrt{n})^n| \leq e^{-\beta n}$ for some $\beta > 0$. Hence, up to an error that is exponentially small in n, $p_n(x)$ equals

$$\frac{1}{2\pi\sqrt{n}} \int_{-\epsilon\sqrt{n}}^{\epsilon\sqrt{n}} e^{-i(x/\sqrt{n})s}\, e^{-s^2/4}\, e^{g(s,n)}\, ds.$$

We now use

$$|e^{g(s,n)} - 1| \leq \begin{cases} c\,s^4/n, & |s| \leq n^{1/4} \\ e^{s^2/8}, & n^{1/4} < |s| \leq \epsilon\sqrt{n} \end{cases}$$

to bound the error term as follows:

$$\left| \frac{1}{2\pi\sqrt{n}} \int_{-\epsilon\sqrt{n}}^{\epsilon\sqrt{n}} e^{-i(x/\sqrt{n})s}\, e^{-s^2/4}\, [e^{g(s,n)} - 1]\, ds \right|$$
$$\leq \frac{c}{\sqrt{n}} \int_{-\epsilon\sqrt{n}}^{\epsilon\sqrt{n}} e^{-s^2/4}\, |e^{g(s,n)} - 1|\, ds,$$

$$\int_{-n^{-1/4}}^{n^{1/4}} e^{-s^2/4}\, |e^{g(s,n)} - 1|\, ds \leq \frac{c}{n} \int_{-\infty}^{\infty} s^4\, e^{-s^2/4}\, ds \leq \frac{c}{n},$$

$$\int_{n^{1/4} \leq |s| \leq \epsilon\sqrt{n}} e^{-s^2/4}\, |e^{g(s,n)} - 1|\, ds \leq \int_{|s| \geq n^{1/4}} e^{-s^2/8}\, ds = o(n^{-1}).$$

Hence, we have

$$p_n(x) = O\left(\frac{1}{n^{3/2}}\right) + \frac{1}{2\pi\sqrt{n}} \int_{-\epsilon\sqrt{n}}^{\epsilon\sqrt{n}} e^{-i(x/\sqrt{n})s}\, e^{-s^2/4}\, ds$$
$$= O\left(\frac{1}{n^{3/2}}\right) + \frac{1}{2\pi\sqrt{n}} \int_{-\infty}^{\infty} e^{-i(x/\sqrt{n})s}\, e^{-s^2/4}\, ds.$$

The last term equals $\overline{p}_n(x)$, see (2.2), and so we have shown that

$$p_n(x) = \overline{p}_n(x) + O\left(\frac{1}{n^{3/2}}\right).$$

We will follow this basic line of proof for theorems in this subsection. Before proceeding, it will be useful to outline the main steps.

- Expand $\log \phi(\theta)$ in a neighborhood $|\theta| < \epsilon$ about the origin.
- Use this expansion to approximate $[\phi(\theta/\sqrt{n})]^n$, which is the characteristic function of $S_n/\sqrt{n}$.
- Use the inversion formula to get an exact expression for the probability and do a change of variables $s = \theta\sqrt{n}$ to yield an integral over $[-\pi\sqrt{n}, \pi\sqrt{n}]^d$. Use Lemma 2.3.2 to show that the integral over $|\theta| \geq \epsilon\sqrt{n}$ is exponentially small.
- Use the approximation of $[\phi(\theta/\sqrt{n})]^n$ to compute the dominant term and to give an expression for the error term that needs to be estimated.
- Estimate the error term.

Our first lemma discusses the approximation of the characteristic function of $S_n/\sqrt{n}$ by an exponential. We state the lemma for all $p \in \mathcal{P}'_d$, and then give examples to show how to get sharper results if one makes stronger assumptions on the moments.

Lemma 2.3.3 *Suppose that $p \in \mathcal{P}'_d$ with covariance matrix Γ and characteristic function ϕ that we write as*

$$\phi(\theta) = 1 - \frac{\theta \cdot \Gamma\theta}{2} + h(\theta),$$

where $h(\theta) = o(|\theta|^2)$ as $\theta \to 0$. There exist $\epsilon > 0, c < \infty$ such that for all positive integers n and all $|\theta| \leq \epsilon\sqrt{n}$, we can write

$$\left[\phi\left(\frac{\theta}{\sqrt{n}}\right)\right]^n = \exp\left\{-\frac{\theta \cdot \Gamma\theta}{2} + g(\theta, n)\right\} = e^{-\frac{\theta\cdot\Gamma\theta}{2}}\,[1 + F_n(\theta)], \qquad (2.15)$$

where $F_n(\theta) = e^{g(\theta,n)} - 1$ and

$$|g(\theta, n)| \leq \min\left\{\frac{\theta \cdot \Gamma\theta}{4},\ n\left|h\left(\frac{\theta}{\sqrt{n}}\right)\right| + \frac{c|\theta|^4}{n}\right\}. \qquad (2.16)$$

In particular,

$$|F_n(\theta)| \leq e^{\frac{\theta\cdot\Gamma\theta}{4}} + 1.$$

Proof Choose $\delta > 0$ such that

$$|\phi(\theta) - 1| \leq \frac{1}{2}, \quad |\theta| \leq \delta.$$

For $|\theta| \le \delta$, we can write

$$\log\phi(\theta) = -\frac{\theta\cdot\Gamma\theta}{2} + h(\theta) - \frac{(\theta\cdot\Gamma\theta)^2}{8} + O(|h(\theta)|\,|\theta|^2) + O(|\theta|^6). \quad (2.17)$$

Define $g(\theta,n)$ by

$$n\,\log\phi\left(\frac{\theta}{\sqrt{n}}\right) = -\frac{\theta\cdot\Gamma\theta}{2} + g(\theta,n),$$

so that (2.15) holds. Note that

$$|g(\theta,n)| \le n\left|h\left(\frac{\theta}{\sqrt{n}}\right)\right| + O\left(\frac{|\theta|^4}{n}\right).$$

Since $n\,h(\theta/\sqrt{n}) = o(|\theta|^2)$, we can find $0 < \epsilon \le \delta$ such that for $|\theta| \le \epsilon\sqrt{n}$,

$$|g(\theta,n)| \le \frac{\theta\cdot\Gamma\theta}{4}.$$

□

♣ The proofs will require estimates for $F_n(\theta)$. The inequality $|e^z - 1| \le O(|z|)$ is valid if z is restricted to a bounded set. Hence, the basic strategy is to find $c_1, r(n) \le O(n^{1/4})$ such that

$$n\left|h\left(\frac{\theta}{\sqrt{n}}\right)\right| \le c_1, \qquad |\theta| \le r(n)$$

Since $O(|\theta|^4/n) \le O(1)$ for $|\theta| \le n^{1/4}$, (2.16) implies that

$$|F_n(\theta)| = \left|e^{g(\theta,n)} - 1\right| \le c\,|g(\theta,n)| \le c\left[n\left|h\left(\frac{\theta}{\sqrt{n}}\right)\right| + \frac{|\theta|^4}{n^2}\right], \quad |\theta| \le r(n),$$

$$|F_n(\theta)| \le e^{\frac{\theta\cdot\Gamma\theta}{4}} + 1, \qquad r(n) \le |\theta| \le \epsilon\sqrt{n}.$$

Examples

We give some examples with different moment assumptions. In the discussion below, ϵ is as in Lemma 2.3.3 and θ is restricted to $|\theta| \le \epsilon\sqrt{n}$.

- If $\mathbb{E}[|X_1|^4] < \infty$, then by (2.11),

$$h(\theta) = q_3(\theta) + O(|\theta|^4),$$

and

$$\log\phi(\theta) = -\frac{\theta\cdot\Gamma\theta}{2} + f_3(\theta) + O(|\theta|^4),$$

where $f_3 = q_3$ is a homogeneous polynomial of degree three. In this case,

$$g(\theta, n) = nf_3\left(\frac{\theta}{\sqrt{n}}\right) + \frac{O(|\theta|^4)}{n}, \tag{2.18}$$

and there exists $c < \infty$ such that

$$|g(\theta, n)| \leq \min\left\{\frac{\theta \cdot \Gamma\theta}{4}, \frac{c\,|\theta|^3}{\sqrt{n}}\right\}.$$

We use here, and below, the fact that $|\theta|^3/\sqrt{n} \geq |\theta|^4/n$ for $|\theta| \leq \epsilon\sqrt{n}$.

- If $\mathbb{E}[X_1|^6] < \infty$ and all the third and fifth moments of X_1 vanish, then

$$h(\theta) = q_4(\theta) + O(|\theta|^6),$$
$$\log\phi(\theta) = -\frac{\theta \cdot \Gamma\theta}{2} + f_4(\theta) + O(|\theta|^6),$$

where $f_4(\theta) = q_4(\theta) - (\theta \cdot \Gamma\theta)^2/8$ is a homogeneous polynomial of degree four. In this case,

$$g(\theta, n) = nf_4\left(\frac{\theta}{\sqrt{n}}\right) + \frac{O(|\theta|^6)}{n^2}, \tag{2.19}$$

and there exists $c < \infty$ such that

$$|g(\theta, n)| \leq \min\left\{\frac{\theta \cdot \Gamma\theta}{4}, \frac{c\,|\theta|^4}{n}\right\}.$$

- More generally, suppose that $k \geq 3$ is a positive integer such that $\mathbb{E}[|X_1|^{k+1}] < \infty$. Then

$$h(\theta) = \sum_{j=3}^{k} q_j(\theta) + O(|\theta|^{k+1}),$$
$$\log\phi(\theta) = -\frac{\theta \cdot \Gamma\theta}{2} + \sum_{j=3}^{k} f_j(\theta) + O(|\theta|^{k+1}),$$

where f_j are homogeneous polynomials of degree j that are determined by $\Gamma, q_3, \ldots, q_k$. In this case,

$$g(\theta, n) = \sum_{j=3}^{k} nf_j\left(\frac{\theta}{\sqrt{n}}\right) + \frac{O(|\theta|^{k+1})}{n^{(k-1)/2}}, \tag{2.20}$$

Moreover, if j is odd and all the odd moments of X of degree less than or equal to j vanish, then $f_j \equiv 0$. Also,

$$|g(\theta, n)| \le \min\left\{\frac{\theta\cdot\Gamma\theta}{4}, \frac{c\,|\theta|^{2+\alpha}}{n^{\alpha/2}}\right\},$$

where $\alpha = 2$ if the third moments vanish and otherwise $\alpha = 1$.

- Suppose that $\mathbb{E}[e^{b\cdot X}] < \infty$ for all b in a real neighborhood of the origin. Then $z \mapsto \phi(z) = e^{iz\cdot X_1}$ is a holomorphic function from a neighborhood of the origin in $\mathbb{C}^n$ to $\mathbb{C}$. Hence, we can choose ϵ so that $\log\phi(z)$ is holomorphic for $|z| < \epsilon$ and hence $z \mapsto g(z,n)$ and $z \mapsto F_n(z)$ are holomorphic for $|z| < \epsilon\sqrt{n}$.

The next lemma computes the dominant term and isolates the integral that needs to be estimated in order to obtain error bounds.

Lemma 2.3.4 *Suppose that $p \in \mathcal{P}'_d$ with covariance matrix Γ. Let ϕ, ϵ, F_n be as in Lemma 2.3.3. There exist $c < \infty, \zeta > 0$ such that for all $0 \le r \le \epsilon\sqrt{n}$, if we define $v_n(x,r)$ by*

$$p_n(x) = \overline{p}_n(x) + v_n(x,r) + \frac{1}{(2\pi)^d\, n^{d/2}} \int_{|\theta|\le r} e^{-\frac{ix\cdot\theta}{\sqrt{n}}}\, e^{-\frac{\theta\cdot\Gamma\theta}{2}}\, F_n(\theta)\, d\theta,$$

then

$$|v_n(x,r)| \le c\, n^{-d/2}\, e^{-\zeta r^2}.$$

Proof The inversion formula (2.12) gives

$$\begin{aligned} p_n(x) &= \frac{1}{(2\pi)^d} \int_{[-\pi,\pi]^d} \phi(\theta)^n\, e^{-ix\cdot\theta}\, d\theta \\ &= \frac{1}{(2\pi)^d\, n^{d/2}} \int_{[-\sqrt{n}\pi,\sqrt{n}\pi]^d} \phi\left(\frac{s}{\sqrt{n}}\right)^n e^{-iz\cdot s}\, ds, \end{aligned}$$

where $z = x/\sqrt{n}$. Lemma 2.3.2 implies that there is a $\beta > 0$ such that $|\phi(\theta)| \le e^{-\beta}$ for $|\theta| \ge \epsilon$. Therefore,

$$\begin{aligned} &\frac{1}{(2\pi)^d\, n^{d/2}} \int_{[-\sqrt{n}\pi,\sqrt{n}\pi]^d} \phi\left(\frac{s}{\sqrt{n}}\right)^n e^{-iz\cdot s}\, ds \\ &\quad = O(e^{-\beta n}) + \frac{1}{(2\pi)^d\, n^{d/2}} \int_{|\theta|\le\epsilon\sqrt{n}} \phi\left(\frac{s}{\sqrt{n}}\right)^n e^{-iz\cdot s}\, ds. \end{aligned}$$

For $|s| \leq \epsilon\sqrt{n}$, we write

$$\phi\left(\frac{s}{\sqrt{n}}\right)^n = e^{-\frac{s\cdot\Gamma s}{2}} + e^{-\frac{s\cdot\Gamma s}{2}} F_n(s).$$

By (2.2) we have

$$\frac{1}{(2\pi)^d\, n^{d/2}} \int_{\mathbb{R}^d} e^{-iz\cdot s}\, e^{-\frac{s\cdot\Gamma s}{2}}\, ds = \overline{p}_n(x). \tag{2.21}$$

Also,

$$\left|\frac{1}{(2\pi)^d\, n^{d/2}} \int_{|s|\geq\epsilon\sqrt{n}} e^{-iz\cdot s}\, e^{-\frac{s\cdot\Gamma s}{2}}\, ds\right|$$
$$\leq \frac{1}{(2\pi)^d\, n^{d/2}} \int_{|s|\geq\epsilon\sqrt{n}} e^{-\frac{s\cdot\Gamma s}{2}}\, ds \leq O(e^{-\beta n}),$$

for perhaps a different β. Therefore,

$$p_n(x) = \overline{p}_n(x) + O(e^{-\beta n}) + \frac{1}{(2\pi)^d\, n^{d/2}} \int_{|\theta|\leq\epsilon\sqrt{n}} e^{\frac{-ix\cdot\theta}{\sqrt{n}}}\, e^{-\frac{\theta\cdot\Gamma\theta}{2}}\, F_n(\theta)\, d\theta.$$

This gives the result for $r = \epsilon\sqrt{n}$. For other values of r, we use the estimate

$$|F_n(\theta)| \leq e^{\frac{\theta\cdot\Gamma\theta}{4}} + 1,$$

to see that

$$\left|\int_{r\leq|\theta|\leq\epsilon\sqrt{n}} e^{\frac{-ix\cdot\theta}{\sqrt{n}}}\, e^{-\frac{\theta\cdot\Gamma\theta}{2}}\, F_n(\theta)\, d\theta\right| \leq 2\int_{|\theta|\geq r} e^{-\frac{\theta\cdot\Gamma\theta}{4}}\, d\theta = O(e^{-\zeta r^2}).$$

□

The next theorem establishes LCLTs for $p \in \mathcal{P}'$ with finite third moment and $p \in \mathcal{P}'$ with finite fourth moment and vanishing third moments. It gives an error term that is uniform over all $x \in \mathbb{Z}^d$. The estimate is good for typical x, but is not very sharp for atypically large x.

Theorem 2.3.5 *Suppose that $p \in \mathcal{P}'$ with $\mathbb{E}[|X_1|^3] < \infty$. Then there exists a $c < \infty$ such that for all n, x,*

$$|p_n(x) - \overline{p}_n(x)| \leq \frac{c}{n^{(d+1)/2}}. \tag{2.22}$$

If $\mathbb{E}[|X_1|^4] < \infty$ *and all the third moments of* X_1 *are zero, then there is a* c *such that for all* n, x,

$$|p_n(x) - \overline{p}_n(x)| \leq \frac{c}{n^{(d+2)/2}}. \tag{2.23}$$

Proof We use the notations of Lemmas 2.3.3 and 2.3.4. Letting $r = n^{1/8}$ in Lemma 2.3.4, we see that

$$p_n(x) = \overline{p}_n(x) + O(e^{-\beta n^{1/4}}) + \frac{1}{(2\pi)^d\, n^{d/2}} \int_{|\theta| \leq n^{1/8}} e^{-\frac{ix\cdot\theta}{\sqrt{n}}}\, e^{-\frac{\theta\cdot\Gamma\theta}{2}}\, F_n(\theta)\, d\theta.$$

Note that $|h(\theta)| = O(|\theta|^{2+\alpha})$ where $\alpha = 1$ under the weaker assumption and $\alpha = 2$ under the stronger assumption. For $|\theta| \leq n^{1/8}$, $|g(\theta, n)| \leq c\, |\theta|^{2+\alpha}/n^{\alpha/2}$, and hence

$$|F_n(\theta)| \leq c\, \frac{|\theta|^{2+\alpha}}{n^{\alpha/2}}.$$

This implies that

$$\left| \int_{|\theta| \leq n^{1/8}} e^{-\frac{ix\cdot\theta}{\sqrt{n}}}\, e^{-\frac{\theta\cdot\Gamma\theta}{2}}\, F_n(\theta)\, d\theta \right| \leq \frac{c}{n^{\alpha/2}} \int_{\mathbb{R}^d} |\theta|^{2+\alpha}\, e^{-\frac{\theta\cdot\Gamma\theta}{2}}\, d\theta \leq \frac{c}{n^{\alpha/2}}.$$

♣ The choice $r = n^{1/8}$ in the proof above was somewhat arbitrary. The value r was chosen sufficiently large so that the error term $v_n(x, r)$ from Lemma 2.3.4 decays faster than any power of n but sufficiently small that $|g(\theta, n)|$ is uniformly bounded for $|\theta| \leq r$. We could just as well have chosen $r(n) = n^\kappa$ for any $0 < \kappa \leq 1/8$.

♣ The constant c in (2.22) and (2.23) depends on the particular p. However, by careful examination of the proof, one can get uniform estimates for all p satisfying certain conditions. The error in the Taylor polynomial approximation of the characteristic function can be bounded in terms of the moments of p. One also needs a uniform bound such as (2.13) which guarantees that the walk is not too close to being a bipartite walk. Such uniform bounds on rates of convergence in CLT or LCLT are often called Berry–Esseen bounds. We will need one such result, see Proposition 2.3.13, but for most of this book, the walk p is fixed and we just allow constants to depend on p.

□

The estimate (2.5) is a special case of (2.23). We have shown that (2.5) holds for any symmetric $p \in \mathcal{P}'$ with $\mathbb{E}[|X_1|^4] < \infty$. One can obtain a difference estimate for $p_n(x)$ from (2.5). However, we will give another proof below that requires only third moments of the increment distribution. This theorem also gives a uniform bound on the error term.

♣ If $\alpha \neq 0$ and

$$f(n) = n^\alpha + O(n^{\alpha-1}), \tag{2.24}$$

then

$$f(n+1) - f(n) = [(n+1)^\alpha - n^\alpha] + [O((n+1)^{\alpha-1}) - O(n^{\alpha-1})].$$

This shows that $f(n+1) - f(n) = O(n^{\alpha-1})$, but the best that we can write about the error terms is $O((n+1)^{\alpha-1}) - O(n^{\alpha-1}) = O(n^{\alpha-1})$, which is as large as the dominant term. Hence, an expression such as (2.24) is not sufficient to give good asymptotics on differences of f. One strategy for proving difference estimates is to go back to the derivation of (2.24) to see if the difference of the errors can be estimated. This is the approach used in the next theorem.

Theorem 2.3.6 *Suppose that $p \in \mathcal{P}'_d$ with $\mathbb{E}[|X_1|^3] < \infty$. Let ∇_y denote the differences in the x variable,*

$$\nabla_y p_n(x) = p_n(x+y) - p_n(x), \qquad \nabla_y \overline{p}_n(x) = \overline{p}_n(x+y) - \overline{p}_n(x),$$

and $\nabla_j = \nabla_{\mathbf{e}_j}$.

- *There exists $c < \infty$ such that for all x, n, y,*

$$\left|\nabla_y p_n(x) - \nabla_y \overline{p}_n(x)\right| \leq \frac{c\,|y|}{n^{(d+2)/2}}.$$

- *If $\mathbb{E}[|X_1|^4] < \infty$ and all the third moments of X_1 vanish, there exists $c < \infty$ such that for all x, n, y,*

$$\left|\nabla_y p_n(x) - \nabla_y \overline{p}_n(x)\right| \leq \frac{c\,|y|}{n^{(d+3)/2}}.$$

Proof By the triangle inequality, it suffices to prove the result for $y = \mathbf{e}_j$, $j = 1, \ldots, d$. Let $\alpha = 1$ under the weaker assumptions and $\alpha = 2$ under the stronger assumptions. As in the proof of Theorem 2.3.5, we see that

$$\nabla_j p_n(x) = \nabla_j \overline{p}_n(x) + O(e^{-\beta n^{1/4}}) + \frac{1}{(2\pi)^d\, n^{d/2}} \int_{|\theta| \leq n^{1/8}} \left[e^{-\frac{i(x+\mathbf{e}_j)\cdot\theta}{\sqrt{n}}} - e^{-\frac{ix\cdot\theta}{\sqrt{n}}} \right] e^{-\frac{\theta\cdot\Gamma\theta}{2}}\, F_n(\theta)\, d\theta.$$

Note that

$$\left| e^{-\frac{i(x+\mathbf{e}_j)\cdot\theta}{\sqrt{n}}} - e^{-\frac{ix\cdot\theta}{\sqrt{n}}} \right| = \left| e^{-\frac{i\mathbf{e}_j\cdot\theta}{\sqrt{n}}} - 1 \right| \leq \frac{|\theta|}{\sqrt{n}},$$

and hence

$$\left| \int_{|\theta|\leq n^{1/8}} \left[e^{-\frac{i(x+\mathbf{e}_j)\cdot\theta}{\sqrt{n}}} - e^{-\frac{ix\cdot\theta}{\sqrt{n}}} \right] e^{-\frac{\theta\cdot\Gamma\theta}{2}} F_n(\theta)\, d\theta \right|$$
$$\leq \frac{1}{\sqrt{n}} \int_{|\theta|\leq n^{1/8}} |\theta|\, e^{-\frac{\theta\cdot\Gamma\theta}{2}} \, |F_n(\theta)|\, d\theta.$$

The estimate

$$\int_{|\theta|\leq n^{1/8}} |\theta|\, e^{-\frac{\theta\cdot\Gamma\theta}{2}} \, |F_n(\theta)|\, d\theta \leq \frac{c}{n^{\alpha/2}},$$

where $\alpha = 1$ under the weaker assumption and $\alpha = 2$ under the stronger assumption is done as in the previous theorem. □

The next theorem improves the LCLT by giving a better error bound for larger x. The basic strategy is to write $F_n(\theta)$ as the sum of a dominant term and an error term. This requires a stronger moment condition. If $\mathbb{E}[|X_1|^j] < \infty$, let f_j be the homogeneous polynomial of degree j defined in (2.20). Let

$$u_j(z) = \frac{1}{(2\pi)^d} \int_{\mathbb{R}^d} e^{-is\cdot z} f_j(s)\, e^{-\frac{s\cdot\Gamma s}{2}}\, ds. \tag{2.25}$$

Using standard properties of Fourier transforms, we can see that

$$u_j(z) = f_j^*(z)\, e^{-(z\cdot\Gamma^{-1}z)/2} = f_j^*(z)\, e^{-\frac{\mathcal{J}^*(z)^2}{2}} \tag{2.26}$$

for some jth degree polynomial f_j^* that depends only on the distribution of X_1.

Theorem 2.3.7 *Suppose that $p \in \mathcal{P}_d'$.*

- *If $\mathbb{E}[|X_1|^4] < \infty$, there exists $c < \infty$ such that*

$$\left| p_n(x) - \overline{p}_n(x) - \frac{u_3(x/\sqrt{n})}{n^{(d+1)/2}} \right| \leq \frac{c}{n^{(d+2)/2}}, \tag{2.27}$$

where u_3 is as defined in (2.25).

- *If* $\mathbb{E}[|X_1|^5] < \infty$ *and the third moments of* X_1 *vanish, there exists* $c < \infty$ *such that*

$$\left| p_n(x) - \overline{p}_n(x) - \frac{u_4(x/\sqrt{n})}{n^{(d+2)/2}} \right| \leq \frac{c}{n^{(d+3)/2}}, \tag{2.28}$$

where u_4 *is as defined in (2.25).*

If $k \geq 3$ *is a positive integer such that* $\mathbb{E}[|X_1|^k] < \infty$ *and* u_k *is as defined in (2.25), then there is a* $c(k)$ *such that*

$$|u_k(z)| \leq c(k)\,(|z|^k + 1)\, e^{-\frac{\mathcal{J}^*(z)^2}{2}}.$$

Moreover, if j *is a positive integer, there is a* $c(k,j)$ *such that if* D_j *is a* j*th-order derivative,*

$$|D_j u_k(z)| \leq c(k,j)\,|(|z|^{k+j} + 1)\, e^{-\frac{\mathcal{J}^*(z)^2}{2}}. \tag{2.29}$$

Proof Let $\alpha = 1$ under the weaker assumptions and $\alpha = 2$ under the stronger assumptions. As in Theorem 2.3.5,

$$n^{d/2}\,[p_n(x) - \overline{p}_n(x)] = O(e^{-\beta n^{1/4}}) + \frac{1}{(2\pi)^d} \int_{|\theta| \leq n^{1/8}} e^{-\frac{ix\cdot\theta}{\sqrt{n}}}\, e^{-\frac{\theta\cdot\Gamma\theta}{2}}\, F_n(\theta)\, d\theta. \tag{2.30}$$

Recalling (2.20), we can see that for $|\theta| \leq n^{1/8}$,

$$F_n(\theta) = \frac{f_{2+\alpha}(\theta)}{n^{\alpha/2}} + \frac{O(|\theta|^{3+\alpha})}{n^{(\alpha+1)/2}}.$$

Up to an error of $O(e^{-\beta n^{1/4}})$, the right-hand side of (2.30) equals

$$\frac{1}{(2\pi)^d} \int_{\mathbb{R}^d} e^{-\frac{ix\cdot\theta}{\sqrt{n}}}\, e^{-\frac{\theta\cdot\Gamma\theta}{2}}\, \frac{f_{2+\alpha}(\theta)}{n^{(\alpha/2)}}\, d\theta$$
$$+ \frac{1}{(2\pi)^d} \int_{|\theta| \leq n^{1/8}} e^{-\frac{ix\cdot\theta}{\sqrt{n}}}\, e^{-\frac{\theta\cdot\Gamma\theta}{2}} \left[F_n(\theta) - \frac{f_{2+\alpha}(\theta)}{n^{\alpha/2}} \right] d\theta.$$

The second integral can be bounded as before

$$\left| \int_{|\theta| \leq n^{1/8}} e^{-\frac{ix\cdot\theta}{\sqrt{n}}}\, e^{-\frac{\theta\cdot\Gamma\theta}{2}} \left[F_n(\theta) - \frac{f_{2+\alpha}(\theta)}{n^{\alpha/2}} \right] d\theta \right|$$
$$\leq \frac{c}{n^{(\alpha+1)/2}} \int_{\mathbb{R}^d} |\theta|^{3+\alpha}\, e^{-\frac{\theta\cdot\Gamma\theta}{2}}\, d\theta \leq \frac{c}{n^{(\alpha+1)/2}}.$$

The estimates on u_k and $D_j u_k$ follow immediately from (2.26). □

The next theorem is proved in the same way as Theorem 2.3.7 starting with (2.20), and we omit the proof. A special case of this theorem is (2.3). The theorem shows that (2.3) holds for all symmetric $p \in \mathcal{P}'_d$ with $\mathbb{E}[|X_1|^6] < \infty$. The results stated for $n \geq |x|^2$ are just restatements of Theorem 2.3.5.

Theorem 2.3.8 *Suppose that $p \in \mathcal{P}'_d$ and $k \geq 3$ is a positive integer such that $\mathbb{E}[|X_1|^{k+1}] < \infty$. There exists $c = c(k)$ such that*

$$\left| p_n(x) - \overline{p}_n(x) - \sum_{j=3}^{k} \frac{u_j(x/\sqrt{n})}{n^{(d+j-2)/2}} \right| \leq \frac{c}{n^{(d+k-1)/2}}, \tag{2.31}$$

where u_j are as defined in (2.25).

In particular, if $z = x/\sqrt{n}$,

$$\left|p_n(x) - \overline{p}_n(x)\right| \leq \frac{c}{n^{(d+1)/2}} \left[|z|^k \, e^{-\frac{\mathcal{J}^*(z)^2}{2}} + \frac{1}{n^{(k-2)/2}} \right], \quad n \leq |x|^2,$$

$$\left|p_n(x) - \overline{p}_n(x)\right| \leq \frac{c}{n^{(d+1)/2}}, \quad n \geq |x|^2.$$

If the third moments of X_1 vanish (e.g. if p is symmetric) then $u_3 \equiv 0$ and

$$\left|p_n(x) - \overline{p}_n(x)\right| \leq \frac{c}{n^{(d+2)/2}} \left[|z|^k \, e^{-\frac{\mathcal{J}^*(z)^2}{2}} + \frac{1}{n^{(k-3)/2}} \right], \quad n \leq |x|^2,$$

$$\left|p_n(x) - \overline{p}_n(x)\right| \leq \frac{c}{n^{(d+2)/2}}, \quad n \geq |x|^2.$$

Remark Theorem 2.3.8 gives improvements to Theorem 2.3.6. Assuming a sufficient number of moments on the increment distribution, one can estimate $\nabla_y p_n(x)$ up to an error of $O(n^{-(d+k-1)/2})$ by taking ∇_y of all the terms on the left-hand side of (2.31). These terms can be estimated using (2.29). This works for higher order differences as well.

The next theorem is the LCLT assuming only a finite second moment.

Theorem 2.3.9 *Suppose that $p \in \mathcal{P}'_d$. Then there exists a sequence $\delta_n \to 0$ such that for all n, x,*

$$|p_n(x) - \overline{p}_n(x)| \leq \frac{\delta_n}{n^{d/2}}. \tag{2.32}$$

Proof By Lemma 2.3.4, there exist c, ζ such that for all n, x and $r > 0$,

$$n^{d/2}|p_n(x) - \overline{p}_n(x)| \leq c\left[e^{-\zeta r^2} + \int_{\theta \leq r} |F_n(\theta)|\, d\theta\right]$$
$$\leq c\left[e^{-\zeta r^2} + r^d \sup_{\theta \leq r} |F_n(\theta)|\right].$$

We now refer to Lemma 2.3.3. Since $h(\theta) = o(|\theta|^2)$,

$$\lim_{n\to\infty} \sup_{|\theta|\leq r} |g(\theta, n)| = 0,$$

and hence

$$\lim_{n\to\infty} \sup_{|\theta|\leq r} |F_n(\theta)| = 0.$$

In particular, for all n sufficiently large,

$$n^{d/2}\,|p_n(x) - \overline{p}_n(x)| \leq 2\,c\,e^{-\zeta r^2}.$$

□

The next theorem improves on this for $|x|$ larger than $\sqrt{n}$. The proof uses an integration by parts. One advantage of this theorem is that it does not need any extra moment conditions. However, if we impose extra moment conditions we get a stronger result.

Theorem 2.3.10 *Suppose that $p \in \mathcal{P}'_d$. Then there exists a sequence $\delta_n \to 0$ such that for all n, x,*

$$|p_n(x) - \overline{p}_n(x)| \leq \frac{\delta_n}{|x|^2\, n^{(d-2)/2}}. \tag{2.33}$$

Moreover,

- *if $\mathbb{E}[|X_1|^3] < \infty$, then δ_n can be chosen $O(n^{-1/2})$;*
- *if $\mathbb{E}[|X_1|^4] < \infty$ and the third moments of X_1 vanish, then δ_n can be chosen $O(n^{-1})$.*

Proof If ψ_1, ψ_2 are C^2 functions on $\mathbb{R}^d$ with period 2π in each component, then it follows from the Green's theorem (integration by parts) that

$$\int_{[-\pi,\pi]^d} [\Delta\psi_1(\theta)]\;\psi_2(\theta)\,d\theta = \int_{[-\pi,\pi]^d} \psi_1(\theta)\;[\Delta\psi_2(\theta)]\;d\theta$$

(the boundary terms disappear by periodicity). Since $\Delta[e^{ix\cdot\theta}] = -|x|^2 e^{-x\cdot\theta}$, the inversion formula gives

$$p_n(-x) = \frac{1}{(2\pi)^d} \int_{[-\pi,\pi]^d} e^{ix\cdot\theta}\, \psi(\theta)\, d\theta$$
$$= -\frac{1}{|x|^2 (2\pi)^d} \int_{[-\pi,\pi]^d} e^{ix\cdot\theta}\, \Delta\psi(\theta)\, d\theta,$$

where $\psi(\theta) = \phi(\theta)^n$. Therefore,

$$\frac{|x|^2}{n} p_n(-x) = -\frac{1}{(2\pi)^d} \int_{[-\pi,\pi]^d} e^{ix\cdot\theta}\, \phi(\theta)^{n-2}$$
$$\times [(n-1)\,\lambda(\theta) + \phi(\theta)\, \Delta\phi(\theta)]\, d\theta,$$

where

$$\lambda(\theta) = \sum_{j=1}^{d} \left[\partial_j \phi(\theta)\right]^2.$$

The first and second derivatives of ϕ are uniformly bounded. Hence, by (2.14), we can see that

$$\left|\phi(\theta)^{n-2}\, [(n-1)\,\lambda(\theta) + \phi(\theta)\, \Delta\phi(\theta)]\right| \le c\,[1 + n|\theta|^2]\, e^{-n|\theta|^2 b} \le c\, e^{-\beta n|\theta|^2},$$

where $0 < \beta < b$. Hence, we can write

$$\frac{|x|^2}{n} p_n(-x) - O(e^{-\beta r^2}) = -\frac{1}{(2\pi)^d} \int_{|\theta| \le r/\sqrt{n}} e^{ix\cdot\theta}\, \phi(\theta)^{n-2}$$
$$\times [(n-1)\,\lambda(\theta) + \phi(\theta)\, \Delta\phi(\theta)]\, d\theta.$$

The usual change of variables shows that the right-hand side equals

$$-\frac{1}{(2\pi)^d\, n^{d/2}} \int_{|\theta| \le r} e^{iz\cdot\theta}\, \phi\left(\frac{\theta}{\sqrt{n}}\right)^{n-2}$$
$$\times \left[(n-1)\,\lambda\left(\frac{\theta}{\sqrt{n}}\right) + \phi\left(\frac{\theta}{\sqrt{n}}\right)\, \Delta\phi\left(\frac{\theta}{\sqrt{n}}\right)\right] d\theta,$$

where $z = x/\sqrt{n}$.

Note that

$$\Delta\left[e^{-\frac{\theta\cdot\Gamma\theta}{2}}\right] = e^{-\frac{\theta\cdot\Gamma\theta}{2}}\,[|\Gamma\theta|^2 - \mathrm{tr}(\Gamma)].$$

We define $\hat{F}_n(\theta)$ by

$$\begin{aligned}\phi\left(\frac{\theta}{\sqrt{n}}\right)^{n-2}&\left[(n-1)\,\lambda\left(\frac{\theta}{\sqrt{n}}\right)+\phi\left(\frac{\theta}{\sqrt{n}}\right)\,\Delta\phi\left(\frac{\theta}{\sqrt{n}}\right)\right]\\&= e^{-\frac{\theta\cdot\Gamma\theta}{2}}\left[|\Gamma\theta|^2 - \mathrm{tr}(\Gamma) - \hat{F}_n(\theta)\right].\end{aligned}$$

A straightforward calculation using the Green's theorem shows that

$$\begin{aligned}\overline{p}_n(-x) &= \frac{1}{(2\pi)^d\,n^{d/2}}\int_{\mathbb{R}^d} e^{i(x/\sqrt{n})\cdot\theta}\,e^{-\frac{\theta\cdot\Gamma\theta}{2}}\,d\theta\\&= -\frac{n}{|x|^2\,(2\pi)^2}\int_{\mathbb{R}^d} e^{i(x/\sqrt{n})\cdot\theta}\,\Delta[e^{-\frac{\theta\cdot\Gamma\theta}{2}}]\,d\theta.\end{aligned}$$

Therefore (with perhaps a different β),

$$\begin{aligned}\frac{|x|^2}{n}\,p_n(-x) &= \frac{|x|^2}{n}\,\overline{p}_n(-x) + O(e^{-\beta r^2})\\&\quad-\frac{1}{(2\pi)^d\,n^{d/2}}\int_{|\theta|\le r} e^{iz\cdot\theta}\,e^{-\frac{\theta\cdot\Gamma\theta}{2}}\,\hat{F}_n(\theta)\,d\theta. \qquad (2.34)\end{aligned}$$

The remaining task is to estimate $\hat{F}_n(\theta)$. Recalling the definition of $F_n(\theta)$ from Lemma 2.3.3, we can see that

$$\begin{aligned}\phi\left(\frac{\theta}{\sqrt{n}}\right)^{n-2}&\left[(n-1)\,\lambda\left(\frac{\theta}{\sqrt{n}}\right)+\phi\left(\frac{\theta}{\sqrt{n}}\right)\,\Delta\phi\left(\frac{\theta}{\sqrt{n}}\right)\right]\\&= e^{-\frac{\theta\cdot\Gamma\theta}{2}}\,[1+F_n(\theta)]\left[(n-1)\,\frac{\lambda(\theta/\sqrt{n})}{\phi(\theta/\sqrt{n})^2}+\frac{\Delta\phi(\theta/\sqrt{n})}{\phi(\theta/\sqrt{n})}\right].\end{aligned}$$

We make three possible assumptions:

- $p \in \mathcal{P}'_d$;
- $p \in \mathcal{P}'_d$ with $\mathbb{E}[|X_1|^3] < \infty$;
- $p \in \mathcal{P}'_d$ with $\mathbb{E}[|X_1|^4] < \infty$ and vanishing third moments.

We set $\alpha = 0, 1, 2$, respectively, in these three cases. Then we can write

$$\phi(\theta) = 1 - \frac{\theta\cdot\Gamma\theta}{2} + q_{2+\alpha}(\theta) + o(|\theta|^{2+\alpha}),$$

where $q_2 \equiv 0$ and q_3, q_4 are homogeneous polynomials of degrees 3 and 4, respectively. Because ϕ is $C^{2+\alpha}$ and we know the values of the derivatives at the origin, we can write

$$\partial_j \phi(\theta) = -\partial_j \frac{\theta \cdot \Gamma\theta}{2} + \partial_j q_{2+\alpha}(\theta) + o(|\theta|^{1+\alpha}),$$

$$\partial_{jj} \phi(\theta) = -\partial_{jj} \frac{\theta \cdot \Gamma\theta}{2} + \partial_{jj} q_{2+\alpha}(\theta) + o(|\theta|^{\alpha}).$$

Using this, we see that

$$\frac{\sum_{j=1}^{d} \left[\partial_j \phi(\theta)\right]^2}{\phi(\theta)^2} = |\Gamma\theta|^2 + \tilde{q}_{2+\alpha}(\theta) + o(|\theta|^{2+\alpha}),$$

$$\frac{\Delta\phi(\theta)}{\phi(\theta)} = -\mathrm{tr}(\Gamma) + \hat{q}_{\alpha}(\theta) + o(|\theta|^{\alpha}),$$

where $\tilde{q}_{2+\alpha}$ is a homogeneous polyonomial of degree $2+\alpha$ with $\tilde{q}_2 \equiv 0$, and $\hat{q}_\alpha$ is a homogeneous polynomial of degree α with $\hat{q}_0 = 0$. Therefore, for $|\theta| \leq n^{1/8}$,

$$(n-1)\,\frac{\lambda(\theta/\sqrt{n})}{\phi(\theta/\sqrt{n})^2} + \frac{\Delta\phi(\theta/\sqrt{n})}{\phi(\theta/\sqrt{n})}$$
$$= |\Gamma\theta|^2 - \mathrm{tr}(\Gamma) + \frac{\tilde{q}_{2+\alpha}(\theta) + \hat{q}_\alpha(\theta)}{n^{\alpha/2}} + o\left(\frac{|\theta|^{\alpha} + |\theta|^{\alpha+2}}{n^{\alpha/2}}\right),$$

which establishes that for $\alpha = 1, 2$

$$|\hat{F}_n(\theta)| = O\left(\frac{1 + |\theta|^{2+\alpha}}{n^{\alpha/2}}\right), \qquad |\theta| \leq n^{1/16},$$

and for $\alpha = 0$, for each $r < \infty$,

$$\lim_{n\to\infty} \sup_{|\theta|\leq r} |\hat{F}_n(\theta)| = 0.$$

The remainder of the argument follows the proofs of Theorems 2.3.5 and 2.3.9. For $\alpha = 1, 2$, we can choose $r = n^{7/16}$ in (2.34) while for $\alpha = 0$, we choose r independent of n and then let $r \to \infty$.

□

2.3.1 Exponential moments

The estimation of probabilities for atypical values can be done more accurately for random walks whose increment distribution has an exponential moment. In this section we prove the following.

Theorem 2.3.11 *Suppose that $p \in \mathcal{P}'_d$ such that for some $b > 0$,*

$$\mathbb{E}\left[e^{b|X_1|}\right] < \infty. \tag{2.35}$$

Then there exists $\rho > 0$ such that for all $n \geq 0$ and all $x \in \mathbb{Z}^d$ with $|x| < \rho\, n$,

$$p_n(x) = \overline{p}_n(x) \exp\left\{O\left(\frac{1}{\sqrt{n}} + \frac{|x|^3}{n^2}\right)\right\}.$$

Moreover, if all the third moments of X_1 vanish,

$$p_n(x) = \overline{p}_n(x) \exp\left\{O\left(\frac{1}{n} + \frac{|x|^4}{n^3}\right)\right\}.$$

♣ Note that the conclusion of the theorem can be written

$$|p_n(x) - \overline{p}_n(x)| \leq c\,\overline{p}_n(x) \left[\frac{1}{n^{\alpha/2}} + \frac{|x|^{2+\alpha}}{n^{1+\alpha}}\right], \quad |x| \leq n^{\frac{1+\alpha}{2+\alpha}},$$

$$|p_n(x) - \overline{p}_n(x)| \leq \overline{p}_n(x) \exp\left\{O\left(\frac{|x|^{2+\alpha}}{n^{1+\alpha}}\right)\right\}, \quad |x| \geq n^{\frac{1+\alpha}{2+\alpha}}$$

where $\alpha = 2$ if the third moments vanish and $\alpha = 1$ otherwise. In particular, if x_n is a sequence of points in $\mathbb{Z}^d$, then as $n \to \infty$,

$$p_n(x_n) \sim \overline{p}_n(x_n) \quad \text{if} \quad |x_n| = o(n^{\beta}),$$

$$p_n(x_n) \asymp \overline{p}_n(x_n) \quad \text{if} \quad |x_n| = O(n^{\beta}),$$

where $\beta = 2/3$ if $\alpha = 1$ and $\beta = 3/4$ if $\alpha = 2$.

Theorem 2.3.11 will follow from a stronger result (Theorem 2.3.12). Before stating it, we introduce some additional notation and make several preliminary observations. Let $p \in \mathcal{P}'_d$ have characteristic function ϕ and covariance matrix Γ, and assume that p satisfies (2.35). If the third moments of p vanish, we let $\alpha = 2$; otherwise, $\alpha = 1$. Let M denote the moment-generating function,

$$M(b) = \mathbb{E}[e^{b \cdot X}] = \phi(-ib),$$

which by (2.35) is well defined in a neighborhood of the origin in $\mathbb{C}^d$. Moreover, we can find $C < \infty, \epsilon > 0$ such that

$$\mathbb{E}\left[|X|^4 e^{|b \cdot X|}\right] \le C, \quad |b| < \epsilon. \tag{2.36}$$

In particular, there is a uniform bound in this neighborhood on all the derivatives of M of order at most four. (A (finite) number of times in this section we will say that something holds for all b in a neighborhood of the origin. At the end, one should take the intersection of all such neighborhoods.) Let $L(b) = \log M(b), L(i\theta) = \log \phi(\theta)$. Then, in a neighborhood of the origin we have

$$M(b) = 1 + \frac{b \cdot \Gamma b}{2} + O\left(|b|^{\alpha+2}\right), \quad \nabla M(b) = \Gamma b + O\left(|b|^{\alpha+1}\right),$$

$$L(b) = \frac{b \cdot \Gamma b}{2} + O(|b|^{\alpha+2}), \quad \nabla L(b) = \frac{\nabla M(b)}{M(b)} = \Gamma b + O(|b|^{\alpha+1}). \tag{2.37}$$

For $|b| < \epsilon$, let $p_b \in \mathcal{P}_d^*$ be the probability measure

$$p_b(x) = \frac{e^{b \cdot x}\, p(x)}{M(b)}, \tag{2.38}$$

and let $\mathbb{P}_b, \mathbb{E}_b$ denote probabilities and expectations associated to a random walk with increment distribution p_b. Note that

$$\mathbb{P}_b\{S_n = x\} = e^{b \cdot x}\, M(b)^{-n}\, \mathbb{P}\{S_n = x\}. \tag{2.39}$$

The mean of p_b is equal to

$$m_b = \frac{\mathbb{E}[X\, e^{b \cdot X}]}{\mathbb{E}[e^{b \cdot X}]} = \nabla L(b).$$

A standard "large deviations" technique for understanding $\mathbb{P}\{S_n = x\}$ is to study $\mathbb{P}_b\{S_n = x\}$ where b is chosen so that $m_b = x/n$. We will apply this technique in the current context. Since Γ is an invertible matrix, (2.37) implies that $b \mapsto \nabla L(b)$ maps $\{|b| < \epsilon\}$ one-to-one and onto a neighborhood of the origin, where $\epsilon > 0$ is sufficiently small. In particular, there is a $\rho > 0$ such that for all $w \in \mathbb{R}^d$ with $|w| < \rho$, there is a unique $|b_w| < \epsilon$ with $\nabla L(b_w) = w$.

♣ One could think of the "tilting" procedure of (2.38) as "weighting by a martingale." Indeed, it is easy to see that for $|b| < \epsilon$, the process

$$N_n = M(b)^{-n} \exp\{bS_n\}$$

is a martingale with respect to the filtration $\{\mathcal{F}_n\}$ of the random walk. The measure $\mathbb{P}_b$ is obtained by weighting by N_n. More precisely, if E is an $\mathcal{F}_n$-measurable event, then

$$\mathbb{P}_b(E) = \mathbb{E}\,[N_n\, 1_E]\,.$$

The martingale property implies the consistency of this definition. Under the measure $\mathbb{P}_b$, S_n has the distribution of a random walk with increment distribution p_b and mean m_b. For fixed n, x we choose $m_b = x/n$ so that x is a typical value for S_n under $\mathbb{P}_b$. This construction is a random walk analogue of the Girsanov transformation for Brownian motion.

Let ϕ_b denote the characteristic function of p_b which we can write as

$$\phi_b(\theta) = \mathbb{E}_b[e^{i\theta\cdot X}] = \frac{M\,(i\theta + b)}{M\,(b)}. \tag{2.40}$$

Then, there is a neighborhood of the origin such that for all b, θ in the neighborhood, we can expand $\log \phi_b$ as

$$\log \phi_b(\theta) = i\,m_b\cdot\theta - \frac{\theta\cdot\Gamma_b\theta}{2} + f_{3,b}(\theta) + h_{4,b}(\theta). \tag{2.41}$$

Here, Γ_b is the covariance matrix for the increment distribution p_b, and $f_{3,b}(\theta)$ is the homogeneous polynomial of degree three

$$f_{3,b}(\theta) = -\frac{i}{6}\left[\mathbb{E}_b[(\theta\cdot X)^3] + 2\,(\mathbb{E}_b[\theta\cdot X])^3\right].$$

Due to (2.36), the coefficients for the third-order Taylor polynomial of $\log\phi_b$ are all differentiable in b with bounded derivatives in the same neighborhood of the origin. In particular, we conclude that

$$|f_{3,b}(\theta)| \le c\,|b|^{\alpha-1}\,|\theta|^3, \quad |b|, |\theta| < \epsilon.$$

To see this, if $\alpha = 1$, use the boundedness of the first and third moments. If $\alpha = 2$, note that $f_{3,0}(\theta) = 0$, $\theta \in \mathbb{R}^d$, and use the fact that the first and third moments have bounded derivatives as functions of b. Similarly,

$$\Gamma_b = \frac{\mathbb{E}[XX^T e^{b\cdot X}]}{M\,(b)} = \Gamma + O(|b|^\alpha),$$

The error term $h_{4,b}$ is bounded by

$$h_{4,b}(\theta) \le c\,|\theta|^4, \quad |b|, |\theta| < \epsilon.$$

Note that due to (2.37) (and invertibility of Γ) we have both $|b_w| = O(|w|)$ and $|w| = O(|b_w|)$. Combining this with the above observations, we can conclude that

$$\begin{aligned} m_b &= \Gamma b + O(|w|^{1+\alpha}), \\ b_w &= \Gamma^{-1} w + O(|w|^{1+\alpha}), &(2.42) \\ \det \Gamma_{b_w} &= \det \Gamma + O(|w|^{\alpha}) = \det \Gamma + O(|b_w|^{\alpha}), &(2.43) \\ L(b_w) &= \frac{b_w \cdot \Gamma b_w}{2} + O(|b_w|^{2+\alpha}) = \frac{w \cdot \Gamma^{-1} w}{2} + O(|w|^{2+\alpha}). &(2.44) \end{aligned}$$

By examining the proof of (2.13) one can find a (perhaps different) $\delta > 0$, such that for all $|b| \leq \epsilon$ and all $\theta \in [-\pi, \pi]^d$,

$$|e^{-im_b \cdot \theta}\, \phi_b(\theta)| \leq 1 - \delta\, |\theta|^2.$$

(For small θ use the expansion of ϕ near 0; otherwise consider $\max_{\theta,b} |e^{-im_b \cdot \theta}\, \phi_b(\theta)|$ where the maximum is taken over all such $\theta \in \{z \in [-\pi, \pi]^d : |z| \geq \epsilon\}$ and all $|b| \leq \epsilon$.)

Theorem 2.3.12 *Suppose that p satisfies the assumptions of Theorem 2.3.11, and let L, b_w be defined as above. Then, there exists $c < \infty$ and $\rho > 0$ such that the following holds. Suppose that $x \in \mathbb{Z}^d$ with $|x| \leq \rho n$ and $b = b_{x/n}$. Then*

$$\left|(2\pi \det \Gamma_b)^{d/2}\, n^{d/2}\, \mathbb{P}_b\{S_n = x\} - 1\right| \leq \frac{c\,(|x|^{\alpha-1} + \sqrt{n})}{n^{(\alpha+1)/2}}. \tag{2.45}$$

In particular,

$$p_n(x) = \mathbb{P}\{S_n = x\} = \overline{p}_n(x) \exp\left\{O\left(\frac{1}{n^{\alpha/2}} + \frac{|x|^{2+\alpha}}{n^{1+\alpha}}\right)\right\}. \tag{2.46}$$

Proof of ((2.46) given (2.45)) We can write (2.45) as

$$\mathbb{P}_b\{S_n = x\} = \frac{1}{(2\pi \det \Gamma_b)^{d/2}\, n^{d/2}} \left[1 + O\left(\frac{|x|^{\alpha-1}}{n^{(\alpha+1)/2}} + \frac{1}{n^{\alpha/2}}\right)\right].$$

By (2.39),

$$\begin{aligned} p_n(x) &= \mathbb{P}\{S_n = x\} \\ &= M\,(b)^n\, e^{-b \cdot x}\, \mathbb{P}_b\{S_n = x\} = \exp\{n\, L(b) - b \cdot x\}\, \mathbb{P}_b\{S_n = x\}. \end{aligned}$$

From (2.43), we see that

$$(\det \Gamma_b)^{-d/2} = (\det \Gamma)^{-d/2} \left[1 + O\left(\frac{|x|^\alpha}{n^\alpha}\right)\right],$$

and due to (2.44), we have

$$\left| n\, L(b) - \frac{x \cdot \Gamma^{-1} x}{2n} \right| \le c\, \frac{|x|^{2+\alpha}}{n^{1+\alpha}}.$$

Applying (2.42), we see that

$$b \cdot x = \left[\Gamma^{-1}\left(\frac{x}{n}\right) + O\left(\frac{|x|^{\alpha+1}}{n^{\alpha+1}}\right)\right] \cdot x = \frac{x \cdot \Gamma^{-1} x}{n} + O\left(\frac{|x|^{\alpha+2}}{n^{\alpha+1}}\right).$$

Therefore,

$$\exp\{n\, L(b) - b \cdot x\} = \exp\left\{-\frac{x \cdot \Gamma^{-1} x}{2n}\right\} \exp\left\{O\left(\frac{|x|^{2+\alpha}}{n^{1+\alpha}}\right)\right\}.$$

Combining these and recalling the definition of $\overline{p}_n(x)$ we get,

$$p_n(x) = \overline{p}_n(x) \exp\left\{O\left(\frac{|x|^{2+\alpha}}{n^{1+\alpha}}\right)\right\}.$$

□

Therefore, it suffices to prove (2.45). The argument uses an LCLT for probability distributions on $\mathbb{Z}^d$ with nonzero mean. Suppose that $K < \infty$ and X is a random variable in $\mathbb{Z}^d$ with mean $m \in \mathbb{R}^d$, covariance matrix Γ, and $\mathbb{E}[|X|^4] \le K$. Let ψ be the characteristic function of X. Then, there exist ϵ, C, depending only on K, such that for $|\theta| < \epsilon$,

$$\left| \log \psi(\theta) - \left[im \cdot \theta + \frac{\theta \cdot \Gamma\theta}{2} + f_3(\theta) \right] \right| \le C\, |\theta|^4. \tag{2.47}$$

where the term $f_3(\theta)$ is a homogeneous polynomial of degree 3. Let us write K_3 for the smallest number such that $|f_3(\theta)| \le K_3 |\theta|^3$ for all θ. Note that there exist uniform bounds for m, Γ and K_3 in terms of K. Moreover, if $\alpha = 2$ and f_3 corresponds to p_b, then $|K_3| \le c\, |b|$. The next proposition is proved in the same way as Theorem 2.3.5, taking some extra care in obtaining uniform bounds. The relation (2.45) then follows from this proposition and the bound $K_3 \le c\, (|x|/n)^{\alpha-1}$.

Proposition 2.3.13 *For every $\delta > 0, K < \infty$, there is a c such that the following holds. Let p be a probability distribution on $\mathbb{Z}^d$ with $\mathbb{E}[|X|^4] \leq K$. Let $m, \Gamma, C, \epsilon, \psi, K_3$ be as in the previous paragraph. Moreover, assume that*

$$|e^{-im\cdot\theta}\,\psi(\theta)| \leq 1 - \delta\,|\theta|^2, \quad \theta \in [-\pi,\pi]^d.$$

Suppose that $X_1, X_2, \ldots$ are independent random variables with distribution p and $S_n = X_1 + \cdots + X_n$. Then, if $nm \in \mathbb{Z}^d$,

$$\left|(2\pi n \det \Gamma)^{d/2}\,\mathbb{P}\{S_n = nm\} - 1\right| \leq \frac{c[K_3\sqrt{n} + 1]}{n}.$$

Remark The error term indicates the existence of two different regimes: $K_3 \leq n^{-1/2}$ and $K_3 \geq n^{-1/2}$.

Proof We fix δ, K and allow all constants in this proof to depend only on δ and K. Proposition 2.2.2 implies that

$$\mathbb{P}\{S_n = nm\} = \frac{1}{(2\pi)^d}\int_{[-\pi,\pi]^d} [e^{-im\cdot\theta}\,\psi(\theta)]^n\,d\theta.$$

The uniform upper bound on $\mathbb{E}[|X|^4]$ implies uniform upper bounds on the lower moments. In particular, $\det\Gamma$ is uniformly bounded and hence it suffices to find n_0 such that the result holds for $n \geq n_0$. Also, observe that (2.47) holds with a uniform C from which we conclude

$$\left|n \log\psi\left(\frac{\theta}{\sqrt{n}}\right) - i\sqrt{n}\,(m\cdot\theta) - \frac{\theta\cdot\Gamma\theta}{2} - nf_3\left(\frac{\theta}{\sqrt{n}}\right)\right| \leq C\,\frac{|\theta|^4}{n}.$$

In addition, we have $|nf_3(\theta/\sqrt{n})| \leq K_3\,|\theta|^3/\sqrt{n}$. The proof proceeds as does the proof of Theorem 2.3.5; the details are left to the reader. □

2.4 Some corollaries of the LCLT

Proposition 2.4.1 *If $p \in \mathcal{P}'$ with bounded support, there is a c such that*

$$\sum_{z\in\mathbb{Z}^d} |p_n(z) - p_n(z+y)| \leq c\,|y|\,n^{-1/2}.$$

Proof By the triangle inequality, it suffices to prove the result for $y = \mathbf{e} = \mathbf{e}_j$. Let $\delta = 1/2d$. By Theorem 2.3.6,

$$p_n(z+\mathbf{e}) - p_n(z) = \nabla_j \bar{p}_n(z) + O\left(\frac{1}{n^{(d+2)/2}}\right).$$

Also, Corollary A.2.7 shows that

$$\sum_{|z| \geq n^{(1/2)+\delta}} |p_n(z) - p_n(z+\mathbf{e})| \leq \sum_{|z| \geq n^{(1/2)+\delta}} [p_n(z) + p_n(z+\mathbf{e})] = o(n^{-1/2}).$$

But,

$$\sum_{|z| \leq n^{(1/2)+\delta}} |p_n(z) - p_n(z+\mathbf{e})|$$

$$\leq \sum_{z \in \mathbb{Z}^d} |\overline{p}_n(z) - \overline{p}_n(z+\mathbf{e})| + \sum_{|z| \leq n^{(1/2)+\delta}} O\left(\frac{1}{n^{(d+2)/2}}\right)$$

$$\leq O(n^{-1/2}) + \sum_{z \in \mathbb{Z}^d} |\nabla_j \overline{p}_n(z)|.$$

A straightforward estimate which we omit gives

$$\sum_{z \in \mathbb{Z}^d} |\nabla_j \overline{p}_n(z)| = O(n^{-1/2}).$$

□

The last proposition holds with much weaker assumptions on the random walk. Recall that $\mathcal{P}^*$ is the set of increment distributions p with the property that for each $x \in \mathbb{Z}^d$, there is an N_x such that $p_n(x) > 0$ for all $n \geq N_x$.

Proposition 2.4.2 *If $p \in \mathcal{P}^*$, there is a c such that*

$$\sum_{z \in \mathbb{Z}^d} |p_n(z) - p_n(z+y)| \leq c\,|y|\,n^{-1/2}.$$

Proof In Exercise 1.3 it was shown that we can write

$$p = \epsilon\, q + (1-\epsilon) q',$$

where $q \in \mathcal{P}'$ with bounded support and $q' \in \mathcal{P}^*$. By considering the process of first choosing q or q' and then doing the jump, we can see that

$$p_n(x) = \sum_{j=0}^{n} \binom{n}{j} \epsilon^j\,(1-\epsilon)^{n-j} \sum_{z \in \mathbb{Z}^d} q_j(x-z)\, q'_{n-j}(z). \tag{2.48}$$

Therefore,

$$\sum_{x\in\mathbb{Z}^d} |p_n(x) - p_n(x+y)|$$
$$\le \sum_{j=0}^{n} \binom{n}{j} \epsilon^j \, (1-\epsilon)^{n-j} \sum_{x\in\mathbb{Z}^d} q'_{n-j}(x) \sum_{z\in\mathbb{Z}^d} |q_j(x-z) - q_j(x+y-z)|.$$

We split the first sum into the sum over $j < (\epsilon/2)n$ and $j \ge (\epsilon/2)n$. Standard exponential estimates for the binomial (see Lemma A.2.8) give

$$\sum_{j<(\epsilon/2)n} \binom{n}{j} \epsilon^j \, (1-\epsilon)^{n-j} \sum_{x\in\mathbb{Z}^d} q'_{n-j}(x) \sum_{z\in\mathbb{Z}^d} |q_j(x-z) - q_j(x+y-z)|$$
$$\le 2 \sum_{j<(\epsilon/2)n} \binom{n}{j} \epsilon^j \, (1-\epsilon)^{n-j} = O(e^{-\alpha n}),$$

for some $\alpha = \alpha(\epsilon) > 0$. By Proposition 2.4.1,

$$\sum_{j\ge(\epsilon/2)n} \binom{n}{j} \epsilon^j \, (1-\epsilon)^{n-j} \sum_{x\in\mathbb{Z}^d} q'_{n-j}(x) \sum_{z\in\mathbb{Z}^d} |q_j(x-z) - q_j(x+y-z)|$$
$$\le c\, n^{-1/2} \, |y| \sum_{j\ge(\epsilon/2)n} \binom{n}{j} \epsilon^j \, (1-\epsilon)^{n-j} \sum_{x\in\mathbb{Z}^d} q'_{n-j}(x) \le c\, n^{-1/2} \, |y|.$$

□

The last proposition has the following useful lemma as a corollary. Since this is essentially a result about Markov chains in general, we leave the proof to the appendix; see Theorem A.4.5.

Lemma 2.4.3 *Suppose that $p \in \mathcal{P}_d^*$. There is a $c < \infty$ such that if $x, y \in \mathbb{Z}^d$, we can define S_n, S_n^* on the same probability space such that: S_n has the distribution of a random walk with increment p with $S_0 = x$; S_n^* has the distribution of a random walk with increment p with $S_0 = y$; and for all n,*

$$\mathbb{P}\{S_m = S_m^* \text{ for all } m \ge n\} \ge 1 - \frac{c\,|x-y|}{\sqrt{n}}.$$

♣ While the proof of this last lemma is somewhat messy to write out in detail, there really is not a lot of content to it once we have Proposition 2.4.2.

Suppose that p, q are two probability distributions on $\mathbb{Z}^d$ with

$$\frac{1}{2}\sum_{z\in\mathbb{Z}^d}|p(z)-q(z)|=\epsilon.$$

Then there is an easy way to define random variables X, Y on the same probability space such that X has distribution p, Y has distribution q, and $\mathbb{P}\{X\neq Y\}=\epsilon$. Indeed, if we let $f(z)=\min\{p(z),q(z)\}$ we can let the probability space be $\mathbb{Z}^d\times\mathbb{Z}^d$ and define μ by

$$\mu(z,z)=f(z)$$

and for $x\neq y$,

$$\mu(x,y)=\epsilon^{-1}\,[p(x)-f(x)]\,[q(y)-f(y)].$$

If we let $X(x,y)=x$, $Y(x,y)=y$, it is easy to check that the marginal of X is p, the marginal of Y is q, and $\mathbb{P}\{X=Y\}=1-\epsilon$. The more general fact is not much more complicated than this.

Proposition 2.4.4 *Suppose that $p\in\mathcal{P}_d^*$. There is a $c<\infty$ such that for all n, x,*

$$p_n(x)\leq\frac{c}{n^{d/2}}. \tag{2.49}$$

Proof If $p\in\mathcal{P}_d'$ with bounded support, this follows immediately from (2.22). For general $p\in\mathcal{P}_d^*$, write $p=\epsilon\,q+(1-\epsilon)\,q'$ with $q\in\mathcal{P}_d', q'\in\mathcal{P}_d^*$ as in the proof of Proposition 2.22. Then $p_n(x)$ is as in (2.48). The sum over $j<(\epsilon/2)n$ is $O(e^{-\alpha n})$ and for $j\geq(\epsilon/2)n$, we have the bound $q_j(x-z)\leq c\,n^{-d/2}$. □

The CLT implies that it takes $O(n^2)$ steps to go distance n. This proposition gives some bounds on large deviations for the number of steps.

Proposition 2.4.5 *Suppose that S is a random walk with increment distribution $p\in\mathcal{P}_d$ and let*

$$\tau_n=\min\{k:|S_k|\geq n\},\qquad \xi_n=\min\{k:\mathcal{J}^*(S_k)\geq n\}.$$

There exist $t>0$ and $c<\infty$ such that for all n and all $r>0$,

$$\mathbb{P}\{\tau_n\leq rn^2\}+\mathbb{P}\{\xi_n\leq rn^2\}\leq c\,e^{-t/r}, \tag{2.50}$$

$$\mathbb{P}\{\tau_n\geq rn^2\}+\mathbb{P}\{\xi_n\geq rn^2\}\leq c\,e^{-rt}. \tag{2.51}$$

Proof There exists a $\tilde{c}$ such that $\xi_{\tilde{c}n} \le \tau_n \le \xi_{n/\tilde{c}}$ so it suffices to prove the estimates for τ_n. It also suffices to prove the result for n sufficiently large. The CLT implies that there is an integer k such that for all n sufficiently large,

$$\mathbb{P}\{|S_{kn^2}| \ge 2n\} \ge \frac{1}{2}.$$

By the strong Markov property, this implies that for all l

$$\mathbb{P}\{\tau_n > kn^2 + l \mid \tau_n > l\} \le \frac{1}{2},$$

and hence

$$\mathbb{P}\{\tau_n > jkn^2\} \le (1/2)^j = e^{-j\log 2} = e^{-jk(\log 2/k)}.$$

This gives (2.51). The estimate (2.50) on τ_n can be written as

$$\mathbb{P}\left\{\max_{1\le j\le rn^2} |S_j| \ge n\right\} = \mathbb{P}\left\{\max_{1\le j\le rn^2} |S_j| \ge (1/\sqrt{r})\sqrt{rn^2}\right\} \le c\,e^{-t/r},$$

which follows from (2.7). □

♣ The upper bound (2.51) for τ_n does not need any assumptions on the distribution of the increments other than that they be nontrivial; see Exercise 2.7.

Theorem 2.3.10 implies that for all $p \in \mathcal{P}'_d$, $p_n(x) \le c\,n^{-d/2}\,(\sqrt{n}/|x|)^2$. The next proposition extends this to real $r > 2$ under the assumption that $\mathbb{E}[|X_1|^r] < \infty$.

Proposition 2.4.6 *Suppose that $p \in \mathcal{P}^*_d$. There is a c such that for all n, x,*

$$p_n(x) \le \frac{c}{n^{d/2}} \max_{0\le j\le n} \mathbb{P}\left\{|S_j| \ge |x|/2\right\}.$$

In particular, if $r > 2$, $p \in \mathcal{P}'_d$, and $\mathbb{E}[|X_1|^r] < \infty$, then there exists $c < \infty$ such that for all n, x,

$$p_n(x) \le \frac{c}{n^{d/2}}\left(\frac{\sqrt{n}}{|x|}\right)^r. \tag{2.52}$$

Proof Let $m = n/2$ if n is even and $m = (n+1)/2$ if n is odd. Then,

$$\{S_n = x\} = \{S_n = x, |S_m| \ge |x|/2\} \;\cup\; \{S_n = x, |S_n - S_m| \ge |x|/2\}.$$

Hence, it suffices to estimate the probabilities of the events on the right-hand side. Using (2.49) we get

$$\begin{aligned}\mathbb{P}\{S_n = x, |S_m| \geq |x|/2\} &= \mathbb{P}\{|S_m| \geq |x|/2\}\, \mathbb{P}\{S_n = x \mid |S_m| \geq |x|/2\} \\ &\leq \mathbb{P}\{|S_m| \geq |x|/2\} \left[\sup_y p_{n-m}(y,x)\right] \\ &\leq c\, n^{-d/2}\, \mathbb{P}\{|S_m| \geq |x|/2\}.\end{aligned}$$

The other probability can be estimated similarly since

$$\mathbb{P}\{S_n = x, |S_n - S_m| \geq |x|/2\} = \mathbb{P}\{S_n = x, |S_{n-m}| \geq |x|/2\}.$$

We claim that if $p \in \mathcal{P}_d'$, $r \geq 2$, and $\mathbb{E}[|X_1|^r] < \infty$, then there is a c such that $\mathbb{E}[|S_n|^r] \leq c\, n^{r/2}$. Once we have this, the Chebyshev inequality gives for $m \leq n$,

$$\mathbb{P}\{|S_m| \geq |x|\} \leq \frac{c\, n^{r/2}}{|x|^r}.$$

The claim is easier when r is an even integer (for then we can estimate the expectation by expanding $(X_1 + \cdots + X_n)^r$), but we give a proof for all $r \geq 2$. Without loss of generality, assume that $d = 1$. For a fixed n, define

$$T_1 = \tilde{T}_1 = \min\{j : |S_j| \geq c_1\sqrt{n}\},$$

and for $l > 1$,

$$\tilde{T}_l = \min\left\{j > \tilde{T}_{l-1} : \left|S_j - S_{\tilde{T}_{l-1}}\right| \geq c_1\sqrt{n}\right\}, \qquad T_l = \tilde{T}_l - \tilde{T}_{l-1},$$

where c_1 is chosen sufficiently large that

$$\mathbb{P}\{T_1 > n\} \geq \frac{1}{2}.$$

The existence of such a c_1 follows from (2.6) applied with $k = 1$.

Let $Y_1 = \left|S_{T_1}\right|$ and for $l > 1$, $Y_l = \left|S_{\tilde{T}_l} - S_{\tilde{T}_{l-1}}\right|$. Note that (T_1, Y_1), $(T_2, Y_2), (T_3, Y_3), \ldots$ are independent, identically distributed random variables taking values in $\{1, 2, \ldots\} \times \mathbb{R}$. Let ξ be the smallest $l \geq 1$ such that $T_l > n$.

Then, one can readily check from the triangle inequality that

$$\begin{aligned}|S_n| &\le Y_1 + Y_2 + \cdots + Y_{\xi-1} + c_1\sqrt{n} \\ &= c_1\sqrt{n} + \sum_{l=1}^{\infty} \hat{Y}_l,\end{aligned}$$

where $\hat{Y}_l = Y_l\, 1\{T_l \le n\}\, 1\{\xi > l-1\}$. Note that

$$\begin{aligned}\mathbb{P}\{Y_1 \ge c_1\sqrt{n} + t; T_1 \le n\} &\le \mathbb{P}\{|X_j| \ge t \text{ for some } 1 \le j \le n\} \\ &\le n\,\mathbb{P}\{|X_1| \ge t\}.\end{aligned}$$

Letting $Z = |X_1|$, we get

$$\begin{aligned}\mathbb{E}[\hat{Y}_1^r] = \mathbb{E}[Y_1^r; T_l \le n] &= c\int_0^\infty s^{r-1}\,\mathbb{P}\{Y_1 \ge s; T_1 \le n\}\,ds \\ &\le c\left[n^{r/2} + \int_{(c_1+1)\sqrt{n}}^\infty s^{r-1}\,n\,\mathbb{P}\{Z \ge s - c_1\sqrt{n}\}\,ds\right] \\ &= c\left[n^{r/2} + \int_{\sqrt{n}}^\infty (s+\sqrt{n})^{r-1}\,n\,\mathbb{P}\{Z \ge s\}\,ds\right] \\ &\le c\left[n^{r/2} + 2^{r-1}\int_{\sqrt{n}}^\infty s^{r-1}\,n\,\mathbb{P}\{Z \ge s\}\,ds\right] \\ &\le c\left[n^{r/2} + 2^{r-1}\,n\,\mathbb{E}[Z^r]\right] \le c\,n^{r/2}.\end{aligned}$$

For $l > 1$,

$$\mathbb{E}[\hat{Y}_l^r] \le \mathbb{P}\{\xi > l-1\}\,\mathbb{E}[Y_l^r\,1\{T_l \ge n\} \mid \xi > l-1] = \left(\frac{1}{2}\right)^{l-1}\mathbb{E}[\hat{Y}_1^r].$$

Therefore,

$$\begin{aligned}\mathbb{E}\left[(\hat{Y}_1 + \hat{Y}_2 + \cdots)^r\right] &= \lim_{l\to\infty}\mathbb{E}\left[(\hat{Y}_1 + \cdots + \hat{Y}_l)^r\right] \\ &\le \lim_{l\to\infty}\left[\mathbb{E}[\hat{Y}_1^r]^{1/r} + \cdots + \mathbb{E}[\hat{Y}_l^r]^{1/r}\right]^r \\ &= \mathbb{E}[\hat{Y}_1^r]\left[\sum_{l=1}^\infty \left(\frac{1}{2}\right)^{(l-1)/r}\right]^r = c\,\mathbb{E}[\hat{Y}_1^r].\end{aligned}$$

□

2.5 LCLT – combinatorial approach

In this section, we give another proof of an LCLT with estimates for one-dimensional simple random walk, both discrete and continuous time, using an elementary combinatorial approach. Our results are no stronger than those derived earlier, and this section is not needed for the remainder of the book, but it is interesting to see how much can be derived by simple counting methods. While we focus on simple random walk, extensions to $p \in \mathcal{P}_d$ are straightforward using (1.2). Although the arguments are relatively elementary, they do require a lot of calculation and estimation. Here is a basic outline:

- Establish the result for discrete time random walk by exact counting of paths. Along the way we will prove Stirling's formula.
- Prove an LCLT for Poisson random variables and use it to derive the result for one-dimensional continuous-time walks. (A result for d-dimensional continuous-time simple random walks follows immediately.)

We could continue this approach and prove an LCLT for multinomial random variables and use it to derive the result for discrete-time d-dimensional simple random walk, but we have chosen to omit this.

2.5.1 Stirling's formula and one-dimensional walks

Suppose that S_n is a simple one-dimensional random walk starting at the origin. Determining the distribution of S_n reduces to an easy counting problem. In order for $X_1 + \cdots + X_{2n}$ to equal $2k$, exactly $n + k$ of the X_j must equal $+1$. Since all 2^{-2n} sequences of ± 1 are equally likely,

$$p_{2n}(2k) = \mathbb{P}\{S_{2n} = 2k\} = 2^{-2n} \binom{2n}{n+k} = 2^{-2n} \frac{(2n)!}{(n+k)!\,(n-k)!}. \qquad (2.53)$$

We will use Stirling's formula, which we now derive, to estimate the factorial. In the proof, we will use some standard estimates about the logarithm; see Section A.1.2.

Theorem 2.5.1 (Stirling's formula) *As $n \to \infty$,*

$$n! \sim \sqrt{2\pi}\, n^{n+(1/2)}\, e^{-n}.$$

In fact,

$$\frac{n!}{\sqrt{2\pi}\, n^{n+(1/2)}\, e^{-n}} = 1 + \frac{1}{12\,n} + O\left(\frac{1}{n^2}\right).$$

Proof Let $b_n = n^{n+(1/2)} e^{-n}/n!$. Then, (A.5) and Taylor's theorem imply that

$$\begin{aligned}
\frac{b_{n+1}}{b_n} &= \frac{1}{e}\left(1+\frac{1}{n}\right)^n \left(1+\frac{1}{n}\right)^{1/2} \\
&= \left[1 - \frac{1}{2n} + \frac{11}{24n^2} + O\left(\frac{1}{n^3}\right)\right]\left[1 + \frac{1}{2n} - \frac{1}{8n^2} + O\left(\frac{1}{n^3}\right)\right] \\
&= 1 + \frac{1}{12n^2} + O\left(\frac{1}{n^3}\right).
\end{aligned}$$

Therefore,

$$\lim_{m\to\infty} \frac{b_m}{b_n} = \prod_{l=n}^{\infty}\left[1 + \frac{1}{12l^2} + O\left(\frac{1}{l^3}\right)\right] = 1 + \frac{1}{12n} + O\left(\frac{1}{n^2}\right).$$

The second equality is obtained by

$$\begin{aligned}
\log \prod_{l=n}^{\infty}\left[1 + \frac{1}{12l^2} + O\left(\frac{1}{l^3}\right)\right] &= \sum_{l=n}^{\infty} \log\left[1 + \frac{1}{12l^2} + O\left(\frac{1}{l^3}\right)\right] \\
&= \sum_{l=n}^{\infty} \frac{1}{12l^2} + \sum_{l=n}^{\infty} O\left(\frac{1}{l^3}\right) \\
&= \frac{1}{12n} + O\left(\frac{1}{n^2}\right).
\end{aligned}$$

This establishes that the limit

$$C := \left[\lim_{m\to\infty} b_m\right]^{-1}$$

exists and

$$b_n = \frac{1}{C}\left[1 - \frac{1}{12n} + O\left(\frac{1}{n^2}\right)\right],$$

$$n! = C\, n^{n+(1/2)}\, e^{-n}\left[1 + \frac{1}{12n} + O\left(\frac{1}{n^2}\right)\right].$$

There are a number of ways to determine the constant C. For example, if S_n denotes a one-dimensional simple random walk, then

$$\mathbb{P}\{|S_{2n}| \leq \sqrt{2n}\,\log n\} = \sum_{|2k| \leq \sqrt{2n}\,\log n} 4^{-n} \binom{2n}{n+k}$$
$$= \sum_{|2k| \leq \sqrt{2n}\,\log n} 4^{-n} \frac{(2n)!}{(n+k)!(n-k)!}.$$

Using (A.3), we see that as $n \to \infty$, if $|2k| \leq \sqrt{2n}\,\log n$,

$$4^{-n} \frac{(2n)!}{(n+k)!(n-k)!} \sim \frac{\sqrt{2}}{C\sqrt{n}} \left(1 + \frac{k}{n}\right)^{-(n+k)} \left(1 - \frac{k}{n}\right)^{-(n-k)}$$
$$= \frac{\sqrt{2}}{C\sqrt{n}} \left(1 - \frac{k^2/n}{n}\right)^{-n} \left(1 + \frac{k^2/n}{k}\right)^{-k} \left(1 - \frac{k^2/n}{k}\right)^{k}$$
$$\sim \frac{\sqrt{2}}{C\sqrt{n}}\, e^{-k^2/n}.$$

Therefore,

$$\lim_{n\to\infty} \mathbb{P}\{|S_{2n}| \leq \sqrt{2n}\,\log n\} = \lim_{n\to\infty} \sum_{|k| \leq \sqrt{n/2}\,\log n} \frac{\sqrt{2}}{C\sqrt{n}}\, e^{-k^2/n}$$
$$= \frac{\sqrt{2}}{C} \int_{-\infty}^{\infty} e^{-x^2}\, dx = \frac{\sqrt{2\pi}}{C}.$$

However, Chebyshev's inequality shows that

$$\mathbb{P}\{|S_{2n}| \geq \sqrt{2n}\,\log n\} \leq \frac{\operatorname{Var}[S_{2n}]}{2n \log^2 n} = \frac{1}{\log^2 n} \longrightarrow 0.$$

Therefore, $C = \sqrt{2\pi}$. □

♣ By adapting this proof, it is easy to see that one can find $r_1 = 1/12$, $r_2, r_3, \ldots$ such that for each positive integer k,

$$n! = \sqrt{2\pi}\, n^{n+(1/2)}\, e^{-n} \left[1 + \frac{r_1}{n} + \frac{r_2}{n^2} + \cdots + \frac{r_k}{n^k} + O\left(\frac{1}{n^{k+1}}\right)\right]. \tag{2.54}$$

We will now prove Theorem 2.1.1 and some difference estimates in the special case of simple random walk in one dimension by using (2.53) and Stirling's formula. As a warmup, we start with the probability of being at the origin.

Proposition 2.5.2 *For simple random walk in* $\mathbb{Z}$, *if* n *is a positive integer, then*

$$\mathbb{P}\{S_{2n} = 0\} = \frac{1}{\sqrt{\pi n}} \left[1 - \frac{1}{8n} + O\left(\frac{1}{n^2}\right)\right].$$

Proof The probability is exactly

$$2^{-2n}\binom{2n}{n} = \frac{(2n)!}{4^n\,(n!)^2}.$$

By plugging into Stirling's formula, we see that the right-hand side equals

$$\frac{1}{\sqrt{\pi n}}\,\frac{1 + (24n)^{-1} + O(n^{-2})}{[1 + (12n)^{-1} + O(n^{-2})]^2} = \frac{1}{\sqrt{\pi n}} \left[1 - \frac{1}{8n} + O\left(\frac{1}{n^2}\right)\right].$$

□

In the last proof, we just plugged into Stirling's formula and evaluated. We will now do the same thing to prove a version of the LCLT for one-dimensional simple random walk.

Proposition 2.5.3 *For simple random walk in* $\mathbb{Z}$, *if* n *is a positive integer and* k *is an integer with* $|k| \le n$,

$$p_{2n}(2k) = \mathbb{P}\{S_{2n} = 2k\} = \frac{1}{\sqrt{\pi n}}\, e^{-k^2/n} \exp\left\{O\left(\frac{1}{n} + \frac{k^4}{n^3}\right)\right\}.$$

In particular, if $|k| \le n^{3/4}$, *then*

$$\mathbb{P}\{S_{2n} = 2k\} = \frac{1}{\sqrt{\pi n}}\, e^{-k^2/n} \left[1 + O\left(\frac{1}{n} + \frac{k^4}{n^3}\right)\right].$$

♣ Note that for one-dimensional simple random walk,

$$2\,\overline{p}_{2n}(2k) = 2\,\frac{1}{\sqrt{(2\pi)\,(2n)}} \exp\left\{-\frac{(2k)^2}{2\,(2n)}\right\} = \frac{1}{\sqrt{\pi n}}\, e^{-k^2/n}.$$

♣ While the theorem is stated for all $|k| \le n$, it is not a very strong statement when k is of order n. For example, for $n/2 \le |k| \le n$, we can rewrite the conclusion as

$$p_{2n}(2k) = \frac{1}{\sqrt{\pi n}}\, e^{-k^2/n}\, e^{O(n)} = e^{O(n)},$$

which only tells us that there exists α such that

$$e^{-\alpha n} \le p_{2n}(2k) \le e^{\alpha n}.$$

In fact, $2\overline{p}_{2n}(2k)$ is not a very good approximation of $p_{2n}(2k)$ for large n. As an extreme example, note that

$$p_{2n}(2n) = 4^{-n}, \qquad 2\,\overline{p}_{2n}(2n) = \frac{1}{\sqrt{\pi n}}\, e^{-n}.$$

Proof If $n/2 \le |k| \le n$, the result is immediate using only the estimate $2^{-2n} \le \mathbb{P}\{S_{2n} = 2k\} \le 1$. Hence, we may assume that $|k| \le n/2$. As noted before,

$$\mathbb{P}\{S_{2n} = 2k\} = 2^{-2n}\binom{2n}{n+k} = \frac{(2n)!}{2^{2n}(n+k)!(n-k)!}.$$

If we restrict to $|k| \le n/2$, we can use Stirling's formula (Lemma 2.5.1) to see that

$$\mathbb{P}\{S_{2n} = 2k\} = \left[1 + O\left(\frac{1}{n}\right)\right] \frac{1}{\sqrt{\pi n}} \left(1 - \frac{k^2}{n^2}\right)^{-1/2}$$
$$\times \left(1 - \frac{k^2}{n^2}\right)^{-n} \left(1 - \frac{2k}{n+k}\right)^{k}.$$

The last two terms approach exponential functions. We need to be careful with the error terms. Using (A.3) we get,

$$\left(1 - \frac{k^2}{n^2}\right)^{n} = e^{-k^2/n} \exp\left\{O\left(\frac{k^4}{n^3}\right)\right\}.$$

$$\left(1 - \frac{2k}{n+k}\right)^{k} = e^{-2k^2/(n+k)} \exp\left\{-\frac{2k^3}{(n+k)^2} + O\left(\frac{k^4}{n^3}\right)\right\}$$
$$= e^{-2k^2/(n+k)} \exp\left\{-\frac{2k^3}{n^2} + O\left(\frac{k^4}{n^3}\right)\right\},$$

$$e^{-2k^2/(n+k)} = e^{-2k^2/n} \exp\left\{\frac{2k^3}{n^2} + O\left(\frac{k^4}{n^3}\right)\right\}.$$

Also, using $k^2/n^2 \le \max\{(1/n), (k^4/n^3)\}$, we can see that

$$\left(1 - \frac{k^2}{n^2}\right)^{-1/2} = \exp\left\{O\left(\frac{1}{n} + \frac{k^4}{n^3}\right)\right\}.$$

Combining all of this gives the theorem. □

♣ We could also prove "difference estimates" by using the equalities

$$p_{2n}(2k+2) = \frac{n-k}{n+k+1}\, p_{2n}(2k),$$

$$p_{2(n+1)}(2k) = p_{2n}(2k)\, 4^{-1}\, \frac{(2n+1)(2n+2)}{(n+k+1)(n-k+1)}.$$

Corollary 2.5.4 *If S_n is simple random walk, then for all positive integers n and all $|k| < n$,*

$$\mathbb{P}\{S_{2n+1} = 2k+1\} = \frac{1}{\sqrt{\pi n}} \exp\left\{-\frac{(k+\frac{1}{2})^2}{n}\right\} \exp\left\{O\left(\frac{1}{n} + \frac{k^4}{n^3}\right)\right\}. \tag{2.55}$$

Proof Note that

$$\mathbb{P}\{S_{2n+1} = 2k+1\} = \frac{1}{2}\,\mathbb{P}\{S_{2n} = 2k\} + \frac{1}{2}\,\mathbb{P}\{S_{2n} = 2(k+1)\}.$$

Hence,

$$\mathbb{P}\{S_{2n+1} = 2k+1\} = \frac{1}{2\sqrt{\pi n}}[e^{-k^2/n} + e^{-(k+1)^2/n}] \exp\left\{O\left(\frac{1}{n} + \frac{k^4}{n^3}\right)\right\}.$$

But,

$$\exp\left\{-\frac{(k+\frac{1}{2})^2}{n}\right\} = e^{-k^2/n}\left[1 - \frac{k}{n} + O\left(\frac{k^2}{n^2}\right)\right],$$

$$\exp\left\{-\frac{(k+1)^2}{n}\right\} = e^{-k^2/n}\left[1 - \frac{2k}{n} + O\left(\frac{k^2}{n^2}\right)\right],$$

which implies that

$$\frac{1}{2}\left[e^{-k^2/n} + e^{-(k+1)^2/n}\right] = \exp\left\{-\frac{(k+\frac{1}{2})^2}{n}\right\}\left[1 + O\left(\frac{k^2}{n^2}\right)\right].$$

Using $k^2/n^2 \le \max\{(1/n), (k^4/n^3)\}$, we get (2.55). □

♣ One might think that we should replace n in (2.55) with $n+(1/2)$. However,

$$\frac{1}{n+(1/2)} = \frac{1}{n}\left[1+O\left(\frac{1}{n}\right)\right].$$

Hence, the same statement with $n+(1/2)$ replacing n is also true.

2.5.2 LCLT for Poisson and continuous-time walks

The next proposition establishes the strong LCLT for Poisson random variables. This will be used for comparing discrete-time and continuous-time random walks with the same p. If N_t is a Poisson random variable with parameter t, then $\mathbb{E}[N_t] = t$, $\mathrm{Var}[N_t] = t$. The CLT implies that as $t \to \infty$, the distribution of $(N_t - t)/\sqrt{t}$ approaches that of a standard normal. Hence, we might conjecture that

$$\mathbb{P}\{N_t = m\} = \mathbb{P}\left\{\frac{m-t}{\sqrt{t}} \le \frac{N_t - t}{\sqrt{t}} < \frac{m+1-t}{\sqrt{t}}\right\}$$
$$\approx \int_{(m-t)/\sqrt{t}}^{(m+1-t)/\sqrt{t}} \frac{1}{\sqrt{2\pi}}\, e^{-\frac{x^2}{2}}\, dx \approx \frac{1}{\sqrt{2\pi t}}\, e^{-\frac{(m-t)^2}{2t}}.$$

In the next proposition, we use a straightforward combinatorial argument to justify this approximation.

Proposition 2.5.5 *Suppose that N_t is a Poisson random variable with parameter t, and m is an integer with $|m-t| \le t/2$. Then*

$$\mathbb{P}\{N_t = m\} = \frac{1}{\sqrt{2\pi t}}\, e^{-\frac{(m-t)^2}{2t}}\, \exp\left\{O\left(\frac{1}{\sqrt{t}} + \frac{|m-t|^3}{t^2}\right)\right\}.$$

Proof For notational ease, we will first consider the case where $t = n$ is an integer, and we let $m = n + k$. Let

$$q(n,k) = \mathbb{P}\{N_n = n+k\} = e^{-n}\, \frac{n^{n+k}}{(n+k)!},$$

and note the recursion formula

$$q(n,k) = \frac{n}{n+k}\, q(n,k-1).$$

Stirling's formula (Theorem 2.5.1) gives

$$q(n,0) = \frac{e^{-n}\, n^n}{n!} = \frac{1}{\sqrt{2\pi n}}\left[1+O\left(\frac{1}{n}\right)\right]. \tag{2.56}$$

By the recursion formula, if $k \le n/2$,

$$q(n,k) = q(n,0)\left[\left(1+\frac{1}{n}\right)\left(1+\frac{2}{n}\right)\cdots\left(1+\frac{k}{n}\right)\right]^{-1},$$

and,

$$\begin{aligned}\log\prod_{j=1}^{k}\left(1+\frac{j}{n}\right) &= \sum_{j=1}^{k}\log\left(1+\frac{j}{n}\right)\\ &= \sum_{j=1}^{k}\left[\frac{j}{n}+O\left(\frac{j^2}{n^2}\right)\right] = \frac{k^2}{2n}+\frac{k}{2n}+O\left(\frac{k^3}{n^2}\right)\\ &= \frac{k^2}{2n}+O\left(\frac{1}{\sqrt{n}}+\frac{k^3}{n^2}\right).\end{aligned}$$

The last equality uses the inequality

$$\frac{k}{n} \le \max\left\{\frac{1}{\sqrt{n}},\frac{k^3}{n^2}\right\},$$

which will also be used in other estimates in this proof. Using (2.56), we get

$$\log q(n,k) = -\log\sqrt{2\pi n} - \frac{k^2}{2n} + O\left(\frac{1}{\sqrt{n}}+\frac{k^3}{n^2}\right),$$

and the result for $k \ge 0$ follows by exponentiating.

Similarly,

$$q(n,-k) = q(n,0)\left(1-\frac{1}{n}\right)\left(1-\frac{2}{n}\right)\cdots\left(1-\frac{k-1}{n}\right)$$

and

$$\begin{aligned}\log q(n,-k) &= -\log\sqrt{2\pi n} + \log\prod_{j=1}^{k-1}\left(1-\frac{j}{n}\right)\\ &= -\log\sqrt{2\pi n} - \frac{k^2}{2n} + O\left(\frac{1}{\sqrt{n}}+\frac{k^3}{n^2}\right).\end{aligned}$$

The proposition for integer n follows by exponentiating. For general t, let $n = \lfloor t \rfloor$ and note that

$$\begin{aligned}
\mathbb{P}\{N_t = n+k\} &= \mathbb{P}\{N_n = n+k\}\, e^{-(t-n)} \left(1 + \frac{t-n}{n}\right)^{n+k} \\
&= \mathbb{P}\{N_n = n+k\} \left(1 + \frac{t-n}{n}\right)^{k} \left[1 + O\left(\frac{1}{n}\right)\right] \\
&= \mathbb{P}\{N_n = n+k\} \left[1 + O\left(\frac{|k|+1}{n}\right)\right] \\
&= (2\pi n)^{-1/2}\, e^{-k^2/(2n)} \exp\left\{O\left(\frac{1}{\sqrt{n}} + \frac{k^3}{n^2}\right)\right\} \\
&= (2\pi t)^{-1/2}\, e^{-(k+n-t)^2/(2t)} \exp\left\{O\left(\frac{1}{\sqrt{t}} + \frac{|n+k-t|^3}{t^2}\right)\right\}.
\end{aligned}$$

The last step uses the estimates

$$\frac{1}{\sqrt{t}} = \frac{1}{\sqrt{n}}\left[1 + O\left(\frac{1}{t}\right)\right], \quad e^{-\frac{k^2}{2t}} = e^{-\frac{k^2}{2n}} \exp\left\{O\left(\frac{k^2}{t^2}\right)\right\}.$$

□

We will use this to prove a version of the LCLT for one-dimensional, continuous-time simple random walk.

Theorem 2.5.6 *If $\tilde{S}_t$ is continuous-time one-dimensional simple random walk, then if $|x| \leq t/2$,*

$$\tilde{p}_t(x) = \frac{1}{\sqrt{2\pi t}}\, e^{-\frac{x^2}{2t}} \exp\left\{O\left(\frac{1}{\sqrt{t}} + \frac{|x|^3}{t^2}\right)\right\}.$$

Proof We will assume that $x = 2k$ is even; the odd case is done similarly. We know that

$$\tilde{p}_t(2k) = \sum_{m=0}^{\infty} \mathbb{P}\{N_t = 2m\}\, p_{2m}(2k).$$

Standard exponential estimates, see (A.12), show that for every $\epsilon > 0$, there exist c, β such that $\mathbb{P}\{|N_t - t| \geq \epsilon t\} \leq c\, e^{-\beta t}$. Hence,

$$\tilde{p}_t(2k) = \sum_{m=0}^{\infty} \mathbb{P}\{N_t = 2m\}\, p_{2m}(2k)$$
$$= O(e^{-\beta t}) + \sum \mathbb{P}\{N_t = 2m\}\, p_{2m}(2k), \tag{2.57}$$

where here and for the remainder of this proof, we write just $\sum$ to denote the sum over all integers m with $|t - 2m| \leq \epsilon t$. We will show that there is an ϵ such that

$$\sum \mathbb{P}\{N_t = 2m\}\, p_{2m}(2k) = \frac{1}{\sqrt{2\pi t}}\, e^{-\frac{x^2}{2t}} \exp\left\{O\left(\frac{1}{\sqrt{t}} + \frac{|x|^3}{t^2}\right)\right\}.$$

A little thought shows that this and (2.57) imply the theorem.

By Proposition 2.5.3 we know that

$$p_{2m}(2k) = \mathbb{P}\{S_{2m} = 2k\} = \frac{1}{\sqrt{\pi m}}\, e^{-\frac{k^2}{m}} \exp\left\{O\left(\frac{1}{m} + \frac{k^4}{m^3}\right)\right\},$$

and by Proposition 2.5.5 we know that

$$\mathbb{P}\{N_t = 2m\} = \frac{1}{\sqrt{2\pi t}}\, e^{-\frac{(2m-t)^2}{2t}} \exp\left\{O\left(\frac{1}{\sqrt{t}} + \frac{|2m-t|^3}{t^2}\right)\right\}.$$

Also, we have

$$\frac{1}{2m} = \frac{1}{t}\left[1 + O\left(\frac{|2m-t|}{t}\right)\right], \quad \frac{1}{\sqrt{2m}} = \frac{1}{\sqrt{t}}\left[1 + O\left(\frac{|2m-t|}{t}\right)\right],$$

which implies that

$$e^{-\frac{k^2}{m}} = e^{-\frac{2k^2}{t}} \exp\left\{O\left(\frac{k^2\,|2m-t|}{t^2}\right)\right\}.$$

Combining all of this, we can see that the sum in (2.57) can be written as

$$\frac{1}{\sqrt{2\pi t}}\, e^{-\frac{x^2}{2t}} \exp\left\{O\left(\frac{1}{\sqrt{t}} + \frac{|x|^3}{t^2}\right)\right\}$$
$$\times \sum \frac{2}{\sqrt{2\pi t}}\, e^{-\frac{(2m-t)^2}{2t}} \exp\left\{O\left(\frac{|2m-t|^3}{t^2}\right)\right\}.$$

We now choose ϵ so that $|O(|2m-t|^3/t^2)| \le (2m-t)^2/(4t)$ for all $|2m-t| \le \epsilon t$. We will now show that

$$\sum \frac{2}{\sqrt{2\pi t}} e^{-\frac{(2m-t)^2}{2t}} \exp\left\{O\left(\frac{|2m-t|^3}{t^2}\right)\right\} = 1 + O\left(\frac{1}{\sqrt{t}}\right),$$

which will complete the argument. Since

$$e^{-\frac{(2m-t)^2}{2t}} \exp\left\{O\left(\frac{|2m-t|^3}{t^2}\right)\right\} \le e^{-\frac{(2m-t)^2}{4t}},$$

it is easy to see that the sum over $|2m-t| > t^{2/3}$ decays faster than any power of t. For $|2m-t| \le t^{2/3}$ we write

$$\exp\left\{O\left(\frac{|2m-t|^3}{t^2}\right)\right\} = 1 + O\left(\frac{|2m-t|^3}{t^2}\right).$$

The estimate

$$\begin{aligned}
&\sum_{|2m-t|\le t^{2/3}} \frac{2}{\sqrt{2\pi t}} e^{-\frac{(2m-t)^2}{2t}} \\
&= \sum_{|m|\le t^{2/3}/2} \frac{2}{\sqrt{2\pi t}} e^{-2(m/\sqrt{t})^2} \\
&= O\left(\frac{1}{\sqrt{t}}\right) + 2\int_{-\infty}^{\infty} \frac{1}{\sqrt{2\pi}} e^{-2y^2}\, dy = 1 + O\left(\frac{1}{\sqrt{t}}\right)
\end{aligned}$$

is a standard approximation of an integral by a sum. Similarly,

$$\begin{aligned}
&\sum_{|2m-t|\le t^{2/3}} O\left(\frac{|2m-t|^3}{t^2}\right) \frac{2}{\sqrt{2\pi t}} e^{-\frac{(2m-t)^2}{2t}} \\
&\le \frac{c}{\sqrt{t}} \int_{-\infty}^{\infty} \frac{|y|^3}{\sqrt{2\pi}} e^{-2y^2}\, dy = O\left(\frac{1}{\sqrt{t}}\right).
\end{aligned}$$

□

Exercises

Exercise 2.1 Suppose that $p \in \mathcal{P}_d'$, $\epsilon \in (0,1)$, and $\mathbb{E}[|X_1|^{2+\epsilon}] < \infty$. Show that the characteristic function has the expansion

$$\phi(\theta) = 1 - \frac{\theta \cdot \Gamma\theta}{2} + o(|\theta|^{2+\epsilon}), \quad \theta \to 0.$$

Show that the δ_n in (2.32) can be chosen so that $n^{\epsilon/2}\,\delta_n \to 0$.

Exercise 2.2 Show that if $p \in \mathcal{P}_d^*$, there exists a c such that for all $x \in \mathbb{Z}^d$ and all positive integers n,

$$|p_n(x) - p_n(0)| \leq c \, \frac{|x|}{n^{(d+1)/2}}.$$

(Hint: first show the estimate for $p \in \mathcal{P}'_d$ with bounded support and then use (2.48). Alternatively, one can use Lemma 2.4.3 at time $n/2$, the Markov property, and (2.49).)

Exercise 2.3 Show that Lemma 2.3.2 holds for $p \in \mathcal{P}^*$.

Exercise 2.4 Suppose that $p \in \mathcal{P}'_d$ with $\mathbb{E}[|X|^3] < \infty$. Show that there is a $c < \infty$ such that for all $|y| = 1$,

$$|p_n(0) - p_n(y)| \leq \frac{c}{n^{(d+2)/2}}.$$

Exercise 2.5 Suppose that $p \in \mathcal{P}_d^*$. Let $A \subset \mathbb{Z}^d$ and

$$h(x) = \mathbb{P}^x\{S_n \in A \text{ i.o.}\}.$$

Show that if $h(x) > 0$ for some $x \in \mathbb{Z}^d$, then $h(x) = 1$ for all $x \in \mathbb{Z}^d$.

Exercise 2.6 Suppose that S_n is a random walk with increment distribution $p \in \mathcal{P}_d$. Show that there exists a $b > 0$ such that

$$\sup_{n>1} \mathbb{E}\left[\exp\left\{\frac{b|S_n|^2}{n}\right\}\right] < \infty.$$

Exercise 2.7 Suppose that $X_1, X_2, \ldots$ are independent, identically distributed random variables in $\mathbb{Z}^d$ with $\mathbb{P}\{X_1 = 0\} < 1$ and let $S_n = X_1 + \cdots + X_n$.

(i) Show that there exists an r such that for all n

$$\mathbb{P}\{|S_{rn^2}| \geq n\} \geq \frac{1}{2}.$$

(ii) Show that there exist c, t such that for all $b > 0$,

$$\mathbb{P}\left\{\max_{1 \leq j \leq n^2} |S_j| \leq bn\right\} \leq ce^{-t/b}.$$

Exercise 2.8 Find r_2, r_3 in (2.54).

Exercise 2.9 Let S_n denote one-dimensional simple random walk. In this exercise we will prove without using Stirling's formula that there exists a constant C such that

$$p_{2n}(0) = \frac{C}{\sqrt{n}} \left[1 - \frac{1}{8n} + O\left(\frac{1}{n^2}\right)\right].$$

(i) Show that if $n \geq 1$,

$$p_{2(n+1)} = \left(1 + \frac{1}{2n}\right) \left(1 + \frac{1}{n}\right)^{-1} p_{2n}.$$

(ii) Let $b_n = \sqrt{n}\, p_{2n}(0)$. Show that $b_1 = 1/2$ and for $n \geq 1$,

$$\frac{b_{n+1}}{b_n} = 1 + \frac{1}{8n^2} + O\left(\frac{1}{n^3}\right).$$

(iii) Use this to show that $b_\infty = \lim b_n$ exists and is positive. Moreover,

$$b_n = b_\infty \left[1 - \frac{1}{8n} + O\left(\frac{1}{n^2}\right)\right].$$

Exercise 2.10 Show that if $p \in \mathcal{P}_d'$ with $\mathbb{E}[|X_1|^3] < \infty$, then

$$\nabla_j^2 p_n(x) = \nabla_j^2 \overline{p}_n(x) + O(n^{-(d+3)/2}).$$

Exercise 2.11 Suppose that $q : \mathbb{Z}^d \to \mathbb{R}$ has finite support, and k is a positive integer such that for all $l \in \{1, \ldots, k-1\}$ and all $j_1, \ldots, j_l \in \{1, \ldots, d\}$,

$$\sum_{x=(x^1,\ldots,x^d)\in\mathbb{Z}^d} x^{j_1}\, x^{j_2} \ldots x^{j_l}\, q(x) = 0.$$

Then we call the operator

$$\Lambda f(x) := \sum_y f(x+y)\, q(y)$$

a difference operator of order (at least) k. The order of the operator is the largest k for which this is true. Suppose that Λ is a difference operator of order $k \geq 1$.

(i) Suppose that g is a C^∞ function on $\mathbb{R}^d$. Define g_ϵ on $\mathbb{Z}^d$ by $g_\epsilon(x) = g(\epsilon x)$. Show that

$$|\Lambda g_\epsilon(0)| = O(|\epsilon|^k), \quad \epsilon \to 0.$$

(ii) Show that if $p \in \mathcal{P}'_d$ with $\mathbb{E}[|X_1|^3] < \infty$, then

$$\Lambda p_n(x) = \Lambda \overline{p}_n(x) + O(n^{-(d+1+k)/2}).$$

(iii) Show that if $p \in \mathcal{P}'_d$ is symmetric with $\mathbb{E}[|X_1|^4] < \infty$, then

$$\Lambda p_n(x) = \Lambda \overline{p}_n(x) + O(n^{-(d+2+k)/2}).$$

Exercise 2.12 Suppose that $p \in \mathcal{P}' \cup \mathcal{P}'_2$. Show that there is a c such that the following is true. Let S_n be a p-walk and let

$$\tau_n = \inf\{j : |S_j| \geq n\}.$$

If $y \in \mathbb{Z}^2$, let

$$V_n(y) = \sum_{j=0}^{\tau_n - 1} 1\{S_j = y\}$$

denote the number of visits to y before time τ_n. Then, if $0 < |y| < n$,

$$\mathbb{E}\left[V_k(y)\right] \leq c \, \frac{1 + \log n - \log |y|}{n}.$$

(Hint: show that there exist c_1, β such that for each positive integer j,

$$\sum_{jn^2 \leq j < (j+1)n^2} 1\{S_j = y; j < \tau_n\} \leq c_1 \, e^{-\beta j} \, n^{-1}.)$$

3

Approximation by Brownian motion

3.1 Introduction

Suppose that $S_n = X_1 + \cdots + X_n$ is a one-dimensional simple random walk. We make this into a (random) continuous function by linear interpolation,

$$S_t = S_n + (t-n)\,[S_{n+1} - S_n], \qquad n \le t \le n+1.$$

For fixed integer n, the LCLT describes the distribution of S_n. A corollary of LCLT is the usual CLT that states that the distribution of $n^{-1/2}\,S_n$ converges to that of a standard normal random variable. A simple extension of this is the following: suppose that $0 < t_1 < t_2 < \cdots < t_k = 1$. Then, as $n \to \infty$, the distribution of

$$n^{-1/2}\,(S_{t_1 n}, S_{t_2 n}, \ldots, S_{t_k n})$$

converges to that of

$$(Y_1, Y_1 + Y_2, \ldots, Y_1 + Y_2 + \cdots Y_k),$$

where $Y_1, \ldots, Y_k$ are independent mean zero normal random variables with variances $t_1, t_2 - t_1, \ldots, t_k - t_{k-1}$, respectively.

The *functional central limit theorem* (also called the *invariance principle* or *Donsker's theorem*) for random walk extends this result to the random function

$$W_t^{(n)} := n^{-1/2}\,S_{tn}. \tag{3.1}$$

The functional CLT states roughly that as $n \to \infty$, the distribution of this random function converges to the distribution of a random function $t \mapsto B_t$. From what we know about the simple random walk, here are some properties that would be expected of the random function B_t:

- If $s < t$, the distribution of $B_t - B_s$ is $N(0, t - s)$.
- If $0 \le t_0 < t_1 < \cdots < t_k$, then $B_{t_1} - B_{t_0}, \ldots, B_{t_k} - B_{t_{k-1}}$ are independent random variables.

These two properties follow almost immediately from the CLT. The third property is not as obvious.

- The function $t \mapsto B_t$ is continuous.

Although this is not obvious, we can guess this from the heuristic argument:

$$\mathbb{E}[(B_{t+\Delta t} - B_t)^2] \approx \Delta t,$$

which indicates that $|B_{t+\Delta t} - B_t|$ should be of order $\sqrt{\Delta t}$. A process satisfying these assumptions will be called a *Brownian motion* (we will define it more precisely in the next section).

There are a number of ways to make rigorous the idea that $W^{(n)}$ approaches a Brownian motion in the limit. For example, if we are restricted to $0 \le t \le 1$, then $W^{(n)}$ and B are random variables taking values in the metric space $C[0, 1]$ with the supremum norm. There is a well-understood theory of convergence in the distribution of random variables taking values in metric spaces.

We will take a different approach using a method that is often called *strong approximation* of random walk by Brownian motion. We start by defining a Brownian motion B on a probability space and then define the random walk S_n as a function of the Brownian motion; i.e. for each realization of random function B_t, we associate a particular random walk path. We will do this in such a way that the random walk S_n has the distribution of simple random walk. We will then carry out some estimates to show that there exist positive numbers c, a such that if $W_t^{(n)}$ is as defined in (3.1), then for all $r \le n^{1/4}$,

$$\mathbb{P}\{\|B - W^{(n)}\| \ge r\, n^{1/4}\, \sqrt{\log n}\} \le c\, e^{-ra}, \tag{3.2}$$

where $\| \cdot \|$ denotes the supremum norm on $C[0, 1]$. The convergence in distribution follows from the strong estimate (3.2).

♣ There is a general approach here that is worth emphasizing. Suppose that we have a discrete process and we want to show that it converges after some scaling to a continuous process. A good approach for proving such a result is first to study the conjectured limit process and then to show that the scaled discrete process is a small perturbation of the limit process.

We start by establishing (3.2) for one-dimensional simple random walk using *Skorokhod embedding*. We then extend this to continuous-time walks and all increment distributions $p \in \mathcal{P}$. The extension will not be difficult; the hard work is done in the one-dimensional case.

We will not handle the general case of $p \in \mathcal{P}'$ in this book. One can give strong approximations in this case to show that the random walk approaches Brownian motion. However, the rate of convergence depends on the moment assumptions. In particular, the estimate (3.2) will not hold assuming only zero mean and a finite second moment.

3.2 Construction of Brownian motion

A *standard (one-dimensional) Brownian motion with respect to a filtration* $\mathcal{F}_t$ is a collection of random variables $B_t, t \geq 0$ satisfying the following:

(a) $B_0 = 0$;
(b) if $s < t$, then $B_t - B_s$ is an $\mathcal{F}_t$-measurable random variable, independent of $\mathcal{F}_s$, with a $N(0, t-s)$ distribution;
(c) with probability one, $t \mapsto B_t$ is a continuous function.

If the filtration is not given explicitly, then it is assumed to be the natural filtration, $\mathcal{F}_t = \sigma\{B_s : 0 \leq s \leq t\}$. In this section, we will construct a Brownian motion and derive an important estimate on the oscillations of the Brownian motion.

We will show how to construct a Brownian motion. There are technical difficulties involved in defining a collection of random variables $\{B_t\}$ indexed over an uncountable set. However, if we know a priori that the distribution should be supported on continuous functions, then we know that the random function $t \mapsto B_t$ should be determined by its value on a countable, dense subset of times. This observation leads us to a method of constructing Brownian motion: define the process on a countable, dense set of times and then extend the process to all times by continuity.

Suppose that $(\Omega, \mathcal{F}, \mathbb{P})$ is any probability space that is large enough to contain a countable collection of independent $N(0, 1)$ random variables which for ease we will index by

$$N_{n,k}, \quad n = 0, 1, \ldots \quad k = 0, 1, \ldots$$

We will use these random variables to define a Brownian motion on $(\Omega, \mathcal{F}, \mathbb{P})$. Let

$$D_n = \left\{ \frac{k}{2^n} : \ k = 0, 1, \ldots \right\}, \qquad D = \bigcup_{n=0}^{\infty} D_n$$

denote the nonnegative dyadic rationals. Our strategy will be as follows:

- to define B_t for t in D satisfying conditions (a) and (b);
- to derive an estimate on the oscillation of $B_t, t \in D$, that implies that with probability one the paths are uniformly continuous on compact intervals;
- to define B_t for other values of t by continuity.

The first step is straightforward using a basic property of normal random variables. Suppose that X, Y are independent normal random variables, each mean zero and variance $1/2$. Then $Z = X + Y$ is $N(0, 1)$. Moreover, the conditional distribution of X given the value of Z is normal with mean $Z/2$ and variance $1/4$. This can be checked directly using the density of the normals. Alternatively, one can check that if Z, N are independent $N(0, 1)$ random variables, then

$$X := \frac{Z}{2} + \frac{N}{2}, \qquad Y := \frac{Z}{2} - \frac{N}{2}, \tag{3.3}$$

are independent $N(0, 1/2)$ random variables. To verify this, one only notes that (X, Y) has a joint normal distribution with $\mathbb{E}[X] = \mathbb{E}[Y] = 0, \mathbb{E}[X^2] = \mathbb{E}[Y^2] = 1/2, \mathbb{E}[XY] = 0$. (See Corollary A.3.1.) This tells us that in order to define X, Y we can start with independent random variables N, Z and then use (3.3).

We start by defining B_t for $t \in D_0 = \mathbb{N}$ by $B_0 = 0$ and

$$B_j = N_{0,1} + \cdots + N_{0,j}.$$

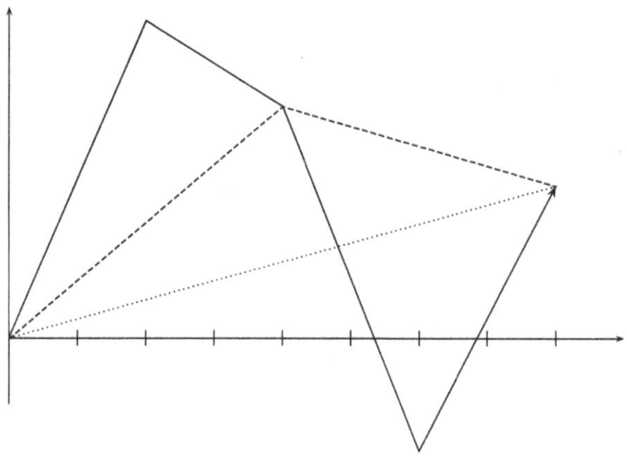

Figure 3.1 The dyadic construction

We then continue recursively using (3.3). Suppose that B_t has been defined for all $t \in D_n$. Then we define B_t for $t \in D_{n+1} \setminus D_n$ by

$$B_{\frac{2k+1}{2^{n+1}}} = B_{\frac{k}{2^n}} + \frac{1}{2}\left[B_{\frac{k+1}{2^n}} - B_{\frac{k}{2^n}}\right] + 2^{-(n+2)/2}\, N_{2k+1,n+1}.$$

By induction, one can check that for each n the collection of random variables $Z_{k,n} := B_{k/2^n} - B_{(k-1)/2^n}$ are independent, each with a $N(0, 2^{-n})$ distribution. Since this is true for each n, we can see that (a) and (b) hold (with the natural filtration) provided that we restrict to $t \in D$. The scaling property for normal random variables shows that for each integer n, the random variables

$$2^{n/2}\, B_{t/2^n}, \quad t \in D,$$

have the same joint distribution as the random variables

$$B_t, \quad t \in D.$$

We define the *oscillation* of B_t (restricted to $t \in D$) by

$$\operatorname{osc}(B; \delta, T) = \sup\{|B_t - B_s| : s, t \in D; 0 \le s, t \le T; |s - t| \le \delta\}.$$

For fixed δ, T, this is an $\mathcal{F}_T$-measurable random variable. We write $\operatorname{osc}(B; \delta)$ for $\operatorname{osc}(B; \delta, 1)$. Let

$$M_n = \max_{0 \le k < 2^n} \sup\left\{|B_{t+k2^{-n}} - B_{k2^{-n}}| : t \in D; 0 \le t \le 2^{-n}\right\}.$$

The random variable M_n is similar to $\operatorname{osc}(B; 2^{-n})$ but is easier to analyze. Note that if $r \le 2^{-n}$,

$$\operatorname{osc}(B; r) \le \operatorname{osc}(B; 2^{-n}) \le 3\, M_n. \tag{3.4}$$

To see this, suppose that $\delta \le 2^{-n}, 0 < s < t \le s + \delta \le 1$, and $|B_s - B_t| \ge \epsilon$. Then there exists a k such that either $k2^{-n} \le s < t \le (k+1)2^{-n}$ or $(k-1)2^{-n} \le s \le k2^{-n} < t \le (k+1)2^{-n}$. In either case, the triangle inequality tells us that $M_n \ge \epsilon/3$. We will prove a proposition that bounds the probability of large values of $\operatorname{osc}(B; \delta, T)$. We start with a lemma which gives a similar bound for M_n.

Lemma 3.2.1 *For every integer n and every $\delta > 0$,*

$$\mathbb{P}\{M_n > \delta\, 2^{-n/2}\} \le 4\sqrt{\frac{2}{\pi}}\, \frac{2^n}{\delta}\, e^{-\delta^2/2}.$$

Proof Note that

$$\mathbb{P}\{M_n > \delta\, 2^{-n/2}\} \leq 2^n\, \mathbb{P}\left\{ \sup_{0\leq t\leq 2^{-n}} |B_t| > \delta\, 2^{-n/2} \right\}$$
$$= 2^n\, \mathbb{P}\left\{ \sup_{0\leq t\leq 1} |B_t| > \delta \right\}.$$

Here the supremums are taken over $t \in D$. Also note that

$$\mathbb{P}\{ \sup\{|B_t| : 0 \leq t \leq 1, t \in D\} > \delta \}$$
$$= \lim_{n\to\infty} \mathbb{P}\{ \max\{|B_{k2^{-n}}| : k = 1, \ldots, 2^n\} > \delta \}$$
$$\leq 2 \lim_{n\to\infty} \mathbb{P}\{ \max\{B_{k2^{-n}} : k = 1, \ldots, 2^n\} > \delta \}.$$

The reflection principle (see Proposition 1.6.2 and the remark following) shows that

$$\mathbb{P}\{ \max\{B_{k2^{-n}} : k = 1, \ldots, 2^n\} > \delta \} \leq 2\, \mathbb{P}\{B_1 > \delta\}$$
$$= 2 \int_\delta^\infty \frac{1}{\sqrt{2\pi}}\, e^{-x^2/2}\, dx$$
$$\leq 2 \int_\delta^\infty \frac{1}{\sqrt{2\pi}}\, e^{-x\delta/2}\, dx$$
$$= 2\sqrt{\frac{2}{\pi}}\, \delta^{-1} e^{-\delta^2/2}.$$

□

Proposition 3.2.2 *There exists a $c > 0$ such that for every $0 < \delta \leq 1$, $r \geq 1$, and positive integer T,*

$$\mathbb{P}\{\mathrm{osc}(B; \delta, T) > c\, r \sqrt{\delta \log(1/\delta)}\} \leq c\, T\, \delta^{r^2}.$$

Proof It suffices to prove the result for $T = 1$ since for general T we can estimate separately the oscillations over the $2T - 1$ intervals $[0, 1]$, $[1/2, 3/2]$, $[1, 2], \ldots, [T - 1, T]$. Also, it suffices to prove the result for $\delta \leq 1/4$. Suppose that $2^{-n-1} \leq \delta \leq 2^{-n}$. Using (3.4), we see that

$$\mathbb{P}\{\mathrm{osc}(B; \delta) > c\, r \sqrt{\delta \log(1/\delta)}\} \leq \mathbb{P}\left\{M_n > \frac{cr}{3\sqrt{2}} \sqrt{2^{-n} \log(1/\delta)}\right\}.$$

By Lemma 3.2.1, if c is chosen sufficiently large, the probability on the right-hand side is bounded by a constant times

$$\exp\left\{-\frac{1}{4}\left(\frac{c^2r^2}{18}\right)\log(1/\delta)\right\},$$

which for c large enough is bounded by a constant times δ^{r^2}. □

Corollary 3.2.3 *With probability one, for every integer $T < \infty$, the function $t \mapsto B_t, t \in D$ is uniformly continuous on* $[0, T]$.

Proof Uniform continuity on $[0, T]$ is equivalent to saying that $\operatorname{osc}(B; 2^{-n}, T) \longrightarrow 0$ as $n \to \infty$. The previous proposition implies that there is a c_1 such that

$$\mathbb{P}\{\operatorname{osc}(B; 2^{-n}, T) > c_1\, 2^{-n/2}\sqrt{n}\} \le c_1\, T\, 2^{-n}.$$

In particular,

$$\sum_{n=1}^{\infty} \mathbb{P}\{\operatorname{osc}(B; 2^{-n}, T) > c_1\, 2^{-n/2}\sqrt{n}\} < \infty,$$

which implies by Borel–Cantelli that with probability one $\operatorname{osc}(B; 2^{-n}, T) \le c_1\, 2^{-n/2}\sqrt{n}$ for all n sufficiently large. □

Given the corollary, we can define B_t for $t \notin D$ by continuity, i.e.

$$B_t = \lim_{t_n \to t} B_{t_n},$$

where $t_n \in D$ with $t_n \to t$. It is not difficult to show that this satisfies the definition of Brownian motion (we omit the details). Moreover, since B_t has continuous paths, we can write

$$\operatorname{osc}(B; \delta, T) = \sup\{|B_t - B_s| : 0 \le s, t \le T; |s - t| \le \delta\}.$$

We restate the estimate and include a fact about scaling of Brownian motion. Note that if B_t is a standard Brownian motion and $a > 0$, then $Y_t := a^{-1/2}\, B_{at}$ is also a standard Brownian motion.

Theorem 3.2.4 (modulus of continuity of Brownian motion) *There is a $c < \infty$ such that if B_t is a standard Brownian motion, $0 < \delta \le 1$, $r \ge c, T \ge 1$,*

$$\mathbb{P}\{\operatorname{osc}(B; \delta, T) > r\sqrt{\delta \log(1/\delta)}\} \le c\, T\, \delta^{(r/c)^2}.$$

Moreover, if $T > 0$, *then* $\mathrm{osc}(B;\delta,T)$ *has the same distribution as* $\sqrt{T}\,\mathrm{osc}(B,\delta/T)$. *In particular, if* $T \geq 1$,

$$\mathbb{P}\{\mathrm{osc}(B;1,T) > c\,r\sqrt{\log T}\}$$
$$= \mathbb{P}\{\mathrm{osc}(B;1/T) > r\sqrt{(1/T)\,\log T}\} \leq c\,T^{-(r/c)^2}. \qquad (3.5)$$

3.3 Skorokhod embedding

We will now define a procedure that takes a Brownian motion path B_t and produces a random walk S_n. The idea is straightforward. Start the Brownian motion and wait until it reaches $+1$ or -1. If it hits $+1$ first, we let $S_1 = 1$; otherwise, we set $S_1 = -1$. Now we wait until the new increment of the Brownian motion reaches $+1$ or -1 and we use this value for the increment of the random walk.

To be more precise, let B_t be a standard one-dimensional Brownian motion, and let

$$\tau = \inf\{t \geq 0 : |B_t| = 1\}.$$

Symmetry tells us that $\mathbb{P}\{B_\tau = 1\} = \mathbb{P}\{B_\tau = -1\} = 1/2$.

Lemma 3.3.1 $\mathbb{E}[\tau] = 1$ *and there exists a* $b < \infty$ *such that* $\mathbb{E}[e^{b\tau}] < \infty$.

Proof Note that for integer n

$$\mathbb{P}\{\tau > n\} \leq \mathbb{P}\{\tau > n-1, |B_n - B_{n-1}| \leq 2\}$$
$$= \mathbb{P}\{\tau > n-1\}\,\mathbb{P}\{|B_n - B_{n-1}| \leq 2\},$$

which implies that for integer n,

$$\mathbb{P}\{\tau > n\} \leq \mathbb{P}\{|B_n - B_{n-1}| \leq 2\}^n = e^{-\rho n},$$

with $\rho > 0$. This implies that $\mathbb{E}[e^{b\tau}] < \infty$ for $b < \rho$. If $s < t$, then $\mathbb{E}[B_t^2 - t \mid \mathcal{F}_s] = B_s^2 - s$ (Exercise 3.1). This shows that $B_t^2 - t$ is a continuous martingale. Also,

$$\mathbb{E}[|B_t^2 - t|; \tau > t] \leq (t+1)\,\mathbb{P}\{\tau > t\} \longrightarrow 0.$$

Therefore, we can use the optional sampling theorem (Theorem A.2.9) to conclude that $\mathbb{E}[B_\tau^2 - \tau] = 0$. Since $\mathbb{E}[B_\tau^2] = 1$, this implies that $\mathbb{E}[\tau] = 1$. □

More generally, let $\tau_0 = 0$ and

$$\tau_n = \inf\{t \geq \tau_{n-1} : |B_t - B_{\tau_{n-1}}| = 1\}.$$

Then $S_n := B_{\tau_n}$ is a simple one-dimensional random walk.[†] Let $T_n = \tau_n - \tau_{n-1}$. The random variables $T_1, T_2, \ldots$ are independent, identically distributed, with mean one satisfying $\mathbb{E}[e^{bT_j}] < \infty$ for some $b > 0$. As before, we define S_t for noninteger t by linear interpolation. Let

$$\Theta(B, S; n) = \max\{|B_t - S_t| : 0 \leq t \leq n\}.$$

In other words, $\Theta(B, S; n)$ is the distance between the continuous functions B and S in $C[0, n]$ using the usual supremum norm. If $j \leq t < j + 1 \leq n$, then

$$|B_t - S_t| \leq |S_j - S_t| + |B_j - B_t| + |B_j - S_j| \leq 1 + \operatorname{osc}(B; 1, n) + |B_j - B_{\tau_j}|.$$

Hence, for integer n,

$$\Theta(B, S; n) \leq 1 + \operatorname{osc}(B; 1, n) + \max\{|B_j - B_{\tau_j}| : j = 1, \ldots, n\}. \tag{3.6}$$

We can estimate the probabilities for the second term with (3.5). We will concentrate on the last term. Before doing the harder estimates, let us consider how large an error we should expect. Since $T_1, T_2, \ldots$ are independent identically distributed random variables with mean one and finite variance, the CLT says roughly that

$$|\tau_n - n| = \left|\sum_{j=1}^{n} [T_j - 1]\right| \approx \sqrt{n}.$$

Hence we would expect that

$$|B_n - B_{\tau_n}| \approx \sqrt{|\tau_n - n|} \approx n^{1/4}.$$

From this reasoning, we can see that we expect $\Theta(B, S; n)$ to be at least of the order $n^{1/4}$. The next theorem shows that it is unlikely that the actual value is much greater than $n^{1/4}$.

† We actually need the *strong Markov property* for Brownian motion to justify this and the next assertion. This is not difficult to prove, but we will not do so here.

Theorem 3.3.2 *There exist $0 < c_1, a < \infty$ such that for all $r \leq n^{1/4}$ and all integers $n \geq 3$*

$$\mathbb{P}\{\Theta(B,S;n) > r\, n^{1/4}\, \sqrt{\log n}\} \leq c_1\, e^{-ar}.$$

Proof It suffices to prove the theorem for $r \geq 9c_*^2$ where c_* is the constant c from Theorem 3.2.4 (if we choose $c_1 \geq e^{9ac_*^2}$, the result holds trivially for $r \leq 9c_*^2$). Suppose that $9c_*^2 \leq r \leq n^{1/4}$. If $|B_n - B_{\tau_n}|$ is large, then either $|n - \tau_n|$ is large or the oscillation of B is large. Using (3.6), we see that the event $\{\Theta(B,S;n) \geq r\, n^{1/4}\, \sqrt{\log n}\}$ is contained in the union of the two events

$$\{\, \text{osc}(B; r\sqrt{n}, 2n) \geq (r/3)\, n^{1/4}\, \sqrt{\log n}\, \},$$

$$\left\{ \max_{1 \leq j \leq n} |\tau_j - j| \geq r\, \sqrt{n} \right\}.$$

Indeed, if $\text{osc}(B; r\sqrt{n}, 2n) \leq (r/3)\, n^{1/4}\, \sqrt{\log n}$ and $|\tau_j - j| \leq r\, \sqrt{n}$ for $j = 1, \ldots, n$, then the three terms on the right-hand side of (3.6) are each bounded by $(r/3)\, n^{1/4}\, \sqrt{\log n}$.

Note that Theorem 3.2.4 gives for $1 \leq r \leq n^{1/4}$,

$$\begin{aligned}
\mathbb{P}\{\, \text{osc}(B; r\sqrt{n}, 2n) &> (r/3)\, n^{1/4}\, \sqrt{\log n}\, \} \\
&\leq 3\, \mathbb{P}\{\, \text{osc}(B; r\sqrt{n}, n) > (r/3)\, n^{1/4}\, \sqrt{\log n}\, \} \\
&= 3\, \mathbb{P}\{\, \text{osc}(B; r\, n^{-1/2}) > (r/3)\, n^{-1/4}\, \sqrt{\log n}\} \\
&\leq 3\, \mathbb{P}\left\{ \text{osc}(B; r\, n^{-1/2}) > (\sqrt{r}/3)\, \sqrt{r\, n^{-1/2}\, \log(n^{1/2}/r)} \right\}.
\end{aligned}$$

If $\sqrt{r}/3 \geq c_*$ and $r \leq n^{1/4}$, we can use Theorem 3.2.4 to conclude that there exist c, a such that

$$\mathbb{P}\left\{ \text{osc}(B; r\, n^{-1/2}) > (\sqrt{r}/3)\, \sqrt{r\, n^{-1/2}\, \log(n^{1/2}/r)} \right\} \leq c\, e^{-ar \log n}.$$

For the second event, consider the martingale

$$M_j = \tau_j - j.$$

Using (A.12) on M_j and $-M_j$, we see that there exist c, a such that

$$\mathbb{P}\left\{ \max_{1 \leq j \leq n} |\tau_j - j| \geq r\, \sqrt{n} \right\} \leq c\, e^{-ar^2}. \tag{3.7}$$

□

♣ The proof actually gives the stronger upper bound of $c\,[e^{-ar^2} + e^{-ar\log n}]$ but we will not need this improvement.

Extending the Skorokhod approximation to continuous time simple random walk $\tilde{S}_t$ is not difficult, although in this case the path $t \mapsto \tilde{S}_t$ is not continuous. Let N_t be a Poisson process with parameter one defined on the same probability space and independent of the Brownian motion B. Then

$$\tilde{S}_t := S_{N_t}$$

has the distribution of the continuous-time simple random walk. Since $N_t - t$ is a martingale, and the Poisson distribution has exponential moments, another application of (A.12) shows that for $r \leq t^{1/4}$,

$$\mathbb{P}\left\{\max_{0\leq s\leq t} |N_s - s| \geq r\sqrt{t}\right\} \leq c\, e^{-ar^2}.$$

Let

$$\Theta(B, \tilde{S}; n) = \sup\{|B_t - \tilde{S}_t| : 0 \leq t \leq n\}.$$

Then the following is proved similarly.

Theorem 3.3.3 *There exist $0 < c, a < \infty$ such that for all $1 \leq r \leq n^{1/4}$ and all positive integers n*

$$\mathbb{P}\{\Theta(B, \tilde{S}; n) \geq r\, n^{1/4}\sqrt{\log n}\} \leq c\, e^{-ar}.$$

3.4 Higher dimensions

It is not difficult to extend Theorems 3.3.2 and 3.3.3 to $p \in \mathcal{P}_d$ for $d > 1$. A d-dimensional Brownian motion with covariance matrix Γ with respect to a filtration $\mathcal{F}_t$ is a collection of random variables $B_t, t \geq 0$ satisfying the following:

- $B_0 = 0$;
- if $s < t$, then $B_t - B_s$ is an $\mathcal{F}_t$-measurable random $\mathbb{R}^d$-valued variable, independent of $\mathcal{F}_s$, whose distribution is joint normal with mean zero and covariance matrix $(t - s)\,\Gamma$.
- with probability one, $t \mapsto B_t$ is a continuous function.

Lemma 3.4.1 *Suppose that $B^{(1)}, \ldots, B^{(l)}$ are independent one-dimensional standard Brownian motions and $v_1, \ldots, v_l \in \mathbb{R}^d$. Then*

$$B_t := B_t^{(1)}\, v_1 + \cdots + B_t^{(l)}\, v_l$$

is a Brownian motion in $\mathbb{R}^d$ with covariance matrix $\Gamma = AA^T$ where $A = [v_1\, v_2, \ldots, v_l]$.

Proof This is straightforward and is left to the reader. □

In particular, a standard d-dimensional Brownian motion is of the form

$$B_t = (B_t^{(1)}, \ldots, B_t^{(d)})$$

where $B^{(1)}, \ldots, B^{(d)}$ are independent one-dimensional Brownian motions. Its covariance matrix is the identity.

The next theorem shows that one can define d-dimensional Brownian motions and d-dimensional random walks on the same probability space so that their paths are close to each other. Although the proof will use Skorokhod embedding, it is not true that the d-dimensional random walk is embedded into the d-dimensional Brownian motion. In fact, it is impossible to have an embedded walk, since for $d > 1$ the probability that a d-dimensional Brownian motion B_t visits the countable set $\mathbb{Z}^d$ after time 0 is zero.

Theorem 3.4.2 *Let $p \in \mathcal{P}_d$ with covariance matrix Γ. There exist c, a and a probability space $(\Omega, \mathcal{F}, \mathbb{P})$ on which are defined a Brownian motion B with covariance matrix Γ; a discrete-time random walk S with increment distribution p; and a continuous-time random walk $\tilde{S}$ with increment distribution p such that for all positive integers n and all $1 \le r \le n^{1/4}$,*

$$\mathbb{P}\{\Theta(B, S; n) \ge r\, n^{1/4}\, \sqrt{\log n}\} \le c\, e^{-ar},$$
$$\mathbb{P}\{\Theta(B, \tilde{S}; n) \ge r\, n^{1/4}\, \sqrt{\log n}\} \le c\, e^{-ar}.$$

Proof Suppose that $v_1, \ldots, v_l$ are the points such that $p(v_j) = p(-v_j) = q_j/2$ and $p(z) = 0$ for all other $z \in \mathbb{Z}^d \setminus \{0\}$. Let $L_n = (L_n^1, \ldots, L_n^l)$ be a multinomial process with parameters $q_1, \ldots, q_l$, and let $B^1, \ldots, B^l$ be independent one-dimensional Brownian motions. Let $S^1, \ldots, S^l$ be the random walks derived from $B^1, \ldots, B^l$ by Skorokhod embedding. As was noted in (1.2),

$$S_n := S^1_{L_n^1}\, v_1 + \cdots + S^l_{L_n^l}\, v_l,$$

has the distribution of a random walk with increment distribution p. Also,

$$B_t := B_t^1 \, v_1 + \cdots + B_t^l \, v_l,$$

is a Brownian motion with covariance matrix Γ. The proof now proceeds as in the previous cases. One fact that is used is that the L_n^j have a binomial distribution and hence we can get an exponential estimate

$$\mathbb{P}\left\{ \max_{1\leq j\leq n} |L_j^i - q^i j| \geq a\sqrt{n} \right\} \leq c\, e^{-a}.$$

□

3.5 An alternative formulation

Here we give a slightly different, but equivalent, form of the strong approximation from which we get (3.2). We will illustrate this in the case of one-dimensional simple random walk. Suppose that B_t is a standard Brownian motion defined on a probability space $(\Omega, \mathcal{F}, \mathbb{P})$. For positive integer n, let $B_t^{(n)}$ denote the Brownian motion

$$B_t^{(n)} = n^{-1/2}\, B_{nt}.$$

Let $S^{(n)}$ denote the simple random walk derived from $B^{(n)}$ using the Skorokhod embedding. Then we know that for all positive integers T,

$$\mathbb{P}\left\{ \max_{0\leq t\leq Tn} |S_t^{(n)} - B_t^{(n)}| \geq c\, r\, (Tn)^{1/4} \sqrt{\log(Tn)} \right\} \leq c\, e^{-ar}.$$

If we let

$$W_t^{(n)} = n^{-1/2}\, S_{tn}^{(n)},$$

then this becomes

$$\mathbb{P}\left\{ \max_{0\leq t\leq T} |W_t^{(n)} - B_t| \geq c\, r\, T^{1/4}\, n^{-1/4} \sqrt{\log(Tn)} \right\} \leq c\, e^{-ar}.$$

In particular, if $r = c_1 \log n$ where $c_1 = c_1(T)$ is chosen sufficiently large,

$$\mathbb{P}\left\{ \max_{0\leq t\leq T} |W_t^{(n)} - B_t| \geq c_1\, n^{-1/4} \log^{3/2} n \right\} \leq c_1\, n^{-2}.$$

By the Borel–Cantelli lemma, with probability one

$$\max_{0 \le t \le T} |W_t^{(n)} - B_t| \le c_1 \, n^{-1/4} \log^{3/2} n$$

for all n sufficiently large. In particular, with probability one $W^{(n)}$ converges to B in the metric space $C[0,T]$.

By using a multinomial process (in the discrete-time case) or a Poisson process (in the continuous-time) case, we can prove the following.

Theorem 3.5.1 *Suppose that $p \in \mathcal{P}_d$ with covariance matrix Γ. There exist $c < \infty, a > 0$ and a probability space $(\Omega, \mathcal{F}, \mathbb{P})$ on which are defined a d-dimensional Brownian motion B_t with covariance matrix Γ; an infinite sequence of discrete-time p-walks, $S^{(1)}, S^{(2)}, \ldots$ and an infinite sequence of continuous time p-walks $\tilde{S}^{(1)}, \tilde{S}^{(2)}, \ldots$ such that the following holds for every $r > 0$, $T \ge 1$. Let*

$$W_t^{(n)} = n^{-1/2} \, S_{nt}^{(n)}, \qquad \tilde{W}_t^{(n)} = n^{-1/2} \, \tilde{S}_{nt}^{(n)}.$$

Then,

$$\mathbb{P}\left\{ \max_{0 \le t \le T} |W_t^{(n)} - B_t| \ge c \, r \, T^{1/4} \, n^{-1/4} \sqrt{\log(Tn)} \right\} \le c \, e^{-ar}.$$

$$\mathbb{P}\left\{ \max_{0 \le t \le T} |\tilde{W}_t^{(n)} - B_t| \ge c \, r \, T^{1/4} \, n^{-1/4} \sqrt{\log(Tn)} \right\} \le c \, e^{-ar}.$$

In particular, with probability one, $W^{(n)} \to B$ and $\tilde{W}^{(n)} \to B$ in the metric space $C^d[0,T]$.

Exercises

Exercise 3.1 Show that if B_t is a standard Brownian motion with respect to the filtration $\mathcal{F}_t$ and $s < t$, then $\mathbb{E}[B_t^2 - t \mid \mathcal{F}_s] = B_s^2 - s$.

Exercise 3.2 Let X be an integer-valued random variable with $\mathbb{P}\{X = 0\} = 0$ and $\mathbb{E}[X] = 0$.

(i) Show that there exist numbers $r_j \in (0, \infty]$,

$$r_1 \le r_2 \le \cdots \qquad r_{-1} \le r_{-2} \le \cdots$$

such that if B_t is a standard Brownian motion and

$$T = \inf\{t : B_t \in \mathbb{Z} \setminus \{0\}, t \le r_{B_t}\},$$

then B_T has the same distribution as X.

(ii) Show that if X has bounded support, then there exists a $b > 0$ with $\mathbb{E}[e^{bT}] < \infty$.

(iii) Show that $\mathbb{E}[T] = \mathbb{E}[X^2]$.
(Hint: you may wish to consider first the cases where X is supported on $\{1, -1\}$, $\{1, 2, -1\}$, and $\{1, 2, -1, -2\}$, respectively.)

Exercise 3.3 Show that there exist $c < \infty, \alpha > 0$ such that the following is true. Suppose that $B_t = (B_t^1, B_t^2)$ is a standard two-dimensional Brownian motion and let $T_R = \inf\{t : |B_t| \ge R\}$. Let U_R denote the unbounded component of the open set $\mathbb{R}^2 \setminus B[0, T_R]$. Then,

$$\mathbb{P}^x\{0 \in U_R\} \le c(|x|/R)^\alpha .$$

(Hint: show there is a $\rho < 1$ such that for all R and all $|x| < R$,

$$\mathbb{P}^x\{0 \in U_{2R} \mid 0 \in U_R\} \le \rho.)$$

Exercise 3.4 Show that there exist $c < \infty, \alpha > 0$ such that the following is true. Suppose that S_n is simple random walk in $\mathbb{Z}^2$ starting at $x \ne 0$, and let $\xi_R = \min\{n : |S_n| \ge R\}$. Then the probability that there is a nearest neighbor path starting at the origin and ending at $\{|z| \ge R\}$ that does intersect $\{S_j : 0 \le j \le \xi_R\}$ is no more than $c(|x|/R)^\alpha$.

(Hint: follow the hint in Exercise 3.3, using the invariance principle to show the existence of a ρ.)

4
The Green's function

4.1 Recurrence and transience

A random walk S_n with increment distribution $p \in \mathcal{P}_d \cup \mathcal{P}_d^*$ is called *recurrent* if $\mathbb{P}\{S_n = 0 \text{ i.o.}\} = 1$. If the walk is not recurrent it is called *transient*. We will also say that p is recurrent or transient. It is easy to see using the Markov property that p is recurrent if and only if for each $x \in \mathbb{Z}^d$,

$$\mathbb{P}^x\{S_n = 0 \text{ for some } n \geq 1\} = 1,$$

and p is transient if and only if the *escape probability*, q, is positive, where q is defined by

$$q = \mathbb{P}\{S_n \neq 0 \text{ for all } n \geq 1\}.$$

Theorem 4.1.1 *If $p \in \mathcal{P}_d'$ with $d = 1, 2$, then p is recurrent. If $p \in \mathcal{P}_d^*$ with $d \geq 3$, then p is transient. For all p,*

$$q = \left[\sum_{n=0}^{\infty} p_n(0)\right]^{-1}, \tag{4.1}$$

where the left-hand side equals zero if the sum is divergent.

Proof Let $Y = \sum_{n=0}^{\infty} 1\{S_n = 0\}$ denote the number of visits to the origin and note that

$$\mathbb{E}[Y] = \sum_{n=0}^{\infty} \mathbb{P}\{S_n = 0\} = \sum_{n=0}^{\infty} p_n(0).$$

If $p \in \mathcal{P}_d'$ with $d = 1, 2$, the LCLT (see Theorem 2.1.1 and Theorem 2.3.9) implies that $p_n(0) \sim c\, n^{-d/2}$ and the sum is infinite. If $p \in \mathcal{P}_d^*$ with $d \geq 3$, then

(2.49) shows that $p_n(0) \le c\, n^{-d/2}$ and hence $\mathbb{E}[Y] < \infty$. We can compute $\mathbb{E}(Y)$ in terms of q. Indeed, the Markov property shows that $\mathbb{P}\{Y = j\} = (1-q)^{j-1}\, q$. Therefore, if $q > 0$,

$$\mathbb{E}[Y] = \sum_{j=0}^{\infty} j\,\mathbb{P}\{Y = j\} = \sum_{j=0}^{\infty} j\,(1-q)^{j-1}\, q = \frac{1}{q}.$$

□

4.2 The Green's generating function

If $p \in \mathcal{P} \cup \mathcal{P}^*$ and $x, y \in \mathbb{Z}^d$, we define the *Green's generating function* to be the power series in ξ:

$$G(x, y; \xi) = \sum_{n=0}^{\infty} \xi^n\, \mathbb{P}^x\{S_n = y\} = \sum_{n=0}^{\infty} \xi^n\, p_n(y - x).$$

Note that the sum is absolutely convergent for $|\xi| < 1$. We write just $G(y; \xi)$ for $G(0, y, \xi)$. If $p \in \mathcal{P}$, then $G(x; \xi) = G(-x; \xi)$.

The generating function is defined for complex ξ, but there is a particular interpretation of the sum for positive $\xi \le 1$. Suppose that T is a random variable independent of the random walk S with a geometric distribution,

$$\mathbb{P}\{T = j\} = \xi^{j-1}\,(1-\xi), \qquad j = 1, 2, \ldots$$

i.e. $\mathbb{P}\{T > j\} = \xi^j$ (if $\xi = 1$, then $T \equiv \infty$). We think of T as a "killing time" for the walk and we will refer to such T as a geometric random variable with *killing rate* $1 - \xi$. At each time j, if the walker has not already been killed, the process is killed with probability $1 - \xi$, where the killing is independent of the walk. If the random walk starts at the origin, then the expected number of visits to x before being killed is given by

$$\mathbb{E}\left[\sum_{j<T} 1\{S_j = x\}\right] = \mathbb{E}\left[\sum_{j=0}^{\infty} 1\{S_j = x; T > j\}\right]$$
$$= \sum_{j=0}^{\infty} \mathbb{P}\{S_j = x; T > j\} = \sum_{j=0}^{\infty} p_j(x)\, \xi^j = G(x; \xi).$$

Theorem 4.1.1 states that a random walk is transient if and only if $G(0; 1) < \infty$, in which case the escape probability is $G(0; 1)^{-1}$. For a transient random walk, we define the *Green's function* to be

$$G(x,y) = G(x,y;1) = \sum_{n=0}^{\infty} p_n(y-x).$$

We write $G(x) = G(0,x)$; if $p \in \mathcal{P}$, then $G(x) = G(-x)$. The strong Markov property implies that

$$G(0,x) = \mathbb{P}\{S_n = x \text{ for some } n \geq 0\}\, G(0,0). \tag{4.2}$$

Similarly, we define

$$\tilde{G}(x,y;\xi) = \int_0^\infty \xi^t \, p_t(x,y)\, dt.$$

For $\xi \in (0,1)$ this is the expected amount of time spent at site y by a continuous-time random walk with increment distribution p before an independent "killing time" that has an exponential distribution with rate $-\log(1-\xi)$. We will now show that if we set $\xi = 1$, we get the same Green's function as that induced by the discrete walk.

Proposition 4.2.1 *If $p \in \mathcal{P}_d^*$ is transient, then*

$$\int_0^\infty \tilde{p}_t(x)\, dt = G(x).$$

Proof Let S_n denote a discrete-time walk with distribution p, let N_t denote an independent Poisson process with parameter one, and let $\tilde{S}_t$ denote the continuous-time walk $\tilde{S}_t = S_{N_t}$. Let

$$Y_x = \sum_{n=0}^{\infty} 1\{S_n = x\}, \qquad \tilde{Y}_x = \int_0^\infty 1\{\tilde{S}_t = x\}\, dt,$$

denote the amount of time spent at x by S and $\tilde{S}$, respectively. Then $G(x) = \mathbb{E}[Y_x]$. If we let $T_n = \inf\{t : N_t = n\}$, then we can write

$$\tilde{Y}_x = \sum_{n=0}^{\infty} 1\{S_n = x\}\,(T_{n+1} - T_n).$$

Independence of S and N implies that

$$\mathbb{E}[1\{S_n = x\}\,(T_{n+1} - T_n)] = \mathbb{P}\{S_n = x\}\,\mathbb{E}[T_{n+1} - T_n] = \mathbb{P}\{S_n = x\}.$$

Hence, $\mathbb{E}[\tilde{Y}_x] = \mathbb{E}[Y_x]$. $\square$

Remark Suppose that p is the increment distribution of a random walk in $\mathbb{Z}^d$. For $\epsilon > 0$, let p_ϵ denote the increment of the "lazy walker" given by

$$p_\epsilon(x) = \begin{cases} (1-\epsilon)\, p(x), & x \neq 0 \\ \epsilon + (1-\epsilon)\, p(0), & x = 0 \end{cases}$$

If p is irreducible and periodic on $\mathbb{Z}^d$, then for each $0 < \epsilon < 1$, p_ϵ is irreducible and aperiodic. Let $\mathcal{L}, \phi$ denote the generator and characteristic function for p, respectively. Then the generator and characteristic function for p_ϵ are

$$\mathcal{L}_\epsilon = (1-\epsilon)\, \mathcal{L}, \qquad \phi_\epsilon(\theta) = \epsilon + (1-\epsilon)\, \phi(\theta). \tag{4.3}$$

If p has mean zero and covariance matrix Γ, then p_ϵ has mean zero and covariance matrix

$$\Gamma_\epsilon = (1-\epsilon)\, \Gamma, \qquad \det \Gamma_\epsilon = (1-\epsilon)^d \det \Gamma. \tag{4.4}$$

If p is transient, and G, G_ϵ denote the Green's function for p, p_ϵ, respectively, then, like the last proposition, we can see that

$$G_\epsilon(x) = \frac{1}{1-\epsilon}\, G(x). \tag{4.5}$$

For some proofs it is convenient to assume that the walk is aperiodic; results for periodic walks can then be derived using these relations.

If $n \geq 1$, let $f_n(x, y)$ denote the probability that a random walk starting at x first visits y at time n (not counting time $n = 0$), i.e.

$$f_n(x, y) = \mathbb{P}^x\{S_n = y; S_1 \neq y, \ldots, S_{n-1} \neq y\} = \mathbb{P}^x\{\tau_y = n\},$$

where

$$\tau_y = \min\{j \geq 1 : S_j = y\}, \qquad \overline{\tau}_y = \min\{j \geq 0 : S_j = y\}.$$

Let $f_n(x) = f_n(0, x)$ and note that

$$\mathbb{P}^x\{\tau_y < \infty\} = \sum_{n=1}^{\infty} f_n(x, y) = \sum_{n=1}^{\infty} f_n(y - x) \leq 1.$$

Define the *first visit generating function* by

$$F(x, y; \xi) = F(y - x; \xi) = \sum_{n=1}^{\infty} \xi^n f_n(y - x).$$

If $\xi \in (0,1)$, then

$$F(x,y;\xi) = \mathbb{P}^x\{\tau_y < T_\xi\},$$

where T_ξ denotes an independent geometric random variable satisfying $\mathbb{P}\{T_\xi > n\} = \xi^n$.

Proposition 4.2.2 *If $n \geq 1$,*

$$p_n(y) = \sum_{j=1}^{n} f_j(y)\, p_{n-j}(0).$$

If $\xi \in \mathbb{C}$,

$$G(y;\xi) = \delta(y) + F(y;\xi)\, G(0;\xi), \tag{4.6}$$

where δ denotes the delta function. In particular, if $|F(0,\xi)| < 1$,

$$G(0;\xi) = \frac{1}{1 - F(0;\xi)}. \tag{4.7}$$

Proof The first equality follows from

$$\mathbb{P}\{S_n = y\} = \sum_{j=1}^{n} \mathbb{P}\{\tau_y = j; S_n - S_j = 0\} = \sum_{j=1}^{n} \mathbb{P}\{\tau_y = j\}\, p_{n-j}(0).$$

The second equality uses

$$\sum_{n=1}^{\infty} p_n(x)\,\xi^n = \left[\sum_{n=1}^{\infty} f_n(x)\,\xi^n\right]\left[\sum_{m=0}^{\infty} p_m(0)\,\xi^m\right],$$

which follows from the first equality. For $\xi \in (0,1]$, there is a probabilistic interpretation of (4.6). If $y \neq 0$, the expected number of visits to y (before time T_ξ) is the product of the probability of reaching y and the expected number of visits to y given that y is reached before time T_ξ. If $y = 0$, we have to add an extra 1 to account for $p_0(y)$. □

♣ If $\xi \in (0,1)$, the identity (4.7) can be considered as a generalization of (4.1). Note that

$$F(0;\xi) = \sum_{j=1}^{\infty} \mathbb{P}\{\tau_0 = j;\, T_\xi > j\} = \mathbb{P}\{\tau_0 < T_\xi\}$$

represents the probability that a random walk killed at rate $1-\xi$ returns to the origin before being killed. Hence, the probability that the walker does not return to the origin before being killed is

$$1 - F(0;\xi) = G(0;\xi)^{-1}. \tag{4.8}$$

The right-hand side is the reciprocal of the expected number of visits before killing. If p is transient, we can plug $\xi = 1$ into this expression and get (4.1).

Proposition 4.2.3 *Suppose that $p \in \mathcal{P}_d \cup \mathcal{P}_d^*$ with characteristic function ϕ. Then, if $x \in \mathbb{Z}^d$, $|\xi| < 1$,*

$$G(x;\xi) = \frac{1}{(2\pi)^d} \int_{[-\pi,\pi]^d} \frac{1}{1-\xi\phi(\theta)}\, e^{-ix\cdot\theta}\, d\theta.$$

If $d \geq 3$, this holds for $\xi = 1$, i.e.

$$G(x) = \frac{1}{(2\pi)^d} \int_{[-\pi,\pi]^d} \frac{1}{1-\phi(\theta)}\, e^{-ix\cdot\theta}\, d\theta.$$

Proof All of the integrals in this proof will be over $[-\pi,\pi]^d$. The formal calculation, using Corollary 2.2.3, is

$$\begin{aligned} G(x;\xi) = \sum_{n=0}^{\infty} \xi^n p_n(x) &= \sum_{n=0}^{\infty} \xi^n \frac{1}{(2\pi)^d} \int \phi(\theta)^n\, e^{-ix\cdot\theta}\, d\theta \\ &= \frac{1}{(2\pi)^d} \int \left[\sum_{n=0}^{\infty} (\xi\,\phi(\theta))^n\right] e^{-ix\cdot\theta}\, d\theta \\ &= \frac{1}{(2\pi)^d} \int \frac{1}{1-\xi\,\phi(\theta)}\, e^{-ix\cdot\theta}\, d\theta. \end{aligned}$$

The interchange of the sum and the integral in the second equality is justified by the dominated convergence theorem as we now describe. For each N,

$$\left|\sum_{n=0}^{N} \xi^n \phi(\theta)^n e^{-ix\cdot\theta}\right| \leq \frac{1}{1-|\xi|\,|\phi(\theta)|}.$$

If $|\xi| < 1$, then the right-hand side is bounded by $1/[1-|\xi|]$. If $p \in \mathcal{P}_d^*$ and $\xi = 1$, then (2.13) shows that the right-hand side is bounded by $c\,|\theta|^{-2}$ for some c. If $d \geq 3$, $|\theta|^{-2}$ is integrable on $[-\pi,\pi]^d$. If $p \in \mathcal{P}_d$ is bipartite, we can use (4.3) and (4.5). $\square$

Some results are easier to prove for geometrically killed random walks than for walks restricted to a fixed number of steps. This is because stopping time arguments work more nicely for such walks. Suppose that S_n is a random walk, τ is a stopping time for the random walk, and T is an independent geometric random variable. Then, on the event $\{T > \tau\}$ the distribution of $T - \tau$ given $S_n, n = 0, \ldots, \tau$ is the same as that of T. This "loss of memory" property for geometric and exponential random variables can be very useful. The next proposition gives an example of a result proved first for geometrically killed walks. The result for fixed length random walks can be deduced from the geometrically killed walk result by using *Tauberian theorems*. Tauberian theorems are one of the major tools for deriving facts about a sequence from its generating functions. We will only use some simple Tauberian theorems; see Section A.5.

Proposition 4.2.4 *Suppose that $p \in \mathcal{P}_d \cup \mathcal{P}'_d$, $d = 1, 2$. Let*

$$q(n) = \mathbb{P}\{S_j \neq 0 : j = 1, \ldots, n\}.$$

Then as $n \to \infty$,

$$q(n) \sim \begin{cases} r\,\pi^{-1}\, n^{-1/2}, & d = 1 \\ r\,(\log n)^{-1}, & d = 2. \end{cases}$$

where $r = (2\pi)^{d/2}\sqrt{\det \Gamma}$.

Proof We will assume that $p \in \mathcal{P}'_d$; it is not difficult to extend this to bipartite $p \in \mathcal{P}_d$. We will establish the corresponding facts about the generating functions for $q(n)$: as $\xi \to 1-$,

$$\sum_{n=0}^{\infty} \xi^n\, q(n) \sim \frac{r}{\mathbf{\Gamma}(1/2)}\, \frac{1}{\sqrt{1-\xi}}, \qquad d = 1, \tag{4.9}$$

$$\sum_{n=0}^{\infty} \xi^n q(n) \sim \frac{r}{1-\xi} \left[\log\left(\frac{1}{1-\xi}\right)\right]^{-1}, \qquad d = 2. \tag{4.10}$$

Here $\mathbf{\Gamma}$ denotes the Gamma function.† Since the sequence $q(n)$ is monotone in n, Propositions A.5.2 and A.5.3 imply the proposition (recall that $\mathbf{\Gamma}(1/2) = \sqrt{\pi}$).

† We use the bold face $\mathbf{\Gamma}$ to denote the Gamma function to distinguish it from the covariance matrix Γ.

Let T be a geometric random variable with killing rate $1-\xi$. Then (4.8) tells us that

$$\mathbb{P}\{S_j \neq 0 : j = 1, \ldots, T-1\} = G(0;\xi)^{-1}.$$

Also,

$$\mathbb{P}\{S_j \neq 0 : j = 1, \ldots, T-1\} = \sum_{n=0}^{\infty} \mathbb{P}\{T = n+1\}\, q(n) = (1-\xi) \sum_{n=0}^{\infty} \xi^n\, q(n).$$

Using (2.32) and Lemma A.5.1, we can see that as $\xi \to 1-$,

$$G(0;\xi) = \sum_{n=0}^{\infty} \xi^n\, p_n(0) = \sum_{n=0}^{\infty} \xi^n \left[\frac{1}{r\, n^{d/2}} + o\left(\frac{1}{n^{d/2}}\right)\right] \sim \frac{1}{r} F\left(\frac{1}{1-\xi}\right),$$

where

$$F(s) = \begin{cases} \Gamma(1/2)\, \sqrt{s}, & d = 1 \\ \log s, & d = 2. \end{cases}$$

This gives (4.9) and (4.10). □

Corollary 4.2.5 *Suppose that S_n is a random walk with increment distribution $p \in \mathcal{P}_d'$ and*

$$\tau = \tau_0 = \min\{j \geq 1 : S_j = 0\}.$$

Then $\mathbb{E}[\tau] = \infty$.

Proof If $d \geq 3$, then transience implies that $\mathbb{P}\{\tau = \infty\} > 0$. For $d = 1, 2$, the result follows from the previous proposition which tells us that

$$\mathbb{P}\{\tau > n\} \geq \begin{cases} c\, n^{-1/2}, & d = 1, \\ c\,(\log n)^{-1}, & d = 2. \end{cases}$$

□

♣ One of the basic ingredients of Proposition 4.2.4 is the fact that the random walk always starts afresh when it returns to the origin. This idea can be extended to returns of a random walk to a set if the set if sufficiently symmetric to look the same at all points. For an example, see Exercise 4.2.

4.3 The Green's function, transient case

In this section, we will study the Green's function for $p \in \mathcal{P}_d, d \geq 3$. The Green's function $G(x,y) = G(y,x) = G(y-x)$ is given by

$$G(x) = \sum_{n=0}^{\infty} p_n(x) = \mathbb{E}\left[\sum_{n=0}^{\infty} 1\{S_n = x\}\right] = \mathbb{E}^x\left[\sum_{n=0}^{\infty} 1\{S_n = 0\}\right].$$

Note that

$$G(x) = 1\{x = 0\} + \sum_y p(x,y)\, \mathbb{E}^y\left[\sum_{n=0}^{\infty} 1\{S_n = 0\}\right] = \delta(x) + \sum_y p(x,y)\, G(y),$$

In other words,

$$\mathcal{L}G(x) = -\delta(x) = \begin{cases} -1, & x = 0, \\ 0, & x \neq 0. \end{cases}$$

Recall from (4.2) that

$$G(x) = \mathbb{P}\{\overline{\tau}_x < \infty\}\, G(0).$$

♣ In the calculations above as well as throughout this section, we use the symmetry of the Green's function, $G(x,y) = G(y,x)$. For nonsymmetric random walks, one must be careful to distinguish between $G(x,y)$ and $G(y,x)$.

The next theorem gives the asymptotics of the Green's function as $|x| \to \infty$. Recall that $\mathcal{J}^*(x)^2 = d\,\mathcal{J}(x)^2 = x \cdot \Gamma^{-1} x$. Since Γ is nonsingular, $\mathcal{J}^*(x) \asymp \mathcal{J}(x) \asymp |x|$.

Theorem 4.3.1 *Suppose that $p \in \mathcal{P}_d$ with $d \geq 3$. As $|x| \to \infty$,*

$$G(x) = \frac{C_d^*}{\mathcal{J}^*(x)^{d-2}} + O\left(\frac{1}{|x|^d}\right) = \frac{C_d}{\mathcal{J}(x)^{d-2}} + O\left(\frac{1}{|x|^d}\right),$$

where

$$C_d^* = d^{(d/2)-1}\, C_d = \frac{\Gamma(\frac{d-2}{2})}{2\,\pi^{d/2}\,\sqrt{\det\Gamma}} = \frac{\Gamma(\frac{d}{2})}{(d-2)\,\pi^{d/2}\,\sqrt{\det\Gamma}}.$$

Here, Γ denotes the covariance matrix and $\mathbf{\Gamma}$ denotes the Gamma function. In particular, for simple random walk,

$$G(x) = \frac{d\,\mathbf{\Gamma}(\frac{d}{2})}{(d-2)\,\pi^{d/2}}\,\frac{1}{|x|^{d-2}} + O\left(\frac{1}{|x|^d}\right).$$

♣ For simple random walk we can write

$$C_d = \frac{2d}{(d-2)\,\omega_d} = \frac{2}{(d-2)\,V_d},$$

where ω_d denotes the surface area of unit $(d-1)$-dimensional sphere and V_d is the volume of the unit ball in $\mathbb{R}^d$. See Exercise 6.18 for a derivation of this relation. More generally,

$$C_d = \frac{2}{(d-2)\,V(\Gamma)}$$

where $V(\Gamma)$ denotes the volume of the ellipsoid $\{x \in \mathbb{R}^d : \mathcal{J}(x) \leq 1\}$.

The last statement of Theorem 4.3.1 follows from the first statement using $\Gamma = d^{-1} I, \mathcal{J}(x) = |x|$ for simple random walk. It suffices to prove the first statement for aperiodic p; the proof for bipartite p follows using (4.4) and (4.5). The proof of the theorem will consist of two estimates:

$$G(x) = \sum_{n=0}^{\infty} p_n(x) = O\left(\frac{1}{|x|^d}\right) + \sum_{n=1}^{\infty} \overline{p}_n(x), \tag{4.11}$$

and

$$\sum_{n=1}^{\infty} \overline{p}_n(x) = \frac{C_d^*}{\mathcal{J}^*(x)^{d-2}} + o\left(\frac{1}{|x|^d}\right).$$

The second estimate uses the next lemma.

Lemma 4.3.2 *Let $b > 1$. Then, as $r \to \infty$,*

$$\sum_{n=1}^{\infty} n^{-b}\, e^{-r/n} = \frac{\mathbf{\Gamma}(b-1)}{r^{b-1}} + O\left(\frac{1}{r^{b+1}}\right)$$

Proof The sum is a Riemann sum approximation of the integral

$$I_r := \int_0^{\infty} t^{-b}\, e^{-r/t}\, dt = \frac{1}{r^{b-1}} \int_0^{\infty} y^{b-2}\, e^{-y}\, dy = \frac{\mathbf{\Gamma}(b-1)}{r^{b-1}}. \tag{4.12}$$

If $f : (0,\infty) \to \mathbb{R}$ is a C^2 function and n is a positive integer, then Lemma A.1.1 gives

$$\left| f(n) - \int_{n-(1/2)}^{n+(1/2)} f(s)\,ds \right| \le \frac{1}{24} \sup\{|f''(t)| : |t-n| \le 1/2\}.$$

Choosing $f(t) = t^{-b}\, e^{-r/t}$, we get

$$\left| n^{-b}\, e^{-r/n} - \int_{n-(1/2)}^{n+(1/2)} t^{-b}\, e^{-r/t}\, dt \right| \le c\, \frac{1}{n^{b+2}} \left[1 + \frac{r^2}{n^2} \right] e^{-r/n}, \qquad n \ge \sqrt{r}.$$

(The restriction $n \ge \sqrt{r}$ is used to guarantee that $e^{-r/(n+(1/2))} \le c\, e^{-r/n}$.) Therefore,

$$\begin{aligned} \sum_{n\ge\sqrt{r}} \left| n^{-b}\, e^{-r/n} - \int_{n-(1/2)}^{n+(1/2)} t^{-b}\, e^{-r/t}\, dt \right| &\le c \sum_{n\ge\sqrt{r}} \frac{1}{n^{b+2}} \left[1 + \frac{r^2}{n^2} \right] e^{-r/n} \\ &\le c \int_0^\infty t^{-(b+2)} \left(1 + \frac{r^2}{t^2} \right) e^{-r/t}\, dt \\ &\le c\, r^{-(b+1)} \end{aligned}$$

The last step uses (4.12). It is easy to check that the sum over $n < \sqrt{r}$ and the integral over $t < \sqrt{r}$ decay faster than any power of r. □

Proof of Theorem 4.3.1. Using Lemma 4.3.2 with $b = d/2, r = \mathcal{J}^*(x)^2/2$, we have

$$\begin{aligned} \sum_{n=1}^{\infty} \overline{p}_n(x) &= \sum_{n=1}^{\infty} \frac{1}{(2\pi n)^{d/2} \sqrt{\det \Gamma}}\, e^{-\mathcal{J}^*(x)^2/(2n)} \\ &= \frac{\Gamma(\frac{d-2}{2})}{2\,\pi^{d/2} \sqrt{\det \Gamma}}\, \frac{1}{\mathcal{J}^*(x)^{(d-2)}} + O\left(\frac{1}{|x|^{d+2}} \right). \end{aligned}$$

Hence, we only need to prove (4.11). A simple estimate shows that

$$\sum_{n<|x|} \overline{p}_n(x)$$

as a function of x decays faster than any power of x. Similarly, using Proposition 2.1.2,

$$\sum_{n<|x|} p_n(x) = o(|x|^{-d}). \tag{4.13}$$

Using (2.5), we see that

$$\sum_{n>|x|^2} |p_n(x) - \bar{p}_n(x)| \le c \sum_{n>|x|^2} n^{-(d+2)/2} = O(|x|^{-d}).$$

Let $k = d + 3$. For $|x| \le n \le |x|^2$, (2.3) implies that there is an r such that

$$|p_n(x) - \bar{p}_n(x)| \le c \left[\left(\frac{|x|}{\sqrt{n}} \right)^k e^{-r|x|^2/n} \frac{1}{n^{(d+2)/2}} + \frac{1}{n^{(d+k-1)/2}} \right]. \tag{4.14}$$

Note that

$$\sum_{n \ge |x|} n^{-(d+k-1)/2} = O(|x|^{-(d+k-3)/2}) = O(|x|^{-d}),$$

and

$$\sum_{n \ge |x|} \left(\frac{|x|}{\sqrt{n}} \right)^k e^{-r\,|x|^2/n} \frac{1}{n^{(d+2)/2}} \le c \int_0^\infty \left(\frac{|x|}{\sqrt{t}} \right)^k \frac{e^{-r|x|^2/t}}{(\sqrt{t})^{d+2}}\, dt \le c\,|x|^{-d}.$$

Remark The error term in this theorem is very small. In order to prove that it is this small we need the sharp estimate (4.14) which uses the fact that the third moments of the increment distribution are zero. If $p \in \mathcal{P}_d'$ with bounded increments but with nonzero third moments, there exists a similar asymptotic expansion for the Green's function except that the error term is $O(|x|^{-(d-1)})$; see Theorem 4.3.5. We have used bounded increments (or at least the existence of sufficiently large moments) in an important way in (4.13). Theorem 4.3.5 proves asymptotics under weaker moment assumptions; however, mean zero, finite variance is not sufficient to conclude that the Green's function is asymptotic to $c\,\mathcal{J}^*(x)^{2-d}$ for $d \ge 4$. See Exercise 4.5.

♣ Often one does not use the full force of these asymptotics. An important thing to remember is that $G(x) \asymp |x|^{2-d}$. There are a number of ways to remember the exponent $2 - d$. For example, the CLT implies that the random walk should visit on the order of R^2 points in the ball of radius R. Since there are R^d points in this ball, the probability that a particular point is visited is of order R^{2-d}. In the case of standard d-dimensional Brownian motion, the Green's function is proportional to $|x|^{2-d}$. This is the unique (up to multiplicative constant) harmonic, radially symmetric function on $\mathbb{R}^d \setminus \{0\}$ that goes to zero as $|x| \to \infty$ (see Exercise 4.4).

Corollary 4.3.3 *If $p \in \mathcal{P}_d$, then*

$$\nabla_j G(x) = \nabla_j \frac{C_d}{\mathcal{J}^*(x)^{d-2}} + O(|x|^{-d}).$$

In particular, $\nabla_j G(x) = O(|x|^{-d+1})$. *Also,*

$$\nabla_j^2 G(x) = O(|x|^{-d}).$$

Remark We could also prove this corollary with improved error terms by using the difference estimates for the LCLT such as Theorem 2.3.6, but in this book we will not need the sharper results. If $p \in \mathcal{P}'_d$ with bounded increments but nonzero third moments, we could also prove difference estimates for the Green's function using Theorem 2.3.6. The starting point is to write

$$\nabla_y G(x) = \sum_{n=0}^{\infty} \nabla_y \overline{p}_n(x) + \sum_{n=0}^{\infty} [\nabla_y p_n(x) - \nabla_y \overline{p}_n(x)].$$

4.3.1 Asymptotics under weaker assumptions

In this section we establish the asymptotics for G for certain $p \in \mathcal{P}'_d, d \geq 3$. We will follow the basic outline of the proof of Theorem 4.3.1. Let $\overline{G}(x) = C_d^*/\mathcal{J}^*(x)^{d-2}$ denote the dominant term in the asymptotics. From that proof we see that

$$G(x) = \overline{G}(x) + o(|x|^{-d}) + \sum_{n=0}^{\infty} [p_n(x) - \overline{p}_n(x)].$$

In the discussion below, we let $\alpha \in \{0, 1, 2\}$. If $\mathbb{E}[|X_1|^4] < \infty$ and the third moments vanish, we set $\alpha = 2$. If this is not the case, but $\mathbb{E}[|X_1|^3] < \infty$, we set $\alpha = 1$. Otherwise, we set $\alpha = 0$. By Theorems 2.3.5 and 2.3.9 we can see that there exists a sequence $\delta_n \to 0$ such that

$$\sum_{n \geq |x|^2} |p_n(x) - \overline{p}_n(x)| \leq c \sum_{n \geq |x|^2} \frac{\delta_n + \alpha}{|x|^{(d+\alpha)/2}} = \begin{cases} o(|x|^{2-d}), & \alpha = 0 \\ O(|x|^{2-d-\alpha}), & \alpha = 1, 2. \end{cases}$$

This is the order of magnitude that we will try to show for the error term, so this estimate suffices for this sum. The sum that is more difficult to handle and which in some cases requires additional moment conditions is

$$\sum_{n < |x|^2} [p_n(x) - \overline{p}_n(x)].$$

Theorem 4.3.4 *Suppose that $p \in \mathcal{P}_3'$. Then*

$$G(x) = \overline{G}(x) + o\left(\frac{1}{|x|}\right).$$

If $\mathbb{E}[|X_1|^3] < \infty$ we can write

$$G(x) = \overline{G}(x) + O\left(\frac{\log |x|}{|x|^2}\right).$$

If $\mathbb{E}[|X_1|^4] < \infty$ and the third moments vanish, then

$$G(x) = \overline{G}(x) + O\left(\frac{1}{|x|^2}\right).$$

Proof By Theorem 2.3.10, there exists $\delta_n \to 0$ such that

$$\sum_{n<|x|^2} |p_n(x) - \overline{p}_n(x)| \leq c \sum_{n<|x|^2} \frac{\delta_n + \alpha}{|x|^2\, n^{(1+\alpha)/2}}.$$

□

The next theorem shows that if we assume enough moments of the distribution, then we get the asymptotics as in Theorem 4.3.1. Note that as $d \to \infty$, the number of moments assumed grows.

Theorem 4.3.5 *Suppose that $p \in \mathcal{P}_d'$, $d \geq 3$.*

- *If $\mathbb{E}|X_1|^{d+1}] < \infty$, then*

$$G(x) = \overline{G}(x) + O(|x|^{1-d}).$$

- *If $\mathbb{E}|X_1|^{d+3}] < \infty$ and the third moments vanish,*

$$G(x) = \overline{G}(x) + O(|x|^{-d}).$$

Proof Let $\alpha = 1$ under the weaker assumption and $\alpha = 2$ under the stronger assumption; set $k = d + 2\alpha - 1$ so that $\mathbb{E}[|X_1|^k] < \infty$. As mentioned above, it suffices to show that

$$\sum_{n<|x|^2} [p_n(x) - \overline{p}_n(x)] = O(|x|^{2-d-\alpha}).$$

Let $\epsilon = 2(1+\alpha)/(1+2\alpha)$. As before,

$$\sum_{n<|x|^\epsilon} \overline{p}_n(x)$$

decays faster than any power of $|x|$. Using (2.52), we have

$$\sum_{n<|x|^\epsilon} p_n(x) \le \frac{c}{|x|^k} \sum_{n<|x|^\epsilon} n^{\frac{k-d}{2}} = O(|x|^{2-d-\alpha}).$$

(The value of ϵ was chosen as the largest value for which this holds.) For the range $|x|^\epsilon \le n < |x|^2$, we use the estimate from Theorem 2.3.8:

$$\left|p_n(x) - \overline{p}_n(x)\right| \le \frac{c}{n^{(d+\alpha)/2}} \left[|x/\sqrt{n}|^{k-1}\, e^{-r|x|^2/n} + n^{-(k-2-\alpha)/2}\right].$$

As before,

$$\sum_{n\ge|x|^\epsilon} \frac{|x/\sqrt{n}|^{k-1}}{n^{(d+\alpha)/2}}\, e^{-r|x|^2/n} \le c \int_0^\infty \frac{|x|^{k-1}}{(\sqrt{t})^{k+d+\alpha-1}}\, e^{-r|x|^2/t}\, dt = O(|x|^{2-d-\alpha}).$$

Also,

$$\sum_{n\ge|x|^\epsilon} \frac{1}{n^{\frac{d}{2}+\frac{k}{2}-1}} = O(|x|^{-\epsilon(d+\alpha-\frac{5}{2})}) \le O(|x|^{2-d-\alpha}),$$

provided that

$$\left(d+\alpha-\frac{5}{2}\right)\frac{2(1+\alpha)}{1+2\alpha} \ge d-2+\alpha,$$

which can be readily checked for $\alpha = 1, 2$ if $d \ge 3$.

□

4.4 Potential kernel

4.4.1 Two dimensions

If $p \in \mathcal{P}_2^*$, the *potential kernel* is the function

$$a(x) = \sum_{n=0}^{\infty}[p_n(0) - p_n(x)] = \lim_{N\to\infty}\left[\sum_{n=0}^{N} p_n(0) - \sum_{n=0}^{N} p_n(x)\right]. \qquad (4.15)$$

Exercise 2.2 shows that $|p_n(0) - p_n(x)| \leq c\,|x|\,n^{-3/2}$, so the first sum converges absolutely. However, since $p_n(0) \asymp n^{-1}$, it is not true that

$$a(x) = \left[\sum_{n=0}^{\infty} p_n(0)\right] - \left[\sum_{n=0}^{\infty} p_n(x)\right]. \tag{4.16}$$

If $p \in \mathcal{P}_2$ is bipartite, the potential kernel for $x \in (\mathbb{Z}^2)_e$ is defined in the same way. If $x \in (\mathbb{Z}^2)_o$ we can define $a(x)$ by the second expression in (4.15). Many authors use the term Green's function for a or $-a$. Note that

$$a(0) = 0.$$

♣ If $p \in \mathcal{P}_d^*$ is transient, then (4.16) is valid, and $a(x) = G(0) - G(x)$, where G is the Green's function for p. Since $|p_n(0) - p_n(x)| \leq c\,|x|\,n^{-3/2}$ for all $p \in \mathcal{P}_2^*$, the same argument shows that a exists for such p.

Proposition 4.4.1 *If $p \in \mathcal{P}_2'$, then $2\,a(x)$ is the expected number of visits to x by a random walk starting at x before its first visit to the origin.*

Proof We delay this until the next section; see (4.31). □

Remark Using Proposition 4.4.1, we can see that if p_ϵ is defined as in (4.3), and a_ϵ denotes the potential kernel for p_ϵ, then

$$a_\epsilon(x) = \frac{1}{1-\epsilon}\,a(x). \tag{4.17}$$

Proposition 4.4.2 *If $p \in \mathcal{P}_2$,*

$$\mathcal{L}a(x) = \delta_0(x) = \begin{cases} 1, & x = 0 \\ 0, & x \neq 0. \end{cases}$$

Proof Recall that

$$\mathcal{L}[p_n(0) - p_n(x)] = -\mathcal{L}p_n(x) = p_n(x) - p_{n+1}(x).$$

For fixed x, the sequence $p_n(x) - p_{n+1}(x)$ is absolutely convergent. Hence, we can write

$$\begin{aligned}
\mathcal{L}a(x) = \sum_{n=0}^{\infty} \mathcal{L}[p_n(0) - p_n(x)] &= \lim_{N\to\infty} \sum_{n=0}^{N} \mathcal{L}[p_n(0) - p_n(x)] \\
&= \lim_{N\to\infty} \sum_{n=0}^{N} [p_n(x) - p_{n+1}(x)] \\
&= \lim_{N\to\infty} [p_0(x) - p_{N+1}(x)] \\
&= p_0(x) = \delta_0(x).
\end{aligned}$$

□

Proposition 4.4.3 *If $p \in \mathcal{P}_2^* \cup \mathcal{P}_2$, then*

$$a(x) = \frac{1}{(2\pi)^2} \int_{[-\pi,\pi]^2} \frac{1 - e^{-ix\cdot\theta}}{1 - \phi(\theta)}\, d\theta.$$

Proof By the remark above, it suffices to consider $p \in \mathcal{P}_2^*$. The formal calculation is

$$\begin{aligned}
a(x) = \sum_{n=0}^{\infty} [p_n(0) - p_n(x)] &= \sum_{n=0}^{\infty} \frac{1}{(2\pi)^2} \int \phi(\theta)^n \,[1 - e^{-ix\cdot\theta}]\, d\theta \\
&= \frac{1}{(2\pi)^2} \int \left[\sum_{n=0}^{\infty} \phi(\theta)^n\right] [1 - e^{-ix\cdot\theta}]\, d\theta \\
&= \frac{1}{(2\pi)^2} \int \frac{1 - e^{-ix\cdot\theta}}{1 - \phi(\theta)}\, d\theta.
\end{aligned}$$

All of the integrals are over $[-\pi,\pi]^2$. To justify the interchange of the sum and the integral we use (2.13) to obtain the estimate

$$\left|\sum_{n=0}^{N} \phi(\theta)^n \,[1 - e^{-ix\cdot\theta}]\right| \le \frac{|1 - e^{-ix\cdot\theta}|}{1 - |\phi(\theta)|} \le \frac{c\,|x\theta|}{|\theta|^2} \le \frac{c\,|x|}{|\theta|}.$$

Since $|\theta|^{-1}$ is an integrable function in $[-\pi,\pi]^2$, the dominated convergence theorem may be applied. □

Theorem 4.4.4 *If $p \in \mathcal{P}_2$, there exists a constant $C = C(p)$ such that as $|x| \to \infty$,*

$$a(x) = \frac{1}{\pi\sqrt{\det \Gamma}} \log[\mathcal{J}^*(x)] + C + O(|x|^{-2}).$$

For simple random walk,

$$a(x) = \frac{2}{\pi} \log |x| + \frac{2\gamma + \log 8}{\pi} + O(|x|^{-2}),$$

where γ is Euler's constant.

Proof We will assume that p is aperiodic; the bipartite case is done similarly. We write that

$$a(x) = \sum_{n \le \mathcal{J}^*(x)^2} p_n(0) - \sum_{n \le \mathcal{J}^*(x)^2} p_n(x) + \sum_{n > \mathcal{J}^*(x)^2} [p_n(0) - p_n(x)].$$

We know from (2.23) that

$$p_n(0) = \frac{1}{2\pi\sqrt{\det \Gamma}} \frac{1}{n} + O\left(\frac{1}{n^2}\right).$$

We therefore get

$$\sum_{n \le \mathcal{J}^*(x)^2} p_n(0) = 1 + O(|x|^{-2}) + \sum_{1 \le n \le \mathcal{J}^*(x)^2} \frac{1}{2\pi\sqrt{\det \Gamma}} \frac{1}{n} + \sum_{n=1}^{\infty} \left[p_n(0) - \frac{1}{2\pi\sqrt{\det \Gamma}} \frac{1}{n} \right],$$

where the last sum is absolutely convergent. Also,

$$\sum_{1 \le n \le \mathcal{J}^*(x)^2} \frac{1}{n} = 2 \log[\mathcal{J}^*(x)] + \gamma + O(|x|^{-2}),$$

where γ is Euler's constant (see Lemma A.1.3). Hence,

$$\sum_{n \le \mathcal{J}^*(x)^2} p_n(0) = \frac{1}{\pi\sqrt{\det \Gamma}} \log[\mathcal{J}^*(x)] + c' + O(|x|^{-2})$$

for some constant c'.

Proposition 2.1.2 shows that

$$\sum_{n \le |x|} p_n(x)$$

decays faster than any power of $|x|$. Theorem 2.3.8 implies that there exist c, r such that for $n \le \mathcal{J}^*(x)^2$,

$$\left| p_n(x) - \frac{1}{2\pi\, n\sqrt{\det \Gamma}}\, e^{-\mathcal{J}^*(x)^2/2n} \right| \le c \left[\frac{|x/\sqrt{n}|^5\, e^{-r|x|^2/n}}{n^2} + \frac{1}{n^3} \right]. \tag{4.18}$$

Therefore,

$$\sum_{|x| \le n \le \mathcal{J}^*(x)^2} \left| p_n(x) - \frac{1}{2\pi\, n\sqrt{\det \Gamma}}\, e^{-\mathcal{J}^*(x)^2/2n} \right|$$

$$\le O(|x|^{-2}) + c \sum_{|x| < n \le \mathcal{J}^*(x)^2} \left[\frac{|x/\sqrt{n}|^5\, e^{-r|x|^2/n}}{n^2} + \frac{1}{n^3} \right] \le c\, |x|^{-2}.$$

The last estimate is done as in the final step of the proof of Theorem 4.3.1. Similarly to the proof of Lemma 4.3.2 we can see that

$$\sum_{|x| \le n \le \mathcal{J}^*(x)^2} \frac{1}{n}\, e^{-\mathcal{J}^*(x)^2/2n} = \int_0^{\mathcal{J}^*(x)^2} \frac{1}{t}\, e^{-\mathcal{J}^*(x)^2/2t}\, dt + O(|x|^{-2})$$

$$= \int_1^{\infty} \frac{1}{y}\, e^{-y/2}\, dy + O(|x|^{-2}).$$

The integral contributes a constant. At this point we have shown that

$$\sum_{n \le \mathcal{J}^*(x)^2} [p_n(0) - p_n(x)] = \frac{1}{\pi\sqrt{\det \Gamma}} \log[\mathcal{J}^*(x)] + C' + O(|x|^{-2})$$

for some constant C'. For $n > \mathcal{J}^*(x)^2$, we use Theorem 2.3.8 and Lemma 4.3.2 again to conclude that

$$\sum_{n > \mathcal{J}^*(x)^2} [p_n(0) - p_n(x)] = c \int_0^1 \frac{1}{y} \left[1 - e^{-y/2} \right] dy + O(|x|^{-2}).$$

For simple random walk in two dimensions, it follows that

$$a(x) = \frac{2}{\pi} \log|x| + k + O(|x|^{-2}),$$

for some constant k. To determine k, we use

$$\phi(\theta^1, \theta^2) = 1 - \frac{1}{2}[\cos\theta^1 + \cos\theta^2].$$

Plugging this into Proposition 4.4.3 and doing the integral (details omitted, see Exercise 4.9), we get an exact expression for $x_n = (n, n)$ for integer $n > 0$

$$a(x_n) = \frac{4}{\pi}\left[1 + \frac{1}{3} + \frac{1}{5} + \cdots + \frac{1}{2n-1}\right].$$

However, we also know that

$$a(x_n) = \frac{2}{\pi}\log n + \frac{2}{\pi}\log\sqrt{2} + k + O(n^{-2}).$$

Therefore,

$$k = \lim_{n\to\infty}\left[-\frac{1}{\pi}\log 2 - \frac{2}{\pi}\log n + \frac{4}{\pi}\sum_{j=1}^{n}\frac{1}{2j-1}\right].$$

Using Lemma A.1.3 we can see that as $n \to \infty$,

$$\sum_{j=1}^{n}\frac{1}{2j-1} = \sum_{j=1}^{2n}\frac{1}{j} - \sum_{j=1}^{n}\frac{1}{2j} = \frac{1}{2}\log n + \log 2 + \frac{1}{2}\gamma + o(1).$$

Therefore,

$$k = \frac{3}{\pi}\log 2 + \frac{2}{\pi}\gamma.$$

□

♣ Roughly speaking, $a(x)$ is the difference between the expected number of visits to 0 and the expected number of visits to x by some large time N. Let us consider $N >> |x|^2$. By time $|x|^2$, the random walker has visited the origin about

$$\sum_{n<|x|^2} p_n(0) \sim \sum_{n<|x|^2}\frac{c}{n} \sim 2\,c\log|x|,$$

times where $c = (2\pi\sqrt{\det\Gamma})^{-1}$. It has visited x about $O(1)$ times. From time $|x|^2$ onward, $p_n(x)$ and $p_n(0)$ are roughly the same and the sum of the difference from then on is $O(1)$. This shows why we expect

$$a(x) = 2c \log |x| + O(1).$$

Note that $\log |x| = \log \mathcal{J}^*(x) + O(1)$.

♣ Although we have included the exact value γ_2 for simple random walk, we will never need to use this value.

Corollary 4.4.5 *If $p \in \mathcal{P}_2$,*

$$\nabla_j a(x) = \nabla_j \left[\frac{1}{\pi \sqrt{\det \Gamma}} \log[\mathcal{J}^*(x)] \right] + O(|x|^{-2}).$$

In particular, $\nabla_j a(x) = O(|x|^{-1})$. *Also,*

$$\nabla_j^2 a(x) = O(|x|^{-2}).$$

Remark One can give better estimates for the differences of the potential kernel by starting with Theorem 2.3.6 and then following the proof of Theorem 4.3.1. We give an example of this technique in Theorem 8.1.2.

4.4.2 Asymptotics under weaker assumptions

We can prove asymptotics for the potential kernel under weaker assumptions. Let

$$\bar{a}(x) = [\pi \sqrt{\det \Gamma}]^{-1} \log[\mathcal{J}^*(x)]$$

denote the leading term in the asymptotics.

Theorem 4.4.6 *Suppose that $p \in \mathcal{P}_2'$. Then*

$$a(x) = \bar{a}(x) + o(\log |x|).$$

If $\mathbb{E}[|X_1|^3] < \infty$, then there exists $C < \infty$ such that

$$a(x) = \bar{a}(x) + C + O(|x|^{-1}).$$

If $\mathbb{E}[|X_1|^6] < \infty$ and the third moments vanish, then

$$a(x) = \bar{a}(x) + C + O(|x|^{-2}).$$

Proof Let $\alpha = 0, 1, 2$ under the three possible assumptions, respectively. We start with $\alpha = 1, 2$ for which we can write

$$\sum_{n=0}^{\infty}[p_n(0) - p_n(x)] = \sum_{n=0}^{\infty}[\overline{p}_n(0) - \overline{p}_n(x)] + \sum_{n=0}^{\infty}[p_n(0) - \overline{p}_n(0)] + \sum_{n=0}^{\infty}[\overline{p}_n(x) - p_n(x)] \tag{4.19}$$

The estimate

$$\sum_{n=0}^{\infty}[\overline{p}_n(0) - \overline{p}_n(x)] = \overline{a}(x) + \tilde{C} + O(|x|^{-2})$$

is done as in Theorem 4.4.4. Since $|p_n(0) - \overline{p}_n(0)| \leq c\,n^{-3/2}$, the second sum on the right-hand side of (4.19) converges, and we set

$$C = \tilde{C} + \sum_{n=0}^{\infty}[p_n(0) - \overline{p}_n(0)].$$

We write

$$\left|\sum_{n=0}^{\infty}[\overline{p}_n(x) - p_n(x)]\right| \leq \sum_{n<|x|^2} |\overline{p}_n(x) - p_n(x)| + \sum_{n\geq|x|^2} |\overline{p}_n(x) - p_n(x)|.$$

By Theorem 2.3.5 and and Theorem 2.3.9,

$$\sum_{n\geq|x|^2} |\overline{p}_n(x) - p_n(x)| \leq c \sum_{n\geq|x|^2} n^{-(2+\alpha)/2} = O(|x|^{-\alpha}).$$

For $\alpha = 1$, Theorem 2.3.10 gives

$$\sum_{n<|x|^2} |\overline{p}_n(x) - p_n(x)| \leq \sum_{n<|x|^2} \frac{c}{|x|^2\, n^{1/2}} = O(|x|^{-1}).$$

If $\mathbb{E}[|X_1|^4] < \infty$ and the third moments vanish, a similar argument shows that the sum on the left-hand side is bounded by $O(|x|^{-2} \log |x|)$ which is a little bigger than we want. However, if we also assume that $\mathbb{E}[|X_1|^6] < \infty$, then we get an estimate as in (4.18), and we can show as in Theorem 4.4.4 that this sum is $O(|x|^{-2})$.

If we only assume that $p \in \mathcal{P}_2'$, then we cannot write (4.19) because the second sum on the right-hand side might diverge. Instead, we write

$$\sum_{n=0}^{\infty}[p_n(0) - p_n(x)] = \sum_{n \geq |x|^2} [p_n(0) - p_n(x)] + \sum_{n<|x|^2} [\overline{p}_n(0) - \overline{p}_n(x)]$$
$$+ \sum_{n<|x|^2} [p_n(0) - \overline{p}_n(0)] + \sum_{n<|x|^2} [\overline{p}_n(x) - p_n(x)].$$

As before,

$$\sum_{n<|x|^2} [\overline{p}_n(0) - \overline{p}_n(x)] = \overline{a}(x) + O(1).$$

Also, Exercise 2.2, Theorem 2.3.10, and (2.32), respectively, imply that

$$\sum_{n \geq |x|^2} |p_n(0) - p_n(x)| \leq \sum_{n \geq |x|^2} \frac{c\,|x|}{n^{3/2}} = O(1),$$

$$\sum_{n<|x|^2} |\overline{p}_n(x) - p_n(x)| \leq \sum_{n<|x|^2} \frac{c}{|x|^2} = O(1),$$

$$\sum_{n<|x|^2} |\overline{p}_n(0) - p_n(0)| \leq \sum_{n<|x|^2} o\left(\frac{1}{n}\right) = o(\log |x|).$$

□

4.4.3 One dimension

If $p \in \mathcal{P}_1'$, the potential kernel is defined in the same way

$$a(x) = \lim_{N \to \infty} \left[\sum_{n=0}^{N} p_n(0) - \sum_{n=0}^{N} p_n(x) \right].$$

In this case, the convergence is a little more subtle. We will restrict ourselves to walks satisfying $\mathbb{E}[|X|^3] < \infty$ for which the proof of the next proposition shows that the sum converges absolutely.

Proposition 4.4.7 *Suppose that $p \in \mathcal{P}_1'$ with $\mathbb{E}[|X|^3] < \infty$. Then there is a c such that for all x,*

$$\left| a(x)\,\sigma^2 - |x| \right| \leq c \log |x|.$$

If $\mathbb{E}[|X|^4] < \infty$ *and* $\mathbb{E}[X^3] = 0$*, then there is a* C *such that*

$$a(x) = \frac{|x|}{\sigma^2} + C + O(|x|^{-1}).$$

Proof Assume that $x > 0$. Let $\alpha = 1$ under the weaker assumption and $\alpha = 2$ under the stronger assumption. Theorem 2.3.6 gives

$$p_n(0) - p_n(x) = \overline{p}_n(0) - \overline{p}_n(x) + x\, O(n^{-(2+\alpha)/2}),$$

which shows that

$$\left| \sum_{n \geq x^2} [p_n(0) - p_n(x)] - [\overline{p}_n(0) - \overline{p}_n(x)] \right| \leq c\, x \sum_{n \geq x^2} n^{-(2+\alpha)/2} \leq c x^{1-\alpha}.$$

If $\alpha = 1$, Theorem 2.3.5 gives

$$\sum_{n < x^2} [p_n(0) - \overline{p}_n(0)] \leq c \sum_{n < x^2} n^{-1} = O(\log x).$$

If $\alpha = 2$, Theorem 2.3.5 gives $|p_n(0) - \overline{p}_n(0)| = O(n^{-3/2})$ and hence

$$\sum_{n < x^2} [p_n(0) - \overline{p}_n(0)] = C' + O(|x|^{-1}), \qquad C' := \sum_{n=0}^{\infty} [p_n(0) - \overline{p}_n(0)],$$

In both cases, Theorem 2.3.10 gives

$$\sum_{n < x^2} [p_n(x) - \overline{p}_n(x)] \leq \frac{c}{x^2} \sum_{n < x^2} n^{(1-\alpha)/2} \leq c\, x^{1-\alpha}.$$

Therefore,

$$a(x) = e(x) + \sum_{n=0}^{\infty} [\overline{p}_n(0) - \overline{p}_n(x)] = e(x) + \sum_{n=1}^{\infty} \frac{1}{\sqrt{2\pi\sigma^2 n}} \left[1 - e^{-\frac{x^2}{2\sigma^2 n}} \right],$$

where $e(x) = O(\log x)$ if $\alpha = 1$ and $e(x) = C' + O(x^{-1})$ if $\alpha = 2$. Standard estimates (see Section A.1.1), which we omit, show that there is a C'' such that as $x \to \infty$,

$$\sum_{n=1}^{\infty} \frac{1}{\sqrt{2\pi\sigma^2 n}} = C'' + \int_0^{\infty} \frac{1}{\sqrt{2\pi\sigma^2 t}} \left[1 - e^{-\frac{x^2}{2\sigma^2 t}} \right] dt + o(x^{-1}),$$

and

$$\int_0^\infty \frac{1}{\sqrt{2\pi\sigma^2 t}}\left[1 - e^{-\frac{x^2}{2\sigma^2 t}}\right] dt = \frac{2x}{\sigma^2\sqrt{2\pi}}\int_0^\infty \frac{1}{u^2}\left(1 - e^{-u^2/2}\right) du = \frac{x}{\sigma^2}.$$

Since $C = C' + C''$ is independent of x, the result also holds for $x < 0$. □

Theorem 4.4.8 *If $p \in \mathcal{P}_1$ and $x > 0$,*

$$a(x) = \frac{x}{\sigma^2} + \mathbb{E}^x\left[a(S_T) - \frac{S_T}{\sigma^2}\right], \tag{4.20}$$

where $T = \min\{n : S_n \le 0\}$. There exists $\beta > 0$ such that for $x > 0$,

$$a(x) = \frac{x}{\sigma^2} + C + O(e^{-\beta x}), \quad x \to \infty, \tag{4.21}$$

where

$$C = \lim_{y\to\infty} \mathbb{E}^y\left[a(S_T) - \frac{S_T}{\sigma^2}\right].$$

In particular, for simple random walk, $a(x) = |x|$.

Proof Assume that $y > x$, let $T_y = \min\{n : S_n \le 0 \text{ or } S_n \ge y\}$, and consider the bounded martingale $S_{n\wedge T_y}$. Then, the optional sampling theorem implies that

$$x = \mathbb{E}^x[S_0] = \mathbb{E}^x[S_{T_y}] = \mathbb{E}^x[S_T; T \le T_y] + \mathbb{E}^x[S_{T_y}; T > T_y].$$

If we let $y \to \infty$, we see that

$$\lim_{y\to\infty} \mathbb{E}^x[S_{T_y}; T > T_y] = x - \mathbb{E}^x[S_T].$$

Also, since $\mathbb{E}[S_{T_y} \mid T_y < T] = y + O(1)$, we can see that

$$\lim_{y\to\infty} y\,\mathbb{P}^x\{T_y < T\} = x - \mathbb{E}^x[S_T].$$

We now consider the bounded martingale $M_n = a(S_{n\wedge T_y})$. Then the optional sampling theorem implies that

$$a(x) = \mathbb{E}^x[M_0] = \mathbb{E}^x[M_{T_y}] = \mathbb{E}^x[a(S_T); T \le T_y] + \mathbb{E}^x[a(S_{T_y}); T > T_y].$$

As $y \to \infty$, $\mathbb{E}^x[a(S_T); T < T_y] \to \mathbb{E}^x[a(S_T)]$. Also, as $y \to \infty$,

$$\mathbb{E}^x[a(S_{T_y}); T > T_y] \sim \mathbb{P}^x\{T_y < T\} \left[\frac{y}{\sigma^2} + O(1)\right] \sim \frac{x - \mathbb{E}^x[S_T]}{\sigma^2}.$$

This gives (4.20).

We will sketch the proof of (4.21); we leave it as an exercise (Exercise 4.12) to fill in the details. We will show that there exists a β such that if $0 < x < y < \infty$, then

$$\sum_{j=0}^{\infty} |\mathbb{P}^x\{S_T = -j\} - \mathbb{P}^y\{S_T = -j\}| = O(e^{-\beta x}). \tag{4.22}$$

Even though we have written this as an infinite sum, the terms are nonzero only for j less than the range of the walk. Let $\rho_z = \min\{n \geq 0 : S_n \leq z\}$. Irreducibility and aperiodicity of the random walk can be used to see that there is an $\epsilon > 0$ such that for all $z > 0$, $\mathbb{P}^{z+1}\{S_{\rho_z} = z\} = \mathbb{P}^1\{\rho_0 = 0\} > \epsilon$. Let

$$f(r) = f_{-j}(r) = \sup_{x,y \geq r} |\mathbb{P}^x\{S_T = -j\} - \mathbb{P}^y\{S_T = -j\}|.$$

Then, if R denotes the range of the walk, we can see that

$$f(r+1) \leq (1-\epsilon) f(r-R).$$

Iteration of this inequality gives $f(kR) \leq (1-\epsilon)^{k-1} f(R)$ and this gives (4.22).

□

Remark There is another (perhaps more efficient) proof of this result; see Exercise 4.13. One may note that the proof does not use the symmetry of the walk to establish

$$a(x) = \frac{x}{\sigma^2} + C + O(e^{-\alpha x}), \quad x \to \infty.$$

Hence, this result holds for all mean zero walks with bounded increments. Applying the proof to negative x yields

$$a(-x) = \frac{|x|}{\sigma^2} + C_- + O(e^{-\alpha |x|}).$$

If the third moment of the increment distribution is nonzero, it is possible that $C \neq C_-$; see Exercise 4.14.

♣ The potential kernel in one dimension is not as useful as the potential kernel or the Green's function in higher dimensions. For $d \geq 2$, we use the fact that the potential kernel or the Green's function is harmonic on $\mathbb{Z}^d \setminus \{0\}$ and that we have very good estimates for the asymptotics. For $d = 1$, similar arguments can be done with the function $f(x) = x$ which is obviously harmonic.

4.5 Fundamental solutions

If $p \in \mathcal{P}$, the Green's function G for $d \geq 3$ or the potential kernel a for $d = 2$ is often called the *fundamental solution of the generator* $\mathcal{L}$, since

$$\mathcal{L}G(x) = -\delta(x), \qquad \mathcal{L}a(x) = \delta(x). \tag{4.23}$$

More generally, we write

$$\mathcal{L}_x G(x,y) = \mathcal{L}_x G(y,x) = -\delta(y-x), \quad \mathcal{L}_x a(x,y) = \mathcal{L}_x a(y,x) = \delta(y-x),$$

where $\mathcal{L}_x$ denotes $\mathcal{L}$ applied to the x variable.

Remark Symmetry of walks in $\mathcal{P}$ is necessary to derive (4.23). If $p \in \mathcal{P}^*$ is transient, the Green's function G does not satisfy (4.23). Instead, it satisfies $\mathcal{L}^R G(x) = -\delta_0(x)$ where $\mathcal{L}^R$ denotes the generator of the "backwards random walk" with increment distribution $p^R(x) = p(-x)$. The function $f(x) = G(-x)$ satisfies $\mathcal{L}f(x) = -\delta_0(x)$ and is therefore the fundamental solution of the generator. Similarly, if $p \in \mathcal{P}_2^*$, the fundamental solution of the generator is $f(x) = a(-x)$.

Proposition 4.5.1 *Suppose that $p \in \mathcal{P}_d$ with $d \geq 2$, and $f : \mathbb{Z}^d \to \mathbb{R}$ is a function satisfying $f(0) = 0$, $f(x) = o(|x|)$ as $x \to \infty$, and $\mathcal{L}f(x) = 0$ for $x \neq 0$. Then, there exists $b \in \mathbb{R}$ such that*

$$f(x) = b\,[G(x) - G(0)], \quad d \geq 3,$$
$$f(x) = b\,a(x), \quad d = 2.$$

Proof See Propositions 6.4.6 and 6.4.8. □

Remark The assumption $f(x) = o(|x|)$ is clearly needed since the function $f(x^1, \ldots, x^d) = x^1$ is harmonic.

Suppose that $d \geq 3$. If $f : \mathbb{Z}^d \to \mathbb{R}$ is a function with finite support, we define

$$Gf(x) = \sum_{y \in \mathbb{Z}^d} G(x,y) f(y) = \sum_{y \in \mathbb{Z}^d} G(y-x) f(y). \tag{4.24}$$

Note that if f is supported on A, then $\mathcal{L}Gf(x) = 0$ for $x \notin A$. Also, if $x \in A$,

$$\mathcal{L}Gf(x) = \mathcal{L}_x \sum_{y \in \mathbb{Z}^d} G(x,y) f(y) = \sum_{y \in \mathbb{Z}^d} \mathcal{L}_x G(x,y) f(y) = -f(x). \tag{4.25}$$

In other words $-G = \mathcal{L}^{-1}$. For this reason the Green's function is often called the *inverse of the (negative of the) Laplacian.* Similarly, if $d = 2$ and f has finite support, we define

$$af(x) = \sum_{y \in \mathbb{Z}^d} a(x,y) f(y) = \sum_{y \in \mathbb{Z}^d} a(y-x) f(y). \tag{4.26}$$

In this case we get

$$\mathcal{L}af(x) = f(x),$$

i.e. $a = \mathcal{L}^{-1}$.

4.6 The Green's function for a set

If $A \subset \mathbb{Z}^d$ and S is a random walk with increment distribution p, let

$$\tau_A = \min\{j \geq 1 : S_j \notin A\}, \qquad \overline{\tau}_A = \min\{j \geq 0 : S_j \notin A\}. \tag{4.27}$$

If $A = \mathbb{Z}^d \setminus \{x\}$, we write just $\tau_x, \overline{\tau}_x$, which is consistent with the definition of τ_x given earlier in this chapter. Note that $\tau_A, \overline{\tau}_A$ agree if $S_0 \in A$, but are different if $S_0 \notin A$. If p is transient or A is a proper subset of $\mathbb{Z}^d$, we define

$$G_A(x,y) = \mathbb{E}^x \left[\sum_{n=0}^{\overline{\tau}_A - 1} 1\{S_n = y\} \right] = \sum_{n=0}^{\infty} \mathbb{P}^x\{S_n = y; n < \overline{\tau}_A\}.$$

Lemma 4.6.1 *Suppose that $p \in \mathcal{P}_d$ and A is a proper subset of $\mathbb{Z}^d$.*

(a) $G_A(x,y) = 0$ unless $x, y \in A$.
(b) $G_A(x,y) = G_A(y,x)$ for all x, y.

(c) *For* $x \in A$, $\mathcal{L}_x G_A(x,y) = -\delta(y-x)$. *In particular, if* $f(y) = G_A(x,y)$, *then* f *vanishes on* $\mathbb{Z}^d \setminus A$ *and satisfies* $\mathcal{L}f(y) = -\delta(y-x)$ *on* A.

(d) *For each* $y \in A$,

$$G_A(y,y) = \frac{1}{\mathbb{P}^y\{\tau_A < \tau_y\}} < \infty.$$

(e) *If* $x, y \in A$, *then*

$$G_A(x,y) = \mathbb{P}^x\{\overline{\tau}_y < \tau_A\}\, G_A(y,y).$$

(f) $G_A(x,y) = G_{A-x}(0, y-x)$ *where* $A - x = \{z - x : z \in A\}$.

Proof This is easy and left to the reader. The second assertion may be surprising at first, but symmetry of the random walk implies that for $x, y \in A$,

$$\mathbb{P}^x\{S_n = y; n < \tau_A\} = \mathbb{P}^y\{S_n = x; n < \tau_A\}.$$

Indeed, if $z_o = x, z_1, z_2, \ldots, z_{n-1}, z_n = y \in A$, then

$$\mathbb{P}^x\{S_1 = z_1, S_2 = z_2, \ldots, S_n = y\} = \mathbb{P}^y\{S_1 = z_{n-1}, S_2 = z_{n-2}, \ldots, S_n = x\}.$$

□

The next proposition gives an important relation between the Green's function for a set and the Green's function or the potential kernel.

Proposition 4.6.2 *Suppose that* $p \in \mathcal{P}_d$, $A \subset \mathbb{Z}^d$, $x, y \in \mathbb{Z}^d$.

(a) *If* $d \geq 3$,

$$\begin{aligned} G_A(x,y) &= G(x,y) - \mathbb{E}^x[G(S_{\overline{\tau}_A}, y); \overline{\tau}_A < \infty] \\ &= G(x,y) - \sum_z \mathbb{P}^x\{S_{\overline{\tau}_A} = z\}\, G(z,y). \end{aligned}$$

(b) *If* $d = 1, 2$ *and* A *is finite,*

$$G_A(x,y) = \mathbb{E}^x[a(S_{\overline{\tau}_A}, y)] - a(x,y) = \left[\sum_z \mathbb{P}^x\{S_{\overline{\tau}_A} = z\}\, a(z,y)\right] - a(x,y). \tag{4.28}$$

Proof The result is trivial if $x \notin A$. We will assume that $x \in A$, in which case $\tau_A = \overline{\tau}_A$.

If $d \geq 3$, let $Y_y = \sum_{n=0}^{\infty} 1\{S_n = y\}$ denote the total number of visits to the point y. Then

$$Y_y = \sum_{n=0}^{\tau_A - 1} 1\{S_n = y\} + \sum_{n=\tau_A}^{\infty} 1\{S_n = y\}.$$

If we assume that $S_0 = x$ and take expectations of both sides, we get

$$G(x,y) = G_A(x,y) + \mathbb{E}^x[G(S_{\tau_A}, y)].$$

The $d = 1, 2$ case could be done using a similar approach, but it is easier to use a different argument. If $S_0 = x$ and g is any function, then it is easy to check that

$$M_n = g(S_n) - \sum_{j=0}^{n-1} \mathcal{L}g(S_j)$$

is a martingale. We apply this to $g(z) = a(z,y)$ for which $\mathcal{L}g(z) = \delta(z-y)$. Then,

$$\begin{aligned} a(x,y) &= \mathbb{E}^x[M_0] = \mathbb{E}^x[M_{n\wedge\tau_A}] \\ &= \mathbb{E}^x[a(S_{n\wedge\tau_A}, y)] - \mathbb{E}^x\left[\sum_{j=0}^{(n\wedge\tau_A)-1} 1\{S_j = y\}\right]. \end{aligned}$$

Since A is finite, the dominated convergence theorem implies that

$$\lim_{n\to\infty} \mathbb{E}^x[a(S_{n\wedge\tau_A}, y)] = \mathbb{E}^x[a(S_{\tau_A}, y)]. \tag{4.29}$$

The monotone convergence theorem implies that

$$\lim_{n\to\infty} \mathbb{E}^x\left[\sum_{j=0}^{(n\wedge\tau_A)-1} 1\{S_j = y\}\right] = \mathbb{E}^x\left[\sum_{j=0}^{\tau_A - 1} 1\{S_j = y\}\right] = G_A(x,y).$$

□

The finiteness assumption on A was used in (4.29). The next proposition generalizes this to all proper subsets A of $\mathbb{Z}^d, d = 1, 2$. Recall that

$\mathcal{B}_n = \{x \in \mathbb{Z}^d : |x| < n\}$. Define a function F_A by

$$F_A(x) = \lim_{n\to\infty} \frac{\log n}{\pi \sqrt{\det \Gamma}} \, \mathbb{P}^x\{\tau_{\mathcal{B}_n} < \overline{\tau}_A\}, \qquad d = 2,$$

$$F_A(x) = \lim_{n\to\infty} \frac{n}{\sigma^2} \, \mathbb{P}^x\{\tau_{\mathcal{B}_n} < \overline{\tau}_A\}, \qquad d = 1.$$

The existence of these limits is established in the next proposition. Note that $F_A \equiv 0$ on $\mathbb{Z}^d \setminus A$ since $\mathbb{P}^x\{\overline{\tau}_A = 0\} = 1$ for $x \in \mathbb{Z}^d \setminus A$.

Proposition 4.6.3 *Suppose that $p \in \mathcal{P}_d, d = 1, 2$ and A is a proper subset of $\mathbb{Z}^d$. Then, if $x, y \in \mathbb{Z}^d$,*

$$G_A(x, y) = \mathbb{E}^x[a(S_{\overline{\tau}_A}, y)] - a(x, y) + F_A(x).$$

Proof The result is trivial if $x \notin A$ so we will suppose that $x \in A$. Choose $n > |x|, |y|$ and let $A_n = A \cap \{|z| < n\}$. Using (4.28), we have

$$G_{A_n}(x, y) = \mathbb{E}^x[a(S_{\tau_{A_n}}, y)] - a(x, y).$$

Note also that

$$\mathbb{E}^x[a(S_{\tau_{A_n}}, y)] = \mathbb{E}^x[a(S_{\tau_A}, y); \tau_A \le \tau_{\mathcal{B}_n}] + \mathbb{E}^x[a(S_{\tau_{\mathcal{B}_n}}, y); \tau_A > \tau_{\mathcal{B}_n}].$$

The monotone convergence theorem implies that as $n \to \infty$,

$$G_{A_n}(x, y) \longrightarrow G_A(x, y), \qquad \mathbb{E}^x[a(S_{\tau_A}, y); \tau_A \le \tau_{\mathcal{B}_n}] \longrightarrow \mathbb{E}^x[a(S_{\tau_A}, y)].$$

Since $G_A(x, y) < \infty$, this implies that

$$\lim_{n\to\infty} \mathbb{E}^x[a(S_{\tau_{\mathcal{B}_n}}, y); \tau_A > \tau_{\mathcal{B}_n}] = G_A(x, y) + a(x, y) - \mathbb{E}^x[a(S_{\tau_A}, y)].$$

However, $n \le |S_{\tau_{\mathcal{B}_n}}| \le n + R$ where R denotes the range of the increment distribution. Hence Theorems 4.4.4 and 4.4.8 show that as $n \to \infty$,

$$\mathbb{E}^x[a(S_{\tau_{\mathcal{B}_n}}, y); \tau_A > \tau_{\mathcal{B}_n}] \sim \mathbb{P}^x\{\tau_A > \tau_{\mathcal{B}_n}\} \, \frac{\log n}{\pi \sqrt{\det \Gamma}}, \qquad d = 2,$$

$$\mathbb{E}^x[a(S_{\tau_{\mathcal{B}_n}}, y); \tau_A > \tau_{\mathcal{B}_n}] \sim \mathbb{P}^x\{\tau_A > \tau_{\mathcal{B}_n}\} \, \frac{n}{\sigma^2}, \qquad d = 1.$$

□

Remark We proved that for $d = 1, 2$,

$$F_A(x) = G_A(x, y) + a(x, y) - \mathbb{E}^x[a(S_{\overline{\tau}_A}, y)]. \tag{4.30}$$

This holds for all y. If we choose $y \in \mathbb{Z}^d \setminus A$, then $G_A(x, y) = 0$, and hence we can write

$$F_A(x) = a(x, y) - \mathbb{E}^x[a(S_{\overline{\tau}_A}, y)].$$

Using this expression, it is easy to see that

$$\mathcal{L}F_A(x) = 0, \qquad x \in A.$$

Also, if $\mathbb{Z}^d \setminus A$ is finite,

$$F_A(x) = a(x) + O_A(1), \quad x \to \infty.$$

In the particular case $A = \mathbb{Z}^d \setminus \{0\}, y = 0$, this gives

$$F_{\mathbb{Z}^d \setminus \{0\}}(x) = a(x).$$

Applying (4.30) with $y = x$, we get

$$G_{\mathbb{Z}^d \setminus \{0\}}(x, x) = F_{\mathbb{Z}^d \setminus \{0\}}(x) + a(0, x) = 2\, a(x). \tag{4.31}$$

The next simple proposition relates the Green's functions to "escape probabilities" from sets. The proof uses a *last-exit decomposition*. Note that the last time a random walk visits a set is a random time that is not a stopping time. If $A \subset A'$, the event $\{\overline{\tau}_{\mathbb{Z}^d \setminus A} < \overline{\tau}_{A'}\}$ is the event that the random walk visits A before leaving A'.

Proposition 4.6.4 (last-exit decomposition) *Suppose that $p \in \mathcal{P}_d$ and $A \subset \mathbb{Z}^d$. Then,*

(a) if A' is a proper subset of $\mathbb{Z}^d$ with $A \subset A'$,

$$\mathbb{P}^x\{\overline{\tau}_{\mathbb{Z}^d \setminus A} < \overline{\tau}_{A'}\} = \sum_{z \in A} G_{A'}(x, z)\, \mathbb{P}^z\{\tau_{\mathbb{Z}^d \setminus A} > \tau_{A'}\};$$

(b) if $\xi \in (0, 1)$ and T_ξ is an independent geometric random variable with killing rate $1 - \xi$, then

$$\mathbb{P}^x\{\overline{\tau}_{\mathbb{Z}^d \setminus A} < T_\xi\} = \sum_{z \in A} G(x, z; \xi)\, \mathbb{P}^z\{\tau_{\mathbb{Z}^d \setminus A} \geq T_\xi\};$$

(c) *if $d \geq 3$ and A is finite,*

$$\mathbb{P}^x\{S_j \in A \text{ for some } j \geq 0\} = \mathbb{P}^x\{\overline{\tau}_{\mathbb{Z}^d \setminus A} < \infty\}$$
$$= \sum_{z \in A} G(x,z)\, \mathbb{P}^z\{\tau_{\mathbb{Z}^d \setminus A} = \infty\}.$$

Proof We will prove the first assertion; the other two are left as Exercise 4.11. We assume that $x \in A'$ (for otherwise the result is trivial). On the event $\{\overline{\tau}_{\mathbb{Z}^d \setminus A} < \overline{\tau}_{A'}\}$, let σ denote the largest $k < \overline{\tau}_{A'}$ such that $S_k \in A$. Then,

$$\mathbb{P}^x\{\overline{\tau}_{\mathbb{Z}^d \setminus A} < \overline{\tau}_{A'}\} = \sum_{k=0}^{\infty} \sum_{z \in A} \mathbb{P}^x\{\sigma = k; S_\sigma = z\}$$
$$= \sum_{z \in A} \sum_{k=0}^{\infty} \mathbb{P}^x\{S_k = z; k < \tau_{A'}; S_j \notin A, j = k+1, \ldots, \tau_{A'}\}.$$

The Markov property implies that

$$\mathbb{P}^x\{S_j \notin A, j = k+1, \ldots, \tau_{A'} \mid S_k = z; k < \tau_{A'}\} = \mathbb{P}^z\{\tau_{A'} < \tau_{\mathbb{Z}^d \setminus A}\}.$$

Therefore,

$$\mathbb{P}^x\{\overline{\tau}_{\mathbb{Z}^d \setminus A} < \overline{\tau}_{A'}\} = \sum_{z \in A} \sum_{k=0}^{\infty} \mathbb{P}^x\{S_k = z; k < \tau_{A'}\}\, \mathbb{P}^z\{\tau_{A'} < \tau_{\mathbb{Z}^d \setminus A}\}$$
$$= \sum_{z \in A} G_{A'}(x,z)\, \mathbb{P}^z\{\tau_{A'} < \tau_{\mathbb{Z}^d \setminus A}\}.$$

□

The next proposition uses a last-exit decomposition to describe the distribution of a random walk conditioned to not return to its starting point before a killing time. The killing time is either geometric or the first exit time from a set.

Proposition 4.6.5 *Suppose that S_n is a p-walk with $p \in \mathcal{P}_d$; $0 \in A \subset \mathbb{Z}^d$; and $\xi \in (0,1)$. Let T_ξ be a geometric random variable independent of the random walk with killing rate $1 - \xi$. Let*

$$\rho = \max\{j \geq 0 : j \leq \tau_A, S_j = 0\}, \quad \rho^* = \max\{j \geq 0 : j < T_\xi, S_j = 0\}.$$

(a) *The distribution of $\{S_j : \rho \leq j \leq \tau_A\}$ is the same as the conditional distribution of $\{S_j : 0 \leq j \leq \tau_A\}$ given that $\rho = 0$.*

(b) The distribution of $\{S_j : \rho^* \le j < T_\xi\}$ *is the same as the conditional distribution of* $\{S_j : 0 \le j < T_\xi\}$ *given that* $\rho^* = 0$.

Proof The usual Markov property implies that for any positive integer j, any $x_1, x_2, \ldots, x_{k-1} \in A \setminus \{0\}$ and any $x_k \in \mathbb{Z}^d \setminus A$,

$$\begin{aligned}\mathbb{P}\{\rho = j, \tau_A = j + k, S_{j+1} = x_1, \ldots, S_{j+k} = x_k\} \\ = \mathbb{P}\{S_j = 0, \tau_A > j, S_{j+1} = x_1, \ldots, S_{j+k} = x_k\} \\ = \mathbb{P}\{S_j = 0, \tau_A > j\}\, \mathbb{P}\{S_1 = x_1, \ldots, S_k = x_k\}.\end{aligned}$$

The first assertion is obtained by summation over j and the other equality is done similarly. □

Exercises

Exercise 4.1 Suppose that $p \in \mathcal{P}_d$ and S_n is a p-walk. Suppose that $A \subset \mathbb{Z}^d$ and that $\mathbb{P}^x\{\overline{\tau}_A = \infty\} > 0$ for some $x \in A$. Show that for every $\epsilon > 0$, there is a y with $\mathbb{P}^y\{\overline{\tau}_A = \infty\} > 1 - \epsilon$.

Exercise 4.2 Suppose that $p \in \mathcal{P}_d \cup \mathcal{P}'_d, d \ge 2$ and let $x \in \mathbb{Z}^d \setminus \{0\}$. Let

$$T = \min\{n > 0 : S_n = jx \text{ for some } j \in \mathbb{Z}\}.$$

Show that there exists $c = c(x)$ such that as $n \to \infty$,

$$\mathbb{P}\{T > n\} \sim \begin{cases} c\,n^{-1/2}, & d = 2 \\ c\,(\log n)^{-1}, & d = 3 \\ c, & d \ge 4. \end{cases}$$

Exercise 4.3 Suppose that $d = 1$. Show that the only function satisfying the conditions of Proposition 4.5.1 is the zero function.

Exercise 4.4 Find all radially symmetric functions f in $\mathbb{R}^d \setminus \{0\}$ satisfying $\Delta f(x) = 0$ for all $x \in \mathbb{R}^d \setminus \{0\}$.

Exercise 4.5 For each positive integer k find positive integer d and $p \in \mathcal{P}'_d$ such that $\mathbb{E}[|X_1|^k] < \infty$ and

$$\limsup_{|x| \to \infty} |x|^{d-2}\, G(x) = \infty.$$

(Hint: consider a sequence of points $z_1, z_2, \ldots$ going to infinity and define $\mathbb{P}\{X_1 = z_j\} = q_j$. Note that $G(z_j) \geq q_j$. Make a good choice of $z_1, z_2, \ldots$ and $q_1, q_2, \ldots$)

Exercise 4.6 Suppose that $X_1, X_2, \ldots$ are independent, identically distributed random variables in $\mathbb{Z}$ with mean zero. Let $S_n = X_1 + \cdots + X_n$ denote the corresponding random walk and let

$$G_n(x) = \sum_{j=0}^{n} \mathbb{P}\{S_j = x\}$$

be the expected number of visits to x in the first n steps of the walk.

(i) Show that $G_n(x) \leq G_n(0)$ for all n.
(ii) Use the law of large numbers to conclude that for all $\epsilon > 0$, there is an N_ϵ such that for $n \geq N_\epsilon$,

$$\sum_{|x| \leq \epsilon n} G_n(x) \geq \frac{n}{2}.$$

(iii) Show that

$$G(0) = \lim_{n \to \infty} G_n(0) = \infty$$

and conclude that the random walk is recurrent.

Exercise 4.7 Suppose that $A \subset \mathbb{Z}^d$ and $x, y \in A$. Show that

$$G_A(x, y) = \lim_{n \to \infty} G_{A_n}(x, y),$$

where $A_n = \{z \in A : |z| < n\}$.

Exercise 4.8 Let S_n denote simple random walk in $\mathbb{Z}^2$ starting at the origin and let $\rho = \min\{j \geq 1 : S_j = 0 \text{ or } \mathbf{e}_1\}$. Show that $\mathbb{P}\{S_\rho = 0\} = 1/2$.

Exercise 4.9 Consider the random walk in $\mathbb{Z}^2$ that moves at each step to one of $(1, 1)$, $(1, -1)$, $(-1, 1)$, $(-1, -1)$ each with probability $1/4$. Although this walk is not irreducible, many of the ideas of this chapter apply to this walk.

(i) Show that $\phi(\theta^1, \theta^2) = 1 - (\cos\theta^1)(\cos\theta^2)$.
(ii) Let a be the potential kernel for this random walk and $\hat{a}$ the potential kernel for simple random walk. Show that for every integer n, $a((n, 0)) = \hat{a}((n, n))$ (see Exercise 1.7).

(iii) Use Proposition 4.4.3 (which is valid for this walk) to show that for all integers $n > 0$,

$$a((n,0)) - a((n-1,0)) = \frac{4}{\pi(2n-1)},$$

$$a((n,0)) = \frac{4}{\pi}\left[1 + \frac{1}{3} + \frac{1}{5} + \cdots + \frac{1}{2n-1}\right].$$

Exercise 4.10 Suppose that $p \in \mathcal{P}_1$ and let $A = \{1, 2, \ldots\}$. Show that

$$F_A(x) = \frac{x - \mathbb{E}^x[S_T]}{\sigma^2},$$

where $T = \min\{j \geq 0 : S_j \leq 0\}$ and F_A is as in (4.30).

Exercise 4.11 Finish the details in Proposition 4.6.4.

Exercise 4.12 Finish the details in Theorem 4.4.8.

Exercise 4.13 Let S_j be a random walk in $\mathbb{Z}$ with increment distribution p satisfying

$$r_1 = \min\{j : p(j) > 0\} < \infty, \qquad r_2 = \max\{j : p(j) > 0\} < \infty,$$

and let $r = r_2 - r_1$.

(i) Show that if $\alpha \in \mathbb{R}$ and k is a nonnegative integer, then $f(x) = \alpha^x x^k$ satisfies $\mathcal{L}f(x) = 0$ for all $x \in \mathbb{R}$ if and only if $(s-\alpha)^{k-1}$ divides the polynomial

$$q(s) = \mathbb{E}\left[s^{X_1}\right].$$

(iii) Show that the set of functions on $\{-r+1, -r+2, \ldots\}$ satisfying $\mathcal{L}f(x) = 0$ for $x \geq 1$ is a vector space of dimension r.

(iii) Suppose that f is a function on $\{-r+1, -r+2, \ldots\}$ satisfying $\mathcal{L}f(x) = 0$ and $f(x) \sim x$ as $x \to \infty$. Show that there exists $c \in \mathbb{R}, c_1, \alpha > 0$ such that

$$|f(x) - x - c| \leq c_1\, e^{-\alpha x}.$$

Exercise 4.14 Find the potential kernel $a(x)$ for the one-dimensional walk with

$$p(-1) = p(-2) = \frac{1}{5}, \qquad p(1) = \frac{3}{5}.$$

5
One-dimensional walks

5.1 Gambler's ruin estimate

We will prove one of the basic estimates for one-dimensional random walks with zero mean and finite variance, often called the *gambler's ruin estimate*. We will not restrict to integer-valued random walks. For this section we assume that $X_1, X_2, \ldots$ are independent, identically distributed (one-dimensional) random variables with $\mathbb{E}[X_1] = 0, \mathbb{E}[X_1^2] = \sigma^2 > 0$. We let $S_n = S_0 + X_1 + \cdots + X_n$ be the corresponding random walk. If $r > 0$, we let

$$\eta_r = \min\{n \geq 0 : S_n \leq 0 \text{ or } S_n \geq r\},$$

$$\eta = \eta_\infty = \min\{n \geq 0 : S_n \leq 0\}.$$

We first consider simple random walk for which the gambler's ruin estimates are identities.

Proposition 5.1.1 *If S_n is one-dimensional simple random walk and $j < k$ are positive integers, then*

$$\mathbb{P}^j\{S_{\eta_k} = k\} = \frac{j}{k}.$$

Proof Since $M_n := S_{n\wedge\eta_k}$ is a bounded martingale, the optional sampling theorem implies that

$$j = \mathbb{E}^j[M_0] = \mathbb{E}^j[M_{\eta_k}] = k\,\mathbb{P}^j\{S_{\eta_k} = k\}.$$

□

Proposition 5.1.2 *If S_n is one-dimensional simple random walk, then for positive integer n,*

$$\mathbb{P}^1\{\eta > 2n\} = \mathbb{P}^1\{S_{2n} > 0\} - \mathbb{P}^1\{S_{2n} < 0\}$$
$$= \mathbb{P}\{S_{2n} = 0\} = \frac{1}{\sqrt{\pi\, n}} + O\left(\frac{1}{n^{3/2}}\right).$$

Proof Symmetry and the Markov property tell us that each $k < 2n$ and each positive integer x,

$$\mathbb{P}^1\{\eta = k, S_{2n} = x\} = \mathbb{P}^1\{\eta = k\}\, p_{2n-k}(x) = \mathbb{P}^1\{\eta = k, S_{2n} = -x\}.$$

Therefore,

$$\mathbb{P}^1\{\eta \le 2n, S_{2n} = x\} = \mathbb{P}^1\{\eta \le 2n, S_{2n} = -x\}.$$

Symmetry also implies that for all x, $\mathbb{P}^1\{S_{2n} = x + 2\} = \mathbb{P}^1\{S_{2n} = -x\}$. Since $\mathbb{P}^1\{\eta > 2n, S_{2n} = -x\} = 0$, for $x \ge 0$, we have

$$\mathbb{P}^1\{\eta > 2n\} = \sum_{x>0} \mathbb{P}\{\eta > 2n; S_{2n} = x\}$$
$$= \sum_{x>0} [p_{2n}(1,x) - p_{2n}(1,-x)]$$
$$= p_{2n}(1,1) + \sum_{x>0} [p_{2n}(1,x+2) - p_{2n}(1,-x)]$$
$$= p_{2n}(0,0) = 4^{-n}\binom{2n}{n} = \frac{1}{\sqrt{\pi\, n}} + O\left(\frac{1}{n^{3/2}}\right).$$

□

The proof of the gambler's ruin estimate for more general walks follows the same idea as that in the proof of Proposition 5.1.1. However, there is a complication arising from the fact that we do not know the exact value of S_{η_k}. Our first lemma shows that the application of the optional sampling theorem is valid. For this we do not need to assume that the variance is finite.

Lemma 5.1.3 *If $X_1, X_2, \ldots$ are independent identically distributed random variables in $\mathbb{R}$ with $\mathbb{E}(X_j) = 0$ and $\mathbb{P}\{X_j > 0\} > 0$, then for every $0 < r < \infty$ and every $x \in \mathbb{R}$,*

$$\mathbb{E}^x[S_{\eta_r}] = x. \tag{5.1}$$

Proof We assume that $0 < x < r$, for otherwise, the result is trivial. We start by showing that $\mathbb{E}^x[|S_{\eta_r}|] < \infty$. Since $\mathbb{P}\{X_j > 0\} > 0$, there exists an integer m and a $\delta > 0$ such that

$$\mathbb{P}\{X_1 + \cdots + X_m > r\} \geq \delta.$$

Therefore, for all x and all positive integers j,

$$\mathbb{P}^x\{\eta_r > jm\} \leq (1 - \delta)^m.$$

In particular, $\mathbb{E}^x[\eta_r] < \infty$. By the Markov property,

$$\begin{aligned}\mathbb{P}^x\{|S_{\eta_r}| \geq r + y; \eta_r = k\} &\leq \mathbb{P}^x\{\eta_r > k - 1; |X_k| \geq y\} \\ &= \mathbb{P}^x\{\eta_r > k - 1\} \, \mathbb{P}\{|X_k| \geq y\}.\end{aligned}$$

Summing over k gives

$$\mathbb{P}^x\{|S_{\eta_r}| \geq r + y\} \leq \mathbb{E}^x[\eta_r] \, \mathbb{P}\{|X_k| \geq y\}.$$

Hence,

$$\begin{aligned}\mathbb{E}^x\left[|S_{\eta_r}|\right] = \int_0^\infty \mathbb{P}^x\{|S_{\eta_r}| \geq y\}\, dy &\leq \mathbb{E}^x[\eta_r]\left[r + \int_0^\infty \mathbb{P}\{|X_k| \geq y\} dy\right] \\ &= \mathbb{E}^x[\eta_r]\left(r + \mathbb{E}\left[|X_j|\right]\right) < \infty.\end{aligned}$$

Since $\mathbb{E}^x[|S_{\eta_r}|] < \infty$, the martingale $M_n := S_{n \wedge \eta_r}$ is dominated by the integrable random variable $r + |S_{\eta_r}|$. Hence, it is a uniformly integrable martingale, and (5.1) follows from the optional sampling theorem (Theorem A.2.3). □

We now prove the estimates assuming a bounded range. We will take some care in showing how the constants in the estimate depend on the range.

Proposition 5.1.4 *For every $\epsilon > 0$ and $K < \infty$, there exist $0 < c_1 < c_2 < \infty$ such that if $\mathbb{P}\{|X_1| > K\} = 0$ and $\mathbb{P}\{X_1 \geq \epsilon\} \geq \epsilon$, then for all $0 < x < r$,*

$$c_1 \, \frac{x + 1}{r} \leq \mathbb{P}^x\{S_{\eta_r} \geq r\} \leq c_2 \, \frac{x + 1}{r}.$$

Proof We fix ϵ, K and allow constants in this proof to depend on ϵ, K. Let m be the smallest integer greater than K/ϵ. The assumption $\mathbb{P}\{X_1 \geq \epsilon\} \geq \epsilon$ implies that for all $x > 0$,

$$\mathbb{P}^x\{S_{\eta_K} \geq K\} \geq \mathbb{P}\{X_1 \geq \epsilon, \ldots, X_m \geq \epsilon\} \geq \epsilon^m.$$

Also note that if $0 \leq x \leq y \leq K$, then translation invariance and monotonicity give $\mathbb{P}^x(S_{\eta_r} \geq r) \leq \mathbb{P}^y(S_{\eta_r} \geq r)$. Therefore, for $0 < x \leq K$,

$$\epsilon^m \, \mathbb{P}^K\{S_{\eta_r} \geq r\} \leq \mathbb{P}^x\{S_{\eta_r} \geq r\} \leq \mathbb{P}^K\{S_{\eta_r} \geq r\}, \tag{5.2}$$

and hence it suffices to show for $K \leq x \leq r$ that

$$\frac{x}{r+K} \leq \mathbb{P}^x\{S_{\eta_r} \geq r\} \leq \frac{x+K}{r}.$$

By the previous lemma, $\mathbb{E}^x[S_{\eta_r}] = x$. If $S_{\eta_r} \geq r$, then $r \leq S_{\eta_r} \leq r+K$. If $S_{\eta_r} \leq 0$, then $-K \leq S_{\eta_r} \leq 0$. Therefore,

$$x = \mathbb{E}^x[S_{\eta_r}] \leq \mathbb{E}^x[S_{\eta_r}; S_{\eta_r} \geq r] \leq \mathbb{P}^x\{S_{\eta_r} \geq r\}\,(r+K),$$

and

$$x = \mathbb{E}^x[S_{\eta_r}] \geq \mathbb{E}^x[S_{\eta_r}; S_{\eta_r} \geq r] - K \geq r\,\mathbb{P}^x\{S_{\eta_r} \geq r\} - K.$$

□

Proposition 5.1.5 *For every $\epsilon > 0$ and $K < \infty$, there exist $0 < c_1 < c_2 < \infty$ such that if $\mathbb{P}\{|X_1| > K\} = 0$ and $\mathbb{P}\{X_1 \geq \epsilon\} \geq \epsilon$, then for all $x > 0, r > 1$,*

$$c_1 \, \frac{x+1}{r} \leq \mathbb{P}^x\{\eta \geq r^2\} \leq c_2 \, \frac{x+1}{r}.$$

Proof For the lower bound, we note that the maximal inequality for martingales (Theorem A.2.5) implies that

$$\mathbb{P}\left\{ \sup_{1 \leq j \leq n^2} |X_1 + \cdots + X_j| \geq 2Kn \right\} \leq \frac{\mathbb{E}[S_{n^2}^2]}{4K^2n^2} \leq \frac{1}{4}.$$

This tells us that if the random walk starts at $z \geq 3Kr$, then the probability that it does not reach the origin in r^2 steps is at least $3/4$. Using this, the strong Markov property, and the last proposition, we get

$$\mathbb{P}^x\{\eta \geq r^2\} \geq \frac{3}{4}\,\mathbb{P}^x\{S_{\eta_{3Kr}} \geq 3Kr\} \geq \frac{c_1\,(x+1)}{r}.$$

For the upper bound, we refer to Lemma 5.1.8 below. In this case, it is just as easy to give the argument for general mean zero, finite variance walks. □

If $p \in \mathcal{P}_d, d \geq 2$, then p induces an infinite family of one-dimensional nonlattice random walks $S_n \cdot \theta$ where $|\theta| = 1$. In Chapter 6, we will need gambler's ruin estimates for these walks that are uniform over all θ. In particular, it will be important that the constant is uniform over all θ.

Proposition 5.1.6 *Suppose that $\overline{S}_n$ is a random walk with increment distribution $p \in \mathcal{P}_d, d \geq 2$. There exist c_1, c_2 such that if $\theta \in \mathbb{R}^d$ with $|\theta| = 1$ and $S_n = \overline{S}_n \cdot \theta$, then the conclusions of Propositions 5.1.4 and 5.1.5 hold with c_1, c_2.*

Proof Clearly, there is a uniform bound on the range. The other condition is satisfied by noting the simple geometric fact that there is an $\epsilon > 0$, independent of θ such that $\mathbb{P}\{\overline{S}_1 \cdot \theta \geq \epsilon\} \geq \epsilon$; see Exercise 1.8. □

5.1.1 General case

We prove the gambler's ruin estimate assuming only mean zero and finite variance. While we will not attempt to get the best values for the constants, we do show that the constants can be chosen uniformly over a wide class of distributions. In this section we fix $K < \infty$, $\delta, b > 0$ and $0 < \rho < 1$, and we let $\mathcal{A}(K, \delta, b, \rho)$ be the collection of distributions on X_1 with $\mathbb{E}[X_1] = 0$,

$$\mathbb{E}[X_1^2] = \sigma^2 \leq K^2,$$

$$\mathbb{P}\{X_1 \geq 1\} \geq \delta,$$

$$\inf_n \mathbb{P}\{S_1, \ldots, S_{n^2} > -n\} \geq b,$$

$$\rho \leq \inf_{n>0} \mathbb{P}\{S_{n^2} \leq -n\}.$$

It is easy to check that for any mean zero, finite nonzero variance random walk S_n, we can find a $t > 0$ and some K, δ, b, ρ such that the estimates above hold for tS_n.

Theorem 5.1.7 (gambler's ruin) *For every K, δ, b, ρ, there exist $0 < c_1 < c_2 < \infty$ such that if $X_1, X_2, \ldots$ are independent identically distributed random variables whose distributions are in $\mathcal{A}(K, \delta, b, \rho)$, then for all $0 < x < r$,*

$$c_1 \frac{x+1}{r} \leq \mathbb{P}^x\{\eta > r^2\} \leq c_2 \frac{x+1}{r},$$

$$c_1 \frac{x+1}{r} \leq \mathbb{P}^x\{S_{\eta_r} \geq r\} \leq c_2 \frac{x+1}{r}.$$

Our argument consists of several steps. We start with the upper bound. Let

$$\eta_r^* = \min\{n > 0 : S_n \leq 0 \text{ or } S_n \geq r\}, \quad \eta^* = \eta_\infty^* = \min\{n > 0 : S_n \leq 0\}.$$

Note that η_r^* differs from η_r in that the minimum is taken over $n > 0$ rather than $n \geq 0$. As before, we write $\mathbb{P}$ for $\mathbb{P}^0$.

Lemma 5.1.8

$$\mathbb{P}\{\eta^* > n\} \leq \frac{4K}{\delta\sqrt{n}}, \qquad \mathbb{P}\{\eta_n^* < \eta^*\} \leq \frac{4K}{b\delta n}.$$

Proof Let $q_n = \mathbb{P}\{\eta^* > n\} = \mathbb{P}\{S_1, \ldots, S_n > 0\}$. Then

$$\mathbb{P}\{S_1, \ldots, S_n \geq 1\} \geq \delta q_{n-1} \geq \delta q_n.$$

Let $J_{k,n}$ be the event

$$J_{k,n} = \{S_{k+1}, \ldots, S_n \geq S_k + 1\}.$$

We will also use $J_{k,n}$ to denote the indicator function of this event. Let $m_n = \min\{S_j : 0 \leq j \leq n\}$, $M_n = \max\{S_j : 0 \leq j \leq n\}$. For each real $x \in [m_n, M_n]$, there is at most one integer k such that $S_k \leq x$ and $S_j > x, k < j \leq n$. On the event $J_{k,n}$, the random set corresponding to the jump from S_k to S_{k+1},

$$\{x : S_k \leq x \text{ and } S_j > x, k < j \leq n\},$$

contains an interval of length at least one. In other words, there are $\sum_k J_{k,n}$ nonoverlapping intervals contained in $[m_n, M_n]$ each of length at least one. Therefore,

$$\sum_{k=0}^{n-1} J_{k,n} \leq M_n - m_n.$$

But, $\mathbb{P}(J_{k,n}) \geq \delta q_{n-k} \geq \delta q_n$. Therefore,

$$n\delta q_n \leq \mathbb{E}[M_n - m_n] \leq 2\,\mathbb{E}[\max\{|S_j| : j \leq n\}].$$

Martingale maximal inequalities (Theorem A.2.5) give

$$\mathbb{P}\left\{ \max\{|S_j| : j \leq n\} \geq t \right\} \leq \frac{\mathbb{E}[S_n^2]}{t^2} \leq \frac{K^2 n}{t^2}.$$

Therefore,

$$\frac{n\delta q_n}{2} \leq \mathbb{E}[\max\{|S_j| : j \leq n\}] = \int_0^\infty \mathbb{P}\left\{\max\{|S_j| : j \leq n\} \geq t\right\} dt$$
$$\leq K\sqrt{n} + \int_{K\sqrt{n}}^\infty K^2 n t^{-2}\, dt = 2K\sqrt{n}.$$

This gives the first inequality. The strong Markov property implies that

$$\mathbb{P}\{\eta^* > n^2 \mid \eta_n^* < \eta^*\} \geq \mathbb{P}\{S_j - S_{\eta_n^*} > -n, 1 \leq j \leq n^2 \mid \eta_n^* < \eta^*\} \geq b.$$

Hence,

$$b\,\mathbb{P}\{\eta_n^* < \eta^*\} \leq \mathbb{P}\{\eta^* > n^2\}, \tag{5.3}$$

which gives the second inequality. □

Lemma 5.1.9 (overshoot lemma I) *For all $x > 0$,*

$$\mathbb{P}^x\{|S_\eta| \geq m\} \leq \frac{1}{\rho}\,\mathbb{E}[X_1^2; |X_1| \geq m]. \tag{5.4}$$

Moreover, if $\alpha > 0$ and $\mathbb{E}[|X_1|^{2+\alpha}] < \infty$, then

$$\mathbb{E}^x\left[|S_\eta|^\alpha\right] \leq \frac{\alpha}{\rho}\,\mathbb{E}[|X_1|^{2+\alpha}].$$

♣ Since $\mathbb{E}^x[\eta] = \infty$, we cannot use the proof from Lemma 5.1.3.

Proof Fix $\epsilon > 0$. For nonnegative integers k, let

$$Y_k = \sum_{n=0}^{\eta} 1\{k\epsilon < S_n \leq (k+1)\epsilon\}$$

be the number of times the random walk visits $(k\epsilon, (k+1)\epsilon]$ before hitting $(-\infty, 0]$, and let

$$g(x,k) = \mathbb{E}^x[Y_k] = \sum_{n=0}^{\infty} \mathbb{P}^x\{k\epsilon < S_n \leq (k+1)\epsilon; \eta > n\}.$$

Note that if $m, x > 0$,

$$\begin{aligned}
\mathbb{P}^x\{|S_\eta| \geq m\} &= \sum_{n=0}^{\infty} \mathbb{P}^x\{|S_\eta| \geq m; \eta = n+1\} \\
&= \sum_{n=0}^{\infty}\sum_{k=0}^{\infty} \mathbb{P}^x\{|S_\eta| \geq m; \eta = n+1; k\epsilon < S_n \leq (k+1)\epsilon\}
\end{aligned}$$

$$\leq \sum_{k=0}^{\infty} \sum_{n=0}^{\infty} \mathbb{P}^x\{\eta > n; k\epsilon < S_n$$
$$\leq (k+1)\epsilon; |S_{n+1} - S_n| \geq m + k\epsilon\}$$
$$= \sum_{k=0}^{\infty} g(x,k)\, \mathbb{P}\{|X_1| \geq m + k\epsilon\}$$
$$= \sum_{k=0}^{\infty} g(x,k) \sum_{l=k}^{\infty} \mathbb{P}\{m + l\epsilon \leq |X_1| < m + (l+1)\epsilon\}$$
$$= \sum_{l=0}^{\infty} \mathbb{P}\{m + l\epsilon \leq |X_1| < m + (l+1)\epsilon\} \sum_{k=0}^{l} g(x,k).$$

Recall that $\mathbb{P}\{S_{n^2} \leq -n\} \geq \rho$ for each n. We claim that for all x, y,

$$\sum_{0 \leq k < y/\epsilon} g(x,k) \leq \frac{y^2}{\rho}. \tag{5.5}$$

To see this, let

$$H_y = \max_{x>0} \sum_{0 \leq k < \lfloor y/\epsilon \rfloor} g(x,k).$$

Note that the maximum is the same if we restrict to $0 < x \leq y$. Then, for any $x \leq y$,

$$\sum_{0 \leq k < \lfloor y/\epsilon \rfloor} g(x,k) \leq y^2 + \mathbb{P}^x\{\eta \geq y^2\} \mathbb{E}\left[\sum_{n \geq y^2} 1\{S_n \leq y; n < \eta\} \,\middle|\, \eta \geq y^2 \right]$$
$$\leq y^2 + (1-\rho)\, H_y. \tag{5.6}$$

By taking the supremum over x we get $H_y \leq y^2 + (1-\rho) H_y$ which gives (5.5). We therefore have

$$\mathbb{P}^x\{|S_\eta| \geq m\} \leq \frac{1}{\rho} \sum_{l=0}^{\infty} \mathbb{P}\{m + l\epsilon \leq |X_1| < m + (l+1)\epsilon\}(l\epsilon + \epsilon)^2$$
$$\leq \frac{1}{\rho} \left(\mathbb{E}[(X_1 - \epsilon)^2; X_1 \leq -m] + \mathbb{E}[(X_1 + \epsilon)^2; X_1 \geq m] \right).$$

Letting $\epsilon \to 0$, we obtain (5.4).

To get the second estimate, let F denote the distribution function of $|X_1|$. Then

$$\begin{aligned}
\mathbb{E}^x[|S_\eta|^\alpha] &= \alpha \int_0^\infty t^{\alpha-1}\, \mathbb{P}^x\{|S_\eta| \geq t\}\, dt \\
&\leq \frac{\alpha}{\rho} \int_0^\infty t^{\alpha-1}\, \mathbb{E}[X_1^2; |X_1| \geq t]\, dt \\
&\leq \frac{\alpha}{\rho} \int_0^\infty \mathbb{E}\left[|X_1|^{1+\alpha}; |X_1| \geq t\right] dt \\
&= \frac{\alpha}{\rho} \int_0^\infty \int_t^\infty x^{1+\alpha}\, dF(x)\, dt \\
&= \frac{\alpha}{\rho} \int_0^\infty \left[\int_0^x dt\right] x^{1+\alpha}\, dF(x) = \frac{\alpha}{\rho}\, \mathbb{E}[|X_1|^{2+\alpha}].
\end{aligned}$$

□

♣ The estimate (5.6) illustrates a useful way to prove upper bounds for Green's functions of a set. If starting at any point y in a set $V \subset U$, there is a probability q of leaving U within N steps, then the expected amount of time spent in V before leaving U starting at any $x \in U$ is bounded above by

$$N + (1-q)\,N + (1-q)^2\,N + \cdots = \frac{N}{q}.$$

♣ The lemma states that the overshoot random variable has two fewer moments than the increment distribution. When the starting point is close to the origin, one might expect that the overshoot would be smaller, since there are fewer chances for the last step before entering $(-\infty, 0]$ to be much larger than a typical step. The next lemma confirms this intuition and shows that one gains one moment if one starts near the origin.

Lemma 5.1.10 (overshoot lemma II) *Let*

$$c' = \frac{32\,K}{b\delta}.$$

Then for all $0 < x \leq 1$,

$$\mathbb{P}^x\{|S_\eta| \geq m\} \leq \frac{c'}{\rho}\, \mathbb{E}[|X_1|; |X_1| \geq m].$$

Moreover, if $\alpha > 0$ *and* $\mathbb{E}[|X_1|^{1+\alpha}] < \infty$, *then*

$$\mathbb{E}^x[|S_\eta|^\alpha] \leq \frac{\alpha\, c'}{\rho}\, \mathbb{E}[|X_1|^{1+\alpha}].$$

Proof The proof proceeds exactly as in Lemma 5.1.9 up to (5.5) which we replace with a stronger estimate that is valid for $0 < x \leq 1$:

$$\sum_{0 \leq k\epsilon < y} g(x,k) \leq \frac{c'\, y}{\rho}. \tag{5.7}$$

To derive this estimate, we note that

$$\sum_{2^{j-1} \leq k\epsilon < 2^j} g(x,k)$$

equals the product of the probability of reaching a value above 2^{j-1} before hitting $(-\infty, 0]$ and the expected number of visits in this range given that event. Due to Lemma 5.1.8, the first probability is no more than $4K/(b\delta 2^{j-1})$ and the conditional expectation, as estimated in (5.5), is less than $2^{2j}/\rho$. Therefore,

$$\sum_{0 \leq k\epsilon < 2^j} g(x,k) \leq \frac{1}{\rho} \sum_{l=1}^{j} \left(\frac{4K}{b\delta 2^{l-1}} \right) 2^{2l} \leq \frac{1}{\rho} \left(\frac{16K}{b\delta} \right) 2^j.$$

For general y we write $2^{j-1} < y \leq 2^j$ and obtain (5.7).

Given this, the same argument gives

$$\mathbb{P}^x\{|S_\eta| \geq m\} \leq \frac{c'}{\rho}\, \mathbb{E}[|X_1|; |X_1| \geq m],$$

and

$$\begin{aligned}
\mathbb{E}[|S_\eta|^\alpha] &= \alpha \int_0^\infty t^{\alpha-1}\, \mathbb{P}\{|S_\eta| \geq t\}\, dt \\
&\leq \frac{\alpha\, c'}{\rho} \int_0^\infty t^{\alpha-1}\, \mathbb{E}[|X_1|; |X_1| \geq t]\, dt \\
&\leq \frac{\alpha\, c'}{\rho} \int_0^\infty \mathbb{E}[X_1^\alpha; |X_1| \geq t]\, dt = \frac{\alpha\, c'}{\rho}\, \mathbb{E}[|X_1|^{1+\alpha}].
\end{aligned}$$

□

♣ The inequalities (5.5) and (5.7) imply that there exists a $c < \infty$ such that for all y, $\mathbb{E}^y[\eta_n^*] < cn^2$, and

$$\mathbb{E}^x[\eta_n^*] \leq cn, \qquad 0 < x \leq 1. \tag{5.8}$$

Lemma 5.1.11

$$\mathbb{P}\{\eta_n^* < \eta^*\} \geq \frac{c^*}{n},$$

where

$$c^* = \frac{\rho\,\delta}{2\,(\rho + 2c'K^2)},$$

and c' is as in Lemma 5.1.10. Also,

$$\mathbb{P}\{\eta^* \geq n\} \geq b\,\mathbb{P}\{\eta^*_{\sqrt{n}} < \eta^*\} \geq \frac{b\,c^*}{\sqrt{n}}.$$

Proof The last assertion follows immediately from the first one and the strong Markov property as in (5.3). Since $\mathbb{P}\{\eta_n^* < \eta^*\} \geq \delta\,\mathbb{P}^1\{\eta_n^* < \eta^*\}$, to establish the first assertion it suffices to prove that

$$\mathbb{P}^1\{\eta_n^* < \eta^*\} \geq \frac{c^*}{\delta\,n}.$$

Using (5.8), we have

$$\begin{aligned}
\mathbb{P}^1\{|S_{\eta_n^*}| \geq s + n\} &\leq \sum_{l=0}^{\infty} \mathbb{P}^1\{\eta_n^* = l+1; |S_{\eta_n^*}| \geq s + n\} \\
&\leq \sum_{l=0}^{\infty} \mathbb{P}^1\{\eta_n^* > l; |X_{l+1}| \geq s\} \\
&\leq \mathbb{P}\{|X_1| \geq s\}\,\mathbb{E}^1[\eta_n^*] \leq \frac{c'\,n}{\rho}\,\mathbb{P}\{|X_1| \geq s\}.
\end{aligned}$$

In particular, if $t > 0$,

$$\begin{aligned}
\mathbb{E}^1\left[|S_{\eta_n^*}|; |S_{\eta_n^*}| \geq (1+t)\,n\right] &= \int_{tn}^{\infty} \mathbb{P}^1\{|S_{\eta_n^*}| \geq s + n\}\,ds \\
&\leq \frac{c'\,n}{\rho} \int_{tn}^{\infty} \mathbb{P}\{|X_1| \geq s\}\,ds
\end{aligned}$$

$$= \frac{c' n}{\rho} \mathbb{E}[|X_1|; |X_1| \geq tn]$$
$$\leq \frac{c'}{\rho t} \mathbb{E}\left[|X_1|^2\right] \leq \frac{c' K^2}{\rho t}. \qquad (5.9)$$

Consider the martingale $M_k = S_{k \wedge \eta_n^*}$. Due to the optional stopping theorem, we have

$$1 = \mathbb{E}^1[M_0] = \mathbb{E}^1[M_\infty] \leq \mathbb{E}^1[S_{\eta_n^*}; S_{\eta_n^*} \geq n].$$

If we let $t_0 = 2c'K^2/\rho$ in (5.9), we obtain

$$\mathbb{E}^1[|S_{\eta_n^*}|; |S_{\eta_n^*}| \geq (1 + t_0)\, n] \leq \frac{1}{2},$$

so it must be

$$\mathbb{E}^1[S_{\eta_n^*}; n \leq S_{\eta_n^*} \leq (1 + t_0)n] \geq \frac{1}{2},$$

which implies that

$$\mathbb{P}^1\{\eta_n^* < \eta^*\} \geq \mathbb{P}^1\{n \leq S_{\eta_n^*} \leq (1 + t_0)n\} \geq \frac{1}{2(1 + t_0)n}.$$

□

Proof of Theorem 5.1.7 Lemmas 5.1.8 and 5.1.11 prove the result for $0 < x \leq 1$. The result is easy if $x \geq r/2$ so we will assume that $1 \leq x \leq r/2$. As already noted, the function $x \mapsto \mathbb{P}^x\{S_{\eta_r} \geq r\}$ is nondecreasing in x. Therefore,

$$\mathbb{P}\{S_{\eta_r^*} \geq r\} = \mathbb{P}\{S_{\eta_x^*} \geq x\}\, \mathbb{P}\{S_{\eta_r^*} \geq r \mid S_{\eta_x^*} \geq x\} \geq \mathbb{P}\{S_{\eta_x^*} \geq x\}\, \mathbb{P}^x\{S_{\eta_r} \geq r\}.$$

Hence, by Lemmas 5.1.8 and 5.1.11,

$$\mathbb{P}^x\{S_{\eta_r} \geq r\} \leq \frac{\mathbb{P}\{S_{\eta_r^*} \geq r\}}{\mathbb{P}\{S_{\eta_x^*} \geq x\}} \leq \frac{4K}{c^* b\delta}\, \frac{x}{r}.$$

For an inequality in the opposite direction, we first show that there is a c_2 such that $\mathbb{E}^x[\eta_r] \leq c_2 xr$. Recall from (5.8) that $\mathbb{E}^y[\eta_r] \leq cr$ for $0 < y \leq 1$. The strong Markov property and monotonicity can be used (Exercise 5.1) to see that

$$\mathbb{E}^x[\eta_r] \leq \mathbb{E}^1[\eta_r] + \mathbb{E}^{x-1}[\eta_r]. \qquad (5.10)$$

Hence, we obtain the claimed bound for general x by induction. As in the previous lemma one can now see that

$$\mathbb{E}^x\left[|S_{\eta_r^*}|; |S_{\eta_r^*}| \geq (1+t)\, r\right] \leq \frac{c_2\, K^2\, x}{t},$$

and hence if $t_0 = 2c_2K^2$,

$$\mathbb{E}^x\left[S_{\eta_r^*}; S_{\eta_r^*} \geq (1+t_0)\, r\right] \leq \frac{x}{2},$$

$$\mathbb{E}^x\left[|S_{\eta_r^*}|; r \leq S_{\eta_r^*} \leq (1+t_0)\, r\right] \geq \frac{x}{2},$$

so that

$$\mathbb{P}^x\left\{r \leq S_{\eta_r^*} \leq (1+t_0)\, r\right\} \geq \frac{x}{2(1+t_0)r}.$$

As we have already shown (see the beginning of the proof of Lemma 5.1.11), this implies that

$$\mathbb{P}^x\{\eta^* \geq r^2\} \geq b\,\frac{x}{2(1+t_0)r}.$$

□

5.2 One-dimensional killed walks

A *symmetric defective increment distribution (on $\mathbb{Z}$)* is a set of nonnegative numbers $\{p_k : k \in \mathbb{Z}\}$ with $\sum p_k < 1$ and $p_{-k} = p_k$ for all k. Given a symmetric defective increment distribution, we have the corresponding symmetric random walk with killing, that we again denote by S. More precisely, S is a Markov chain with state space $\mathbb{Z} \cup \{\infty\}$, where ∞ is an absorbing state, and

$$\mathbb{P}\{S_{j+1} = k + l \mid S_j = k\} = p_l, \qquad \mathbb{P}\{S_{j+1} = \infty \mid S_j = k\} = p_\infty,$$

where $p_\infty = 1 - \sum p_k$. We let

$$T = \min\{j : S_j = \infty\}$$

denote the *killing time* for the random walk. Note that $\mathbb{P}\{T = j\} = p_\infty\,(1 - p_\infty)^{j-1}, j \in \{1, 2, \ldots\}$.

Examples

- Suppose that $p(j)$ is the increment distribution of a symmetric one-dimensional random walk and $s \in [0, 1)$. Then $p_j = s\,p(j)$ is a defective increment distribution corresponding to the random walk with killing rate $1 - s$. Conversely, if p_j is a symmetric defective increment distribution, and $p(j) = p_j/(1 - p_\infty)$, then $p(j)$ is an increment distribution of a symmetric one-dimensional random walk (not necessarily aperiodic or irreducible). If we kill this walk at rate $1 - p_\infty$, we get back p_j.
- Suppose that $\overline{S}_j$ is a symmetric random walk in $\mathbb{Z}^d, d \geq 2$ which we write $\overline{S}_j = (Y_j, Z_j)$ where Y_j is a random walk in $\mathbb{Z}$ and Z_j is a random walk in $\mathbb{Z}^{d-1}$. Suppose that the random walk is killed at rate $1 - s$ and let $\hat{T}$ denote the killing time. Let

$$\tau = \min\{j \geq 1 : Z_j = 0\}, \tag{5.11}$$

$$p_k = \mathbb{P}\{Y_\tau = k; \tau < \hat{T}\}.$$

Note that

$$p_k = \sum_{j=1}^{\infty} \mathbb{P}\{\tau = j; Y_j = k; j < \hat{T}\}$$

$$= \sum_{j=1}^{\infty} s^j\, \mathbb{P}\{\tau = j; Y_j = k\} = \mathbb{E}[s^{\hat{T}}; Y_{\hat{T}} = k; \hat{T} < \infty].$$

If Z is a transient random walk, then $\mathbb{P}\{\tau < \infty\} < 1$ and we can let $s = 1$.
- Suppose that $\overline{S}_j = (Y_j, Z_j)$ and τ are as in the previous example and suppose that $A \subset \mathbb{Z}^{d-1} \setminus \{0\}$. Let

$$\sigma_A = \min\{j : Z_j \in A\},$$

$$p_k = \mathbb{P}\{Y_\tau = k; \tau < \sigma_A\}.$$

If $\mathbb{P}\{Z_j \in A \text{ for some } j\} > 0$, then $\{p_k\}$ is a defective increment distribution.

Given a symmetric defective increment distribution $\{p_k\}$ with corresponding walk S_j and killing time T, define the events

$$V_+ = \{S_j > 0 : j = 1, \ldots, T-1\}, \quad \overline{V}_+ = \{S_j \geq 0 : j = 1, \ldots, T-1\},$$

$$V_- = \{S_j < 0 : j = 1, \ldots, T-1\}, \quad \overline{V}_- = \{S_j \leq 0 : j = 1, \ldots, T-1\}.$$

Symmetry implies that $\mathbb{P}(V_+) = \mathbb{P}(V_-), \mathbb{P}(\overline{V}_+) = \mathbb{P}(\overline{V}_-)$. Note that $V_+ \subset \overline{V}_+, V_- \subset \overline{V}_-$ and

$$\mathbb{P}(V_+ \cap \overline{V}_-) = \mathbb{P}(\overline{V}_+ \cap V_-) = \mathbb{P}\{T = 1\} = p_\infty. \tag{5.12}$$

Define a new defective increment distribution $p_{k,-}$, which is supported on $k = 0, -1, -2, \ldots$, by setting $p_{k,-}$ equal to the probability that the first visit to $\{\ldots, -2, -1, 0\}$ after time 0 occurs at position k and this occurs before the killing time T, i.e.

$$p_{k,-} = \sum_{j=1}^{\infty} \mathbb{P}\{S_j = k \; ; \; j < T \; ; \; S_l > 0, l = 1, \ldots, j-1\}.$$

Define $p_{k,+}$ similarly so that $p_{k,+} = p_{-k,-}$. The strong Markov property implies that

$$\mathbb{P}(\overline{V}_+) = \mathbb{P}(V_+) + p_{0,-}\, \mathbb{P}(\overline{V}_+),$$

and hence

$$\mathbb{P}(V_+) = (1 - p_{0,-})\, \mathbb{P}(\overline{V}_+) = (1 - p_{0,+})\, \mathbb{P}(\overline{V}_+). \tag{5.13}$$

In the next proposition we prove a nonintuitive fact.

Proposition 5.2.1 *The events $\overline{V}_+$ and V_- are independent. In particular,*

$$\mathbb{P}(V_-) = \mathbb{P}(V_+) = (1 - p_{0,+})\, \mathbb{P}(\overline{V}_+) = \sqrt{p_\infty\, (1 - p_{0,+})}. \tag{5.14}$$

Proof Independence is equivalent to the statement $\mathbb{P}(V_- \cap \overline{V}_+) = \mathbb{P}(V_-)\, \mathbb{P}(\overline{V}_+)$. We will prove the equivalent statement $\mathbb{P}(V_- \cap \overline{V}_+^c) = \mathbb{P}(V_-)\, \mathbb{P}(\overline{V}_+^c)$. Note that $V_- \cap \overline{V}_+^c$ is the event that $T > 1$ but no point in $\{0, 1, 2, \ldots\}$ is visited during the times $\{1, \ldots, T-1\}$. In particular, at least one point in $\{\ldots, -2, -1\}$ is visited before time T.

Let

$$\rho = \max\{k \in \mathbb{Z} : S_j = k \text{ for some } j = 1, \ldots, T-1\},$$

$$\xi_k = \max\{j \geq 0 : S_j = k; j < T\}.$$

In words, ρ is the rightmost point visited after time zero, and ξ_k is the last time that k is visited before the walk is killed. Then,

$$\mathbb{P}(V_- \cap \overline{V}_+^c) = \sum_{k=1}^{\infty} \mathbb{P}\{\rho = -k\} = \sum_{k=1}^{\infty} \sum_{j=1}^{\infty} \mathbb{P}\{\rho = -k; \xi_{-k} = j\}.$$

Note that the event $\{\rho = -k; \xi_{-k} = j\}$ is the same as the event

$$\{S_j = -k \; ; \; j < T \; ; \; S_l \le -k, l = 1, \ldots, j-1 \; ; \\ S_l < -k, l = j+1, \ldots, T-1\}.$$

Since,

$$\mathbb{P}\{S_l < -k, l = j+1, \ldots, T-1 \mid S_j = -k \; ; \\ j < T \; ; \; S_l \le -k, l = 1, \ldots, j-1\} = \mathbb{P}(V_-),$$

we have

$$\mathbb{P}\{\rho = -k; \xi_{-k} = j\} = \mathbb{P}\{S_j = -k \; ; \\ j < T \; ; \; S_l \le -k, l = 1, \ldots, j-1\} \, \mathbb{P}(V_-).$$

Due to the symmetry of the random walk, the probability of the path $[x_0 = 0, x_1, \ldots, x_j]$ is the same as the probability of the reversed path $[x_j - x_j, x_{j-1} - x_j, \ldots, x_0 - x_j]$. Note that if $x_j = -k$ and $x_l \le -k$, $l = 1, \ldots, j-1$, then $x_0 - x_j = k$ and $\sum_{i=1}^{l} (x_{j-i} - x_{j-i+1}) = x_{j-l} - x_j \le 0$, for $l = 1, \ldots, j-1$. Therefore, we have

$$\mathbb{P}\{S_j = -k \; ; \; j < T \; ; \; S_l \le -k, l = 1, \ldots, j-1\} = \mathbb{P}\{\eta = j; j < T; S_j = k\},$$

where

$$\eta = \min\{j \ge 1 : S_j > 0\}.$$

Since

$$\sum_{k=1}^{\infty} \sum_{j=1}^{\infty} \mathbb{P}\{\eta = j; j < T; S_j = k\} = \mathbb{P}\{\eta < T\} = \mathbb{P}(\overline{V}_-^c) = \mathbb{P}(\overline{V}_+^c),$$

we obtain the stated independence. The equality (5.14) now follows from

$$p_\infty = \mathbb{P}(V_- \cap \overline{V}_+) = \mathbb{P}(V_-) \, \mathbb{P}(\overline{V}_+) = \frac{\mathbb{P}(V_-) \, \mathbb{P}(V_+)}{1 - p_{0,-}} = \frac{\mathbb{P}(V_+)^2}{1 - p_{0,+}}.$$

□

5.3 Hitting a half-line

We will give an application of Proposition 5.2.1 to walks in $\mathbb{Z}^d$. Suppose that $d \ge 2$ and $\overline{S}_n$ is a random walk with increment distribution $p \in \mathcal{P}_d$. We write $\overline{S}_n = (Y_n, Z_n)$ where Y_n is a one-dimensional walk and Z_n is a $(d-1)$-dimensional walk. Let Γ denote the covariance matrix for $\overline{S}_n$ and let Γ^* denote the covariance matrix for Z_n. Let $T = \min\{j > 0 : Z_j = 0\}$ be the first time that the random walk returns to the line $\{(j,x) \in \mathbb{Z} \times \mathbb{Z}^{d-1} : x = 0\}$. Let $T_+, \overline{T}_+$ denote the corresponding quantities for the (nonpositive and negative) half-line

$$T_+ = \min\left\{n > 0 : S_n \in \{(j,x) \in \mathbb{Z} \times \mathbb{Z}^{d-1} : j \le 0, x = 0\}\right\},$$

$$\overline{T}_+ = \min\left\{n > 0 : S_n \in \{(j,x) \in \mathbb{Z} \times \mathbb{Z}^{d-1} : j < 0, x = 0\}\right\},$$

and finally let

$$p_{0,+} = \mathbb{P}\{Y_{T_+} = 0\}.$$

Proposition 5.3.1 *If $p \in \mathcal{P}_d, d = 2, 3$, there is a C such that as $n \to \infty$,*

$$(1 - p_{0,+})\,\mathbb{P}\{\overline{T}_+ > n\} \sim \mathbb{P}\{T_+ > n\} \sim \begin{cases} C\, n^{-1/4}, & d = 2, \\ C\,(\log n)^{-1/2}, & d = 3. \end{cases}$$

Proof We will prove the second asymptotic relation; a similar argument shows that the first and third terms are asympotic. Let $\sigma = \sigma_\xi$ denote a geometric random variable, independent of the random walk, with killing rate $1 - \xi$, i.e. $\mathbb{P}\{\sigma > k\} = \xi^k$. Let $q_n = \mathbb{P}\{T_+ > n\}, q(\xi) = \mathbb{P}\{T_+ > \sigma\}$. Then,

$$q(\xi) = \mathbb{P}\{T_+ > \sigma\} = \sum_{n=1}^{\infty} \mathbb{P}\{\sigma = n; T_+ > n\}$$

$$= \sum_{n=1}^{\infty} (1 - \xi)\, \xi^{n-1}\, q_n.$$

By Propositions A.5.2 and A.5.3, it suffices to show that $q(\xi) \sim c\,(1-\xi)^{1/4}$ if $d = 2$ and $q(\xi) \sim c\,[-\log(1-\xi)]^{-1/2}$ if $d = 3$.

This is the same situation as the second example of the last subsection (although (τ, T) there corresponds to (T, σ) here). Hence, Proposition 5.2.1 tells us that

$$q(\xi) = \sqrt{p_\infty(\xi)\,(1 - p_{0,+}(\xi))},$$

where $p_\infty(\xi) = \mathbb{P}\{T > \sigma\}$ and $p_{0,+}(\xi) = \mathbb{P}\{T_+ \le \sigma; Y_{T_+} = 0\}$. Clearly, as $\xi \to 1-$, $1 - p_{0,+}(\xi) \to 1 - p_{0,+} > 0$. By applying (4.9) and (4.10) to the random walk Z_n, we can see that

$$\mathbb{P}\{T > \sigma\} \sim c\,(1-\xi)^{1/2}, \quad d = 2,$$

$$\mathbb{P}\{T > \sigma\} \sim c\left[\log\left(\frac{1}{1-\xi}\right)\right]^{-1}, \quad d = 3.$$

□

♣ From the proof, one can see that the constant C can be determined in terms of Γ^* and $p_{0,+}$. We do not need the exact value and the proof is a little easier to follow if we do not try to keep track of this constant. It is generally hard to compute $p_{0,+}$; for simple random walk, see Proposition 9.9.8.

♣ The above proof uses the surprising fact that the events "avoid the positive x^1-axis" and "avoid the negative x^1- axis" are independent up to a multiplicative constant. This idea does not extend to other sets; for example the event "avoid the positive x^1-axis" and "avoid the positive x^2-axis" are not independent up to a multiplicative constant in two dimensions. However, they are in three dimensions (which is a nontrivial fact).

In Section 6.8 we will need some estimates for two-dimensional random walks avoiding a half-line. The argument given below uses the Harnack inequality (Theorem 6.3.9), which will be proved independently of this estimate. In the remainder of this section, let $d = 2$ and let $S_n = (Y_n, Z_n)$ be the random walk. Let

$$\zeta_r = \min\{n > 0 : Y_n \geq r\},$$

$$\rho_r = \min\{n > 0 : S_n \notin (-r,r)\times(-r,r)\},$$

$$\rho_r^* = \min\{n > 0 : S_n \notin \mathbb{Z}\times(-r,r)\}.$$

If $|S_0| < r$, the event $\{\zeta_r = \rho_r\}$ occurs if and only if the first visit of the random walk to the complement of $(-r,r)\times(-r,r)$ is at a point (j,k) with $j \geq r$.

Proposition 5.3.2 *If $p \in \mathcal{P}_2$, then*

$$\mathbb{P}\{T_+ > \rho_r\} \asymp r^{-1/2}. \tag{5.15}$$

Moreover, for all $z \neq 0$,

$$\mathbb{P}^z\{\rho_r < T_+\} \leq c\,|z|^{1/2}\, r^{-1/2}. \tag{5.16}$$

In addition, there is a $c < \infty$ *such that if* $1 \le k \le r$ *and* $A_k = \{j\mathbf{e}_1 : j = -k, -k+1, \ldots\}$, *then*

$$\mathbb{P}\{T_{A_k} > \rho_r\} \le c\, k^{-1/2}\, r^{-1/2}. \tag{5.17}$$

Proof It suffices to show that there exist c_1, c_2 with

$$\mathbb{P}\{T_+ > \rho_r\} \le c_2\, r^{-1/2}, \quad \mathbb{P}\{T_+ > \rho_r^*\} \ge c_1\, r^{-1/2}.$$

The gambler's ruin estimate applied to the second component implies that $\mathbb{P}\{T > \rho_r^*\} \asymp r^{-1}$ and an application of Proposition 5.2.1 gives $\mathbb{P}\{T_+ > \rho_r^*\} \asymp r^{-1/2}$.

Using the invariance principle, it is not difficult to show that there is a c such that for r sufficiently large, $\mathbb{P}\{\zeta_r = \rho_r\} \ge c$. By translation invariance and monotonicity, one can see that for $j \ge 1$,

$$\mathbb{P}^{-j\mathbf{e}_1}\{\zeta_r = \rho_r\} \le \mathbb{P}\{\zeta_r = \rho_r\}.$$

Hence, the strong Markov property implies that $\mathbb{P}\{\zeta_r = \rho_r \mid T_+ < \rho_r\} \le \mathbb{P}\{\zeta_r = \rho_r\}$; therefore, it has to be that $\mathbb{P}\{\zeta_r = \rho_r \mid T_+ > \rho_r\} \ge c$ and

$$\mathbb{P}\{\rho_r < T_+\} \le c\, \mathbb{P}\{\zeta_r = \rho_r < T_+\}. \tag{5.18}$$

Another application of the invariance principle shows that

$$\mathbb{P}\{T_+ > r^2 \mid \zeta_r = \rho_r < T_+\} \ge c,$$

since this conditional probability is bounded below by the probability that a random walk goes no farther than distance $r/2$ in r^2 steps. Hence,

$$\mathbb{P}\{\rho_r < T_+\} \le c\, \mathbb{P}\{\rho_r < T_+, T_+ > r^2\} \le c\, \mathbb{P}\{T_+ > r^2\} \le c\, r^{-1/2}.$$

This gives (5.15).

For the remaining results, we will assume that $|z|$ is an integer greater than the range R of the walk, but one can easily adapt the argument to arbitrary z. Let $h_r(x) = \mathbb{P}^x\{\rho_r < T_+\}$ and let $M = M(r, |z|)$ be the maximum value of $h_r(x)$ over $x \in (-|z| - R, |z| + R) \times (-|z| - R, |z| + R)$. By translation invariance, this is maximized at a point with maximal first component and by the Harnack inequality (Theorem 6.3.9),

$$c_1 M \le h_r(x) \le c_2 M,$$
$$x \in (|z| - R, |z| + R) \times (-|z| - R, |z| + R).$$

Together with a strong Markov property this implies that

$$\mathbb{P}\{\rho_r < T_+\} \le c M\, \mathbb{P}\{\rho_{|z|} < T_+\},$$

and due to (5.18)

$$\mathbb{P}\{\rho_r < T_+\} \ge c M\, \mathbb{P}\{\rho_{|z|} = \zeta_{|z|} < T_+\} \ge c M\, \mathbb{P}\{\rho_{|z|} < T_+\}.$$

Since $\mathbb{P}\{\rho_r < T_+\} \asymp r^{-1/2}$, we conclude that $M \asymp |z|^{1/2}\, r^{-1/2}$, implying (5.16). To prove (5.17), we write

$$\begin{aligned}\mathbb{P}\{T_{A_k} > \rho_r\} &= \mathbb{P}\{T_{A_k} > \rho_k\}\, \mathbb{P}\{T_{A_k} > \rho_r \mid T_{A_k} > \rho_k\}\\ &\le c\, \mathbb{P}\{T_{A_k} > \rho_k\}\, (k/r)^{1/2}.\end{aligned}$$

Therefore, it suffices to show that $\mathbb{P}\{T_{A_k} > \rho_k\} \le ck^{-1}$ This is very close to the gambler's ruin estimate, but it is not exactly the form we have proved thus far, so we will sketch a proof.

Let

$$q(k) = \mathbb{P}\{T_{A_k} > \rho_k\}.$$

Note that for all integers $|j| < k$,

$$\mathbb{P}^{j\mathbf{e}_1}\{T_{A_k} > \rho_k\} \ge q(2k).$$

A last-exit decomposition focusing on the last visit to A_k before time ρ_k shows that

$$1 = \sum_{|j|<k} \tilde{G}_k(0, j\mathbf{e}_1)\, \mathbb{P}^{j\mathbf{e}_1}\{T_{A_k} > \rho_k\} \ge q(2k) \sum_{|j|<k} \tilde{G}_k(0, j\mathbf{e}_1).$$

where $\tilde{G}_k$ denotes the Green's function for the set $\mathbb{Z}^2 \cap [(-k,k) \times (-k,k)]$. Hence, it suffices to prove that

$$\sum_{|j|<k} \tilde{G}_k(0, j\mathbf{e}_1) \geq c\,k.$$

We leave this to the reader (alternatively, see the next chapter for such estimates). □

Exercises

Exercise 5.1 Prove inequality (5.10).

Exercise 5.2 Suppose that $p \in \mathcal{P}_2$ and $x \in \mathbb{Z}^2 \setminus \{0\}$. Let

$$T = \min\{n \geq 1 : S_n = jx \text{ for some } j \in \mathbb{Z}\}.$$

$$T_+ = \min\{n \geq 1 : S_n = jx \text{ for some } j = 0, 1, 2, \ldots\}.$$

(i) Show that there exists c such that

$$\mathbb{P}\{T > n\} \sim c\,n^{-1}.$$

(ii) Show that there exists c_1 such that

$$\mathbb{P}\{T_+ > n\} \sim c_1\,n^{-1/2}.$$

Establish the analog of Proposition 5.3.2 in this setting.

6
Potential theory

6.1 Introduction

There is a close relationship between random walks with increment distribution p and functions that are harmonic with respect to the generator $\mathcal{L} = \mathcal{L}_p$.

We start by setting some notation. We fix $p \in \mathcal{P}$. If $A \subsetneq \mathbb{Z}^d$, we let

$$\partial A = \{x \in \mathbb{Z}^d \setminus A : p(y,x) > 0 \text{ for some } y \in A\}$$

denote the *(outer) boundary* of A and we let $\overline{A} = A \cup \partial A$ be the discrete closure of A. Note that the above definition of $\partial A, \overline{A}$ depends on the choice of p. We omit this dependence from the notation, and hope that this will not confuse the reader. In the case of simple random walk,

$$\partial A = \{x \in \mathbb{Z}^d \setminus A : |y - x| = 1 \text{ for some } y \in A\}.$$

Since p has a finite range, if A is finite, then $\partial A, \overline{A}$ are finite. The *inner boundary* of $A \subset \mathbb{Z}^d$ is defined by

$$\partial_i A = \partial(\mathbb{Z}^d \setminus A) = \{x \in A : p(x,y) > 0 \text{ for some } y \notin A\}.$$

A function $f : \overline{A} \to \mathbb{R}$ is *harmonic (with respect to p) or p-harmonic in* A if

$$\mathcal{L}f(y) := \sum_x p(x)\,[f(y+x) - f(y)] = 0$$

for every $y \in A$. Note that we cannot define $\mathcal{L}f(y)$ for all $y \in A$, unless f is defined on $\overline{A}$.

We say that A is *connected (with respect to p)* if for every $x, y \in A$, there is a finite sequence $x = z_0, z_1, \ldots, z_k = y$ of points in A with $p(z_{j+1} - z_j) > 0, j = 0, \ldots, k-1$.

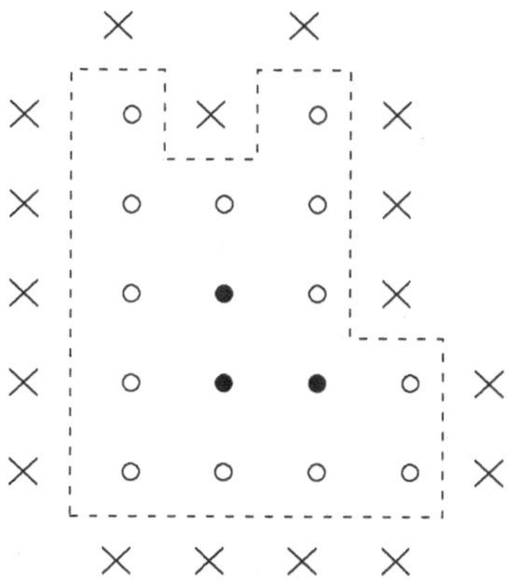

Figure 6.1 Suppose A is the set of lattice points "inside" the dashed curve. Then the points in $A \setminus \partial_i A$, $\partial_i A$ and ∂A are marked by $\bullet$, $\circ$, and $\times$, respectively

---

♣ This chapter contains a number of results about functions on subsets of $\mathbb{Z}^d$. These results have analogues in the continuous setting. The set A corresponds to an open set $D \subset \mathbb{R}^d$, the outer boundary ∂A corresponds to the usual topological boundary ∂D, and $\overline{A}$ corresponds to the closure $\overline{D} = D \cup \partial D$. The term domain is often used for open, connected subsets of $\mathbb{R}^d$. Finiteness assumptions on A correspond to boundedness assumptions on D.

Proposition 6.1.1 *Suppose that S_n is a random walk with increment distribution $p \in \mathcal{P}_d$ starting at $x \in \mathbb{Z}^d$. Suppose that $f : \mathbb{Z}^d \to \mathbb{R}$. Then*

$$M_n := f(S_n) - \sum_{j=0}^{n-1} \mathcal{L}f(S_j)$$

is a martingale. In particular, if f is harmonic on $A \subset \mathbb{Z}^d$, then $Y_n := f(S_{n \wedge \overline{\tau}_A})$ is a martingale, where $\overline{\tau}_A$ is as defined in (4.27).

Proof This is immediate from the definition. □

Proposition 6.1.2 *Suppose that $p \in \mathcal{P}_d$ and $f : \mathbb{Z}^d \to \mathbb{R}$ is bounded and harmonic on $\mathbb{Z}^d$. Then f is constant.*

Proof We may assume that p is aperiodic; if not consider $\hat{p} = (1/2)\,p + (1/2)\delta_0$ and note that f is p-harmonic if and only if it is $\hat{p}$-harmonic. Let $x, y \in \mathbb{Z}^d$. By

Lemma 2.4.3 we can define random walks $S, \hat{S}$ on the same probability space so that S is a random walk starting at x; $\hat{S}$ is a random walk starting at y; and

$$\mathbb{P}\{S_n \neq \hat{S}_n\} \leq c\,|x-y|\,n^{-1/2}.$$

In particular,

$$|\mathbb{E}[f(S_n)] - \mathbb{E}[f(\hat{S}_n)]| \leq 2\,c\,|x-y|\,n^{-1/2}\,\|f\|_\infty \longrightarrow 0.$$

Proposition 6.1.1 implies that $f(x) = \mathbb{E}[f(S_n)], f(y) = \mathbb{E}[f(\hat{S}_n)]$. □

♣ The fact that all bounded harmonic functions are constant is closely related to the fact that a random walk eventually forgets its starting point. Lemma 2.4.3 gives a precise formulation of this loss of memory property. The last proposition is not true for simple random walk on a regular tree.

6.2 Dirichlet problem

The standard *Dirichlet problem* for harmonic functions is to find a harmonic function on a region with specified values on the boundary.

Theorem 6.2.1 (Dirichlet problem I) *Suppose that $p \in \mathcal{P}_d$, and $A \subset \mathbb{Z}^d$ satisfies $\mathbb{P}^x\{\overline{\tau}_A < \infty\} = 1$ for all $x \in A$. Suppose that $F : \partial A \to \mathbb{R}$ is a bounded function. Then there is a unique bounded function $f : \overline{A} \to \mathbb{R}$ satisfying*

$$\mathcal{L}f(x) = 0, \quad x \in A, \tag{6.1}$$

$$f(x) = F(x), \quad x \in \partial A. \tag{6.2}$$

It is given by

$$f(x) = \mathbb{E}^x[F(S_{\overline{\tau}_A})]. \tag{6.3}$$

Proof A simple application of the Markov property shows that f defined by (6.3) satisfies (6.1) and (6.2). Now suppose that f is a bounded function satisfying (6.1) and (6.2). Then $M_n := f(S_{n\wedge\overline{\tau}_A})$ is a bounded martingale. Hence, the optional sampling theorem (Theorem A.2.3) implies that

$$f(x) = \mathbb{E}^x[M_0] = \mathbb{E}^x[M_{\overline{\tau}_A}] = \mathbb{E}^x[F(S_{\overline{\tau}_A})].$$

□

Remark

- If A is finite, then ∂A is also finite and all functions on $\overline{A}$ are bounded. Hence for each F on ∂A, there is a unique function satisfying (6.1) and (6.2). In this case we could prove existence and uniqueness using linear algebra since (6.1) and (6.2) give $\#(A)$ linear equations in $\#(A)$ unknowns. However, algebraic methods do not yield the nice probabilistic form (6.3).
- If A is infinite, there may well be more than one solution to the Dirichlet problem if we allow unbounded solutions. For example, if $d = 1, p$ is simple random walk, $A = \{1, 2, 3, \ldots\}$, and $F(0) = 0$, then there is an infinite number of solutions of the form $f_b(x) = bx$. If $b \neq 0, f_b$ is unbounded.
- Under the conditions of the theorem, it follows that any function f on $\overline{A}$ that is harmonic on A satisfies the *maximum principle*:

$$\sup_{x \in \overline{A}} |f(x)| = \sup_{x \in \partial A} |f(x)|.$$

- If $d = 1, 2$ and A is a proper subset of $\mathbb{Z}^d$, then we know by recurrence that $\mathbb{P}^x\{\overline{\tau}_A < \infty\} = 1$ for all $x \in A$.
- If $d \geq 3$ and $\mathbb{Z}^d \setminus A$ is finite, then there are points $x \in A$ with $\mathbb{P}^x\{\overline{\tau}_A = \infty\} > 0$. The function

$$f(x) = \mathbb{P}^x\{\overline{\tau}_A = \infty\}$$

is a bounded function satisfying (6.1) and (6.2) with $F \equiv 0$ on ∂A. Hence, the condition $\mathbb{P}^x\{\overline{\tau}_A < \infty\} = 1$ is needed to guarantee uniqueness. However, as Proposition 6.2.2 below shows, all solutions with $F \equiv 0$ on ∂A are multiples of f.

Remark This theorem has a well-known continuous analogue. Suppose that $f : \{|z| \in \mathbb{R}^d : |z| \leq 1\} \to \mathbb{R}$ is a continuous function with $\Delta f(x) = 0$ for $|x| < 1$. Then

$$f(x) = \mathbb{E}^x[f(B_T)],$$

where B is a standard d-dimensional Brownian motion and T is the first time t that $|B_t| = 1$. If $|x| < 1$, the distribution of B_T given $B_0 = x$ has a density with respect to surface measure on $\{|z| = 1\}$. This density $h(x, z) = c\,(1 - |x|^2)/|x - z|^d$ is called the *Poisson kernel* and we can write

$$f(x) = c \int_{|z|=1} f(z)\, \frac{1 - |x|^2}{|x - z|^d}\, ds(z), \tag{6.4}$$

where s denotes surface measure. To verify that this is correct, one can check directly that f as defined above is harmonic in the ball and satisfies the boundary condition on the sphere. Two facts follow almost immediately from this integral formula:

- **Derivative estimates** For every k, there is a $c = c(k) < \infty$ such that if f is harmonic in the unit ball and D denotes a kth order derivative, then $|Df(0)| \leq c_k \|f\|_\infty$.
- **Harnack inequality** For every $r < 1$, there is a $c = c_r < \infty$ such that if f is a positive harmonic function on the unit ball, then $f(x) \leq c f(y)$ for $|x|, |y| \leq r$.

An important aspect of these estimates is the fact that the constants do not depend on f. We will prove the analogous results for random walk in Section 6.3.

Proposition 6.2.2 (Dirichlet problem II) *Suppose that $p \in \mathcal{P}_d$ and $A \subsetneq \mathbb{Z}^d$. Suppose that $F : \partial A \to \mathbb{R}$ is a bounded function. Then the only bounded functions $f : \overline{A} \to \mathbb{R}$ satisfying (6.1) and (6.2) are of the form*

$$f(x) = \mathbb{E}^x[F(S_{\overline{\tau}_A}); \overline{\tau}_A < \infty] + b\,\mathbb{P}^x\{\overline{\tau}_A = \infty\}, \tag{6.5}$$

for some $b \in \mathbb{R}$.

Proof We may assume that p is aperiodic. We also assume that $\mathbb{P}^x\{\overline{\tau}_A = \infty\} > 0$ for some $x \in A$; if not, Theorem 6.2.1 applies. Assume that f is a bounded function satisfying (6.1) and (6.2). Since $M_n := f(S_{n\wedge\overline{\tau}_A})$ is a martingale, we know that

$$\begin{aligned} f(x) &= \mathbb{E}^x[M_0] = \mathbb{E}^x[M_n] = \mathbb{E}^x[f(S_{n\wedge\overline{\tau}_A})] \\ &= \mathbb{E}^x[f(S_n)] - \mathbb{E}^x[f(S_n); \overline{\tau}_A < n] + \mathbb{E}^x[F(S_{\overline{\tau}_A}); \overline{\tau}_A < n]. \end{aligned}$$

Using Lemma 2.4.3, we can see that for all x, y,

$$\lim_{n\to\infty} |\mathbb{E}^x[f(S_n)] - \mathbb{E}^y[f(S_n)]| = 0.$$

Therefore,

$$|f(x) - f(y)| \leq 2\,\|f\|_\infty\,[\mathbb{P}^x\{\overline{\tau}_A < \infty\} + \mathbb{P}^y\{\overline{\tau}_A < \infty\}].$$

Let $U_\epsilon = \{z \in \mathbb{Z}^d : \mathbb{P}^z\{\overline{\tau}_A = \infty\} \geq 1 - \epsilon\}$. Since $\mathbb{P}^x\{\overline{\tau}_A = \infty\} > 0$ for some x, one can see (Exercise 4.1) that U_ϵ is nonempty for each $\epsilon \in (0, 1)$. Then,

$$|f(x) - f(y)| \leq 4\epsilon\,\|f\|_\infty, \qquad x, y \in U_\epsilon.$$

Hence, there is a b such that

$$|f(x) - b| \leq 4\epsilon \, \|f\|_\infty, \qquad x \in U_\epsilon.$$

Let ρ_ϵ be the minimum of $\overline{\tau}_A$ and the smallest n such that $S_n \in U_\epsilon$. Then for every $x \in \mathbb{Z}^d$, the optional sampling theorem implies that

$$f(x) = \mathbb{E}^x[f(S_{\rho_\epsilon})] = \mathbb{E}^x[F(S_{\overline{\tau}_A}); \overline{\tau}_A \leq \rho_\epsilon] + \mathbb{E}^x[f(S_{\rho_\epsilon}); \overline{\tau}_A > \rho_\epsilon].$$

(Here, we use the fact that $\mathbb{P}^x\{\tau_A \wedge \rho_\epsilon < \infty\} = 1$ which can be verified easily.) By the dominated convergence theorem,

$$\lim_{\epsilon \to 0} \mathbb{E}^x[F(S_{\overline{\tau}_A}); \overline{\tau}_A \leq \rho_\epsilon] = \mathbb{E}^x[F(S_{\overline{\tau}_A}); \overline{\tau}_A < \infty].$$

Also,

$$\left|\mathbb{E}^x[f(S_{\rho_\epsilon}); \overline{\tau}_A > \rho_\epsilon] - b\,\mathbb{P}^x\{\tau_A > \rho_\epsilon\}\right| \leq 4\epsilon \|f\|_\infty \, \mathbb{P}^x\{\tau_A > \rho_\epsilon\},$$

and since $\rho_\epsilon \to \infty$ as $\epsilon \to 0$,

$$\lim_{\epsilon \to 0} \mathbb{E}^x[f(S_{\rho_\epsilon}); \overline{\tau}_A > \rho_\epsilon] = b\,\mathbb{P}^x\{\tau_A = \infty\}.$$

This gives (6.5). □

Remark We can think of (6.5) as a generalization of (6.3) where we have added a boundary point at infinity. The constant b in the last proposition is the boundary value at infinity and can be written as $F(\infty)$. The fact that there is a single boundary value at infinity is closely related to Proposition 6.1.2.

Definition If $p \in \mathcal{P}_d$ and $A \subset \mathbb{Z}^d$, then the *Poisson kernel* is the function $H : \overline{A} \times \partial A \to [0, 1]$ defined by

$$H_A(x, y) = \mathbb{P}^x\{\overline{\tau}_A < \infty; S_{\overline{\tau}_A} = y\}.$$

As a slight abuse of notation we will also write

$$H_A(x, \infty) = \mathbb{P}^x\{\overline{\tau}_A = \infty\}.$$

Note that

$$\sum_{y \in \partial A} H_A(x, y) = \mathbb{P}^x\{\overline{\tau}_A < \infty\}.$$

For fixed $y \in \partial A, f(x) = H_A(x, y)$ is a function on $\overline{A}$ that is harmonic on A and equals $\delta(\cdot - y)$ on ∂A. If p is recurrent, there is a unique such function. If p is transient, f is the unique such function that tends to 0 as x tends to infinity. We can write (6.3) as

$$f(x) = \mathbb{E}^x[F(S_{\overline{\tau}_A})] = \sum_{y \in \partial A} H_A(x, y)\, F(y), \tag{6.6}$$

and (6.5) as

$$f(x) = \mathbb{E}^x[F(S_{\overline{\tau}_A}); \overline{\tau}_A < \infty] + b\,\mathbb{P}^x\{\overline{\tau}_A = \infty\} = \sum_{y \in \partial A \cup \{\infty\}} H_A(x, y)\, F(y),$$

where $F(\infty) = b$. The expression (6.6) is a random walk analogue of (6.4).

Proposition 6.2.3 *Suppose that $p \in \mathcal{P}_d$ and $A \subsetneq \mathbb{Z}^d$. Let $g : A \to \mathbb{R}$ be a function with finite support. Then, the function*

$$f(x) = \sum_{y \in A} G_A(x, y)\, g(y) = \mathbb{E}^x\left[\sum_{j=0}^{\overline{\tau}_A - 1} g(S_j)\right],$$

is the unique bounded function on $\overline{A}$ that vanishes on ∂A and satisfies

$$\mathcal{L}f(x) = -g(x), \quad x \in A. \tag{6.7}$$

Proof Since g has finite support,

$$|f(x)| \leq \sum_{y \in A} G_A(x, y)\, |g(y)| < \infty,$$

and hence f is bounded. We have already noted in Lemma 4.6.1 that f satisfies (6.7). Now suppose that f is a bounded function vanishing on ∂A satisfying (6.7). Then, Proposition 6.1.1 implies that

$$M_n := f(S_{n \wedge \overline{\tau}_A}) + \sum_{j=0}^{n \wedge \overline{\tau}_A - 1} g(S_j),$$

is a martingale. Note that $|M_n| \leq \|f\|_\infty + Y$ where

$$Y = \sum_{j=0}^{\overline{\tau}_A - 1} |g(S_j)|,$$

and that

$$\mathbb{E}^x[Y] = \sum_y G_A(x,y)\,|g(y)| < \infty.$$

Hence, M_n is dominated by an integrable random variable and we can use the optional sampling theorem (Theorem A.2.3) to conclude that

$$f(x) = \mathbb{E}^x[M_0] = \mathbb{E}^x[M_{\overline{\tau}_A}] = \mathbb{E}^x\left[\sum_{j=0}^{\overline{\tau}_A-1} g(S_j)\right].$$

□

Remark Suppose that $A \subset \mathbb{Z}^d$ is finite with $\#(A) = m$. Then $G_A = [G_A(x,y)]_{x,y\in A}$ is an $m \times m$ symmetric matrix with nonnegative entries. Let $\mathcal{L}^A = [\mathcal{L}^A(x,y)]_{x,y\in A}$ be the $m \times m$ symmetric matrix defined by

$$\mathcal{L}^A(x,y) = p(x,y), \quad x \neq y; \qquad \mathcal{L}^A(x,x) = p(x,x) - 1.$$

If $g : A \to \mathbb{R}$ and $x \in A$, then $\mathcal{L}^A g(x)$ is the same as $\mathcal{L}g(x)$ where g is extended to $\overline{A}$ by setting $g \equiv 0$ on ∂A. The last proposition can be rephrased as $\mathcal{L}^A[G_A g] = -g$, or in other words, $G_A = -(\mathcal{L}^A)^{-1}$.

Corollary 6.2.4 *Suppose that $p \in \mathcal{P}_d$ and $A \subset \mathbb{Z}^d$ is finite. Let $g : A \to \mathbb{R}$, $F : \partial A \to \mathbb{R}$ be given. Then, the function*

$$\begin{aligned} f(x) &= \mathbb{E}^x[F(S_{\overline{\tau}_A})] + \mathbb{E}^x\left[\sum_{j=0}^{\overline{\tau}_A-1} g(S_j)\right] \\ &= \sum_{z\in\partial A} H_A(x,z)\,F(z) + \sum_{y\in A} G_A(x,y)\,g(y), \end{aligned} \tag{6.8}$$

is the unique function on $\overline{A}$ that satisfies

$$\mathcal{L}f(x) = -g(x), \quad x \in A.$$

$$f(x) = F(x), \quad x \in \partial A.$$

In particular, for any $f : \overline{A} \to \mathbb{R}$, $x \in \overline{A}$,

$$f(x) = \mathbb{E}^x[f(S_{\overline{\tau}_A})] - \mathbb{E}^x\left[\sum_{j=0}^{\overline{\tau}_A-1} \mathcal{L}f(S_j)\right]. \tag{6.9}$$

Proof Use the fact that $h(x) := f(x) - \mathbb{E}^x[F(S_{\overline{\tau}_A})]$ satisfies the assumptions in the previous proposition. □

Corollary 6.2.5 *Suppose that $p \in \mathcal{P}_d$ and $A \subset \mathbb{Z}^d$ is finite. Then*

$$f(x) = \mathbb{E}^x[\overline{\tau}_A] = \sum_{y \in A} G_A(x, y)$$

is the unique bounded function $f : \overline{A} \to \mathbb{R}$ that vanishes on ∂A and satisfies

$$\mathcal{L}f(x) = -1, \quad x \in A.$$

Proof This is Proposition 6.2.3 with $g \equiv 1_A$. □

Proposition 6.2.6 *Let $\rho_n = \overline{\tau}_{\mathcal{B}_n} = \inf\{j \geq 0 : |S_j| \geq n\}$. Then if $p \in \mathcal{P}_d$ with range R and $|x| < n$,*

$$[n^2 - |x|^2] \leq (\mathrm{tr}\Gamma)\, \mathbb{E}^x[\rho_n] \leq [(n + R)^2 - |x|^2].$$

Proof In Exercise 1.4 it was shown that $M_j =: |S_{j \wedge \rho_n}|^2 - (\mathrm{tr}\Gamma)(j \wedge \rho_n)$ is a martingale. Also, $\mathbb{E}^x[\rho_n] < \infty$ for each x, so M_j is dominated by the integrable random variable $(n + R)^2 + (\mathrm{tr}\Gamma)\, \rho_n$. Hence,

$$|x|^2 = \mathbb{E}^x[M_0] = \mathbb{E}^x[M_{\rho_n}] = \mathbb{E}^x[|S_{\rho_n}|^2] - (\mathrm{tr}\Gamma)\, \mathbb{E}^x[\rho_n].$$

Moreover, $n \leq |S_{\rho_n}| < (n + R)$. □

6.3 Difference estimates and Harnack inequality

In the next two sections we will prove useful results about random walk and harmonic functions. The main tools in the proofs are the optional sampling theorem and the estimates for the Green's function and the potential kernel. The basic idea in many of the proofs is to define a martingale in terms of the Green's function or potential kernel and then to stop it at a region at which that function is approximately constant. We recall that

$$\mathcal{B}_n = \{z \in \mathbb{Z}^d : |z| < n\}, \quad \mathcal{C}_n = \{z \in \mathbb{Z}^d : \mathcal{J}(z) < n\}.$$

Also, there is a $\delta > 0$ such that

$$\mathcal{C}_{\delta n} \subset \mathcal{B}_n \subset \mathcal{C}_{n/\delta}.$$

We set

$$\xi_n = \tau_{\mathcal{C}_n} = \min\{j \geq 1 : S_j \notin \mathcal{C}_n\}, \qquad \xi_n^* = \tau_{\mathcal{B}_n} = \min\{j \geq 1 : S_j \notin \mathcal{B}_n\}.$$

As the next proposition points out, the Green's function and potential kernel are almost constant on $\partial\mathcal{C}_n$. We recall that Theorems 4.3.1 and 4.4.4 imply that as $x \to \infty$,

$$G(x) = \frac{C_d}{\mathcal{J}(x)^{d-2}} + O\left(\frac{1}{|x|^d}\right), \qquad d \geq 3, \tag{6.10}$$

$$a(x) = C_2 \log \mathcal{J}(x) + \gamma_2 + O\left(\frac{1}{|x|^2}\right). \tag{6.11}$$

Here $C_2 = [\pi\sqrt{\det\Gamma}]^{-1}$ and $\gamma_2 = C + C_2 \log\sqrt{2}$ where C is as in Theorem 4.4.4.

Proposition 6.3.1 *If $p \in \mathcal{P}_d, d \geq 2$ then for $x \in \partial\mathcal{C}_n \cup \partial_i\mathcal{C}_n$,*

$$G(x) = \frac{C_d}{n^{d-2}} + O(n^{1-d}), \qquad d \geq 3,$$

$$a(x) = C_2 \log n + \gamma_2 + O(n^{-1}), \qquad d = 2,$$

where C_d, C_2, γ_2 are as (6.10) and (6.11).

Proof This follows immediately from (6.10) and (6.11) and the estimate

$$\mathcal{J}(x) = n + O(1), \qquad x \in \partial\mathcal{C}_n \cup \partial_i\mathcal{C}_n.$$

Note that the error term $O(n^{1-d})$ comes from the estimates

$$[n + O(1)]^{2-d} = n^{2-d} + O(n^{1-d}), \qquad \log[n + O(1)] = \log n + O(n^{-1}).$$

□

♣ Many of the arguments in this section use $\mathcal{C}_n$ rather than $\mathcal{B}_n$ because we can then use Proposition 6.3.1. We recall that for simple random walk $\mathcal{B}_n = \mathcal{C}_n$.

♣ Proposition 6.3.1 requires the walk to have bounded increments. If the walk does not have bounded increments, then many of the arguments in this chapter still hold. However, one needs to worry about "overshoot" estimates, i.e. giving upper bounds for the probability that the first visit of a random walk to the complement of $\mathcal{C}_n$ is far from $\mathcal{C}_n$. These kinds of estimates can be done in a spirit similar to Lemmas 5.1.9 and 5.1.10, but they complicate the arguments. For this reason, we restrict our attention to walks with bounded increments.

♣ Proposition 6.3.1 gives the estimates for the Green's function or potential kernel on $\partial \mathcal{C}_n$. In order to prove these estimates, it suffices for the error terms in (6.10) and (6.11) to be $O(|x|^{1-d})$ rather than $O(|x|^{-d})$. For this reason, many of the ideas in this section extend to random walks with bounded increments that are not necessarily symmetric (see Theorems 4.3.5 and 4.4.6). However, in this case we would need to deal with the Green's functions for the reversed walk as well as the Green's functions for the forward walk, and this complicates the notation in the arguments. For this reason, we restrict our attention to symmetric walks.

Proposition 6.3.2 *If $p \in \mathcal{P}_d$,*

$$G_{\mathcal{C}_n}(0,0) = G(0,0) - \frac{C_d}{n^{d-2}} + O(n^{1-d}), \quad d \geq 3,$$

$$G_{\mathcal{C}_n}(0,0) = C_2 \log n + \gamma_2 + O(n^{-1}), \quad d = 2. \tag{6.12}$$

where C_d, γ_2 are as defined in Proposition 6.3.1.

Proof Applying Proposition 4.6.2 at $x = y = 0$ gives

$$G_{\mathcal{C}_n}(0,0) = G(0,0) - \mathbb{E}[G(S_{\tau_{\mathcal{C}_n}}, 0)], \quad d \geq 3,$$

$$G_{\mathcal{C}_n}(0,0) = \mathbb{E}[a(S_{\tau_{\mathcal{C}_n}}, 0)], \quad d = 2.$$

We now apply Proposition 6.3.1. □

♣ It follows from Proposition 6.3.2 that

$$G_{\mathcal{B}_n}(0,0) = G(0,0) + O(n^{2-d}), \quad d \geq 3,$$

$$a_{\mathcal{B}_n}(0,0) = C_2 \log n + O(1), \quad d = 2.$$

It can be shown that $G_{\mathcal{B}_n}(0,0) = G(0,0) - \hat{C}_d\, n^{2-d} + o(n^{1-d})$, $a_{\mathcal{B}_n}(0,0) = C_2 \log n + \hat{\gamma}_2 + O(n^{-1})$ where $\hat{C}_d, \hat{\gamma}_2$ are different from C_d, γ_2 but we will not need this in the sequel, hence we omit the argument.

We will now prove difference estimates and a Harnack inequality for harmonic functions. There are different possible approaches to proving these results. One would be to use the result for Brownian motion and approximate.

We will use a different approach where we start with the known difference estimates for the Green's function G and the potential kernel a and work from there. We begin by proving a difference estimate for G_A. We then use this to prove a result on probabilities that is closely related to the gambler's ruin estimate for one-dimensional walks.

Lemma 6.3.3 *If $p \in \mathcal{P}_d, d \geq 2$, then for every $\epsilon > 0$, $r < \infty$, there is a c such that if $\mathcal{B}_{\epsilon n} \subset A \subsetneq \mathbb{Z}^d$, then for every $|x| > \epsilon n$ and every $|y| \leq r$,*

$$|G_A(0,x) - G_A(y,x)| \leq \frac{c}{n^{d-1}}.$$

$$|2\, G_A(0,x) - G_A(y,x) - G_A(-y,x)| \leq \frac{c}{n^d}.$$

Proof It suffices to prove the result for finite A for we can approximate any A by finite sets (see Exercise 4.7). Assume that $x \in A$, for otherwise the result is trivial. By symmetry $G_A(0,x) = G_A(x,0), G_A(y,x) = G_A(x,y)$. By Proposition 4.6.2,

$$G_A(x,0) - G_A(x,y) = G(x,0) - G(x,y) - \sum_{z \in \partial A} H_A(x,z)\,[G(z,0) - G(z,y)],$$

$$d \geq 3,$$

$$G_A(x,y) - G_A(x,0) = a(x,0) - a(x,y) - \sum_{z \in \partial A} H_A(x,z)\,[a(z,0) - a(z,y)],$$

$$d = 2.$$

There are similar expressions for the second differences. The difference estimates for the Green's function and the potential kernel (Corollaries 4.3.3 and 4.4.5) give, provided that $|y| \leq r$ and $|z| \geq (\epsilon/2)\,n$,

$$|G(z) - G(z+y)| \leq c_\epsilon\, n^{1-d}, \quad |2G(z) - G(z+y) - G(z-y)| \leq c_\epsilon\, n^{-d}$$

for $d \geq 3$ and

$$|a(z) - a(z+y)| \leq c_\epsilon\, n^{-1}, \quad |2a(z) - a(z+y) - a(z-y)| \leq c_\epsilon\, n^{-2}$$

for $d = 2$. □

The next lemma is very closely related to the one-dimensional gambler's ruin estimate. This lemma is particularly useful for x on or near the boundary of $\mathcal{C}_n$. For x in $\mathcal{C}_n \setminus \mathcal{C}_{n/2}$ that are away from the boundary, there are sharper estimates. See Propositions 6.4.1 and 6.4.2.

Lemma 6.3.4 *Suppose that $p \in \mathcal{P}_d, d \geq 2$. There exist c_1, c_2 such that for all n sufficiently large and all $x \in \overline{\mathcal{C}_n} \setminus \mathcal{C}_{n/2}$,*

$$\mathbb{P}^x\{S_{\tau_{\mathcal{C}_n \setminus \mathcal{C}_{n/2}}} \in \mathcal{C}_{n/2}\} \geq c_1\, n^{-1}, \tag{6.13}$$

and if $x \in \partial \mathcal{C}_n$,

$$\mathbb{P}^x\{S_{\tau_{\mathcal{C}_n \setminus \mathcal{C}_{n/2}}} \in \mathcal{C}_{n/2}\} \leq c_2\, n^{-1}. \tag{6.14}$$

Proof We will do the proof for $d \geq 3$; the proof for $d = 2$ is almost identical, replacing the Green's function with the potential kernel. It follows from (6.10) that there exist r, c such that for all n sufficiently large and all $y \in \mathcal{C}_{n-r}, z \in \partial \mathcal{C}_n$,

$$G(y) - G(z) \geq \, cn^{1-d}. \tag{6.15}$$

By choosing n sufficiently large, we can assure that $\partial_i \mathcal{C}_{n/2} \cap \mathcal{C}_{n/4} = \emptyset$.

Suppose that $x \in \mathcal{C}_{n-r}$ and let $T = \tau_{\mathcal{C}_n \setminus \mathcal{C}_{n/2}}$. Applying the optional sampling theorem to the bounded martingale $G(S_{j \wedge T})$, we see that

$$G(x) = \mathbb{E}^x[G(S_T)] \leq \mathbb{E}^x[G(S_T); S_T \in \mathcal{C}_{n/2}] + \max_{z \in \partial \mathcal{C}_n} G(z).$$

Therefore, (6.15) implies that

$$\mathbb{E}^x[G(S_T); S_T \in \mathcal{C}_{n/2}] \geq c\, n^{1-d}.$$

For n sufficiently large, $S_T \notin \mathcal{C}_{n/4}$ and hence (6.10) gives

$$\mathbb{E}^x[G(S_T); S_T \in \mathcal{C}_{n/2}] \leq c\, n^{2-d}\, \mathbb{P}^x\{\tau_{\mathcal{C}_n \setminus \mathcal{C}_{n/2}} < \tau_{\mathcal{C}_n}\}.$$

This establishes (6.13) for $x \in \mathcal{C}_{n-r}$.

To prove (6.13) for other x we note the following fact that holds for any $p \in \mathcal{P}_d$: there is an $\epsilon > 0$ such that for all $|x| \geq r$, there is a y with $p(y) \geq \epsilon$ and $\mathcal{J}(x + y) \leq \mathcal{J}(x) - \epsilon$. It follows that there is a $\delta > 0$ such that for all n sufficiently large and all $x \in \mathcal{C}_n$, there is probability at least δ that a random walk starting at x reaches $\mathcal{C}_{n-r}$ before leaving $\mathcal{C}_n$.

Since our random walk has a finite range, it suffices to prove (6.14) for $x \in \mathcal{C}_n \setminus \mathcal{C}_{n-r}$, and any finite r. For such x,

$$G(x) = C_d\, n^{2-d} + O(n^{1-d}).$$

Also,

$$\mathbb{E}^x[G(S_T) \mid S_T \in \mathcal{C}_{n/2}] = C_d\, 2^{d-2}\, n^{2-d} + O(n^{1-d}),$$
$$\mathbb{E}^x[G(S_T) \mid S_T \in \partial\mathcal{C}_n] = C_d\, n^{2-d} + O(n^{1-d}).$$

The optional sampling theorem gives

$$G(x) = \mathbb{E}^x[G(S_T)] = \mathbb{P}^x\{S_T \in \mathcal{C}_{n/2}\}\, \mathbb{E}^x[G(S_T) \mid S_T \in \mathcal{C}_{n/2}] + \mathbb{P}^x\{S_T \in \partial\mathcal{C}_n\}\, \mathbb{E}^x[G(S_T) \mid S_T \in \partial\mathcal{C}_n].$$

The left-hand side equals $C_d\, n^{2-d} + O(n^{1-d})$ and the right-hand side equals

$$C_d\, n^{2-d} + O(n^{1-d}) + C_d\, [2^{d-2} - 1]\, n^{2-d}\, \mathbb{P}^x\{S_T \in \mathcal{C}_{n/2}\}.$$

Therefore, $\mathbb{P}^x\{S_T \in \mathcal{C}_{n/2}\} = O(n^{-1})$. □

Proposition 6.3.5 *If $p \in \mathcal{P}_d$ and $x \in \mathcal{C}_n$,*

$$G_{\mathcal{C}_n}(0,x) = C_d\left[\mathcal{J}(x)^{2-d} - n^{2-d}\right] + O(|x|^{1-d}), \quad d \geq 3,$$
$$G_{\mathcal{C}_n}(0,x) = C_2\,[\log n - \log \mathcal{J}(x)] + O(|x|^{-1}), \quad d = 2.$$

In particular, for every $0 < \epsilon < 1/2$, there exist c_1, c_2 such that for all n sufficiently large,

$$c_1\, n^{2-d} \leq G_{\mathcal{C}_n}(y,x) \leq c_2\, n^{2-d}, \qquad y \in \mathcal{C}_{\epsilon n},\ \ x \in \partial_i \mathcal{C}_{2\epsilon n} \cup \partial \mathcal{C}_{2\epsilon n}.$$

Proof Symmetry and Lemma 4.6.2 tell us that

$$G_{\mathcal{C}_n}(0,x) = G_{\mathcal{C}_n}(x,0) = G(x,0) - \mathbb{E}^x[G(S_{\tau_{\mathcal{C}_n}})], \quad d \geq 3,$$
$$G_{\mathcal{C}_n}(0,x) = G_{\mathcal{C}_n}(x,0) = \mathbb{E}^x[a(S_{\tau_{\mathcal{C}_n}})] - a(x), \quad d = 2. \tag{6.16}$$

Also, (6.10) and (6.11) give

$$G(x) = C_d\, \mathcal{J}(x)^{2-d} + O(|x|^{-d}), \quad d \geq 3,$$
$$a(x) = C_2\, \log[\mathcal{J}(x)] + \gamma_2 + O(|x|^{-2}), \quad d = 2,$$

and Proposition 6.3.1 implies that

$$\mathbb{E}^x[G(S_{\tau_{\mathcal{C}_n}})] = \frac{C_d}{n^{d-2}} + O(n^{1-d}), \quad d \geq 3,$$
$$\mathbb{E}^x[a(S_{\tau_{\mathcal{C}_n}})] = C_2\, \log n + \gamma_2 + O(n^{-1}), \quad d = 2.$$

Since $|x| \leq c\,n$, we can write $O(|x|^{-d}) + O(n^{1-d}) \leq O(|x|^{1-d})$. To get the final assertion we use the estimate

$$G_{\mathcal{C}_{(1-\epsilon)n}}(0, x-y) \leq G_{\mathcal{C}_n}(y, x) \leq G_{\mathcal{C}_{(1+\epsilon)n}}(0, x-y).$$

□

We now focus on $H_{\mathcal{C}_n}$, the distribution of the first visit of a random walker to the complement of $\mathcal{C}_n$. Our first lemma uses the last-exit decomposition.

Lemma 6.3.6 *If $p \in \mathcal{P}_d$, $x \in B \subset A \subsetneq \mathbb{Z}^d$, $y \in \partial A$,*

$$H_A(x,y) = \sum_{z \in B} G_A(x,z)\, \mathbb{P}^z\{S_{\tau_{A \setminus B}} = y\}$$
$$= \sum_{z \in B} G_A(z,x)\, \mathbb{P}^y\{S_{\tau_{A \setminus B}} = z\}.$$

In particular,

$$H_A(x,y) = \sum_{z \in A} G_A(x,z)\, p(z,y) = \sum_{z \in \partial_i A} G_A(x,z)\, p(z,y).$$

Proof In the first display the first equality follows immediately from Proposition 4.6.4, and the second equality uses the symmetry of p. The second display is the particular case $B = A$. □

Lemma 6.3.7 *If $p \in \mathcal{P}_d$, there exist c_1, c_2 such that for all n sufficiently large and all $x \in \mathcal{C}_{n/4}, y \in \partial \mathcal{C}_n$,*

$$\frac{c_1}{n^{d-1}} \leq H_{\mathcal{C}_n}(x,y) \leq \frac{c_2}{n^{d-1}}.$$

♣ We think of $\partial \mathcal{C}_n$ as a $(d-1)$-dimensional subset of $\mathbb{Z}^d$ that contains on the order of n^{d-1} points. This lemma states that the hitting measure is mutually absolutely continuous with respect to the uniform measure on $\partial \mathcal{C}_n$ (with a constant independent of n).

Proof By the previous lemma,

$$H_{\mathcal{C}_n}(x,y) = \sum_{z \in \mathcal{C}_{n/2}} G_{\mathcal{C}_n}(z,x)\, \mathbb{P}^y\{S_{\tau_{\mathcal{C}_n \setminus \mathcal{C}_{n/2}}} = z\}.$$

Using Proposition 6.3.5 we see that for $z \in \partial_i \mathcal{C}_{n/2}, x \in \mathcal{C}_{n/4}, G_{\mathcal{C}_n}(z, x) \asymp n^{2-d}$. Also, Lemma 6.3.4 implies that

$$\sum_{z \in \mathcal{C}_{n/2}} \mathbb{P}^y\{S_{\tau_{\mathcal{C}_n \setminus \mathcal{C}_{n/2}}} = z\} \asymp n^{-1}.$$

□

Theorem 6.3.8 (difference estimates) *If $p \in \mathcal{P}_d$ and $r < \infty$, there exists c such that the following holds for every n sufficiently large.*

- *If $g : \overline{\mathcal{B}}_n \to \mathbb{R}$ is harmonic in $\mathcal{B}_n$ and $|y| \le r$,*

$$|\nabla_y g(0)| \le c \, \|g\|_\infty \, n^{-1}, \tag{6.17}$$

$$|\nabla_y^2 g(0)| \le c \, \|g\|_\infty \, n^{-2}. \tag{6.18}$$

- *If $f : \overline{\mathcal{B}}_n \to [0, \infty)$ is harmonic in $\mathcal{B}_n$ and $|y| \le r$, then*

$$|\nabla_y f(0)| \le c f(0) \, n^{-1}, \tag{6.19}$$

$$|\nabla_y^2 f(0)| \le c f(0) \, n^{-2}. \tag{6.20}$$

Proof Choose $\epsilon > 0$ such that $\mathcal{C}_{2\epsilon n} \subset \mathcal{B}_n$. Choose n sufficiently large so that $\mathcal{B}_r \subset \mathcal{C}_{(\epsilon/2)n}$ and $\partial_i \mathcal{C}_{2\epsilon n} \cap \mathcal{C}_{\epsilon n} = \emptyset$. Let $H(x, z) = H_{\mathcal{C}_{2\epsilon n}}(x, z)$. Then for $|x| \le r$,

$$g(x) = \sum_{z \in \partial \mathcal{C}_{2\epsilon n}} H(x, z) \, g(z),$$

and similarly for f. Hence, to prove the theorem, it suffices to establish (6.19) and (6.20) for $f(x) = H(x, z)$ (with c independent of n, z). Let $\rho = \rho_{n,\epsilon} = \tau_{\mathcal{C}_{2\epsilon n} \setminus \mathcal{C}_{\epsilon n}}$. By Lemma 6.3.6, if $x \in \mathcal{C}_{(\epsilon/2)n}$,

$$f(x) = \sum_{w \in \partial_i \mathcal{C}_{\epsilon n}} G_{\mathcal{C}_{2\epsilon n}}(w, x) \, \mathbb{P}^z\{S_\rho = w\}.$$

Lemma 6.3.7 shows that $f(x) \asymp n^{1-d}$ and in particular that

$$f(z) \le c f(w), \qquad z, w \in \mathcal{C}_{(\epsilon/2)n}. \tag{6.21}$$

There is a $\delta > 0$ such that for n sufficiently large, $|w| \ge \delta n$ for $w \in \partial_i \mathcal{C}_{\epsilon n}$. The estimates (6.19) and (6.20) now follow from Lemma 6.3.3 and Lemma 6.3.4. □

Theorem 6.3.9 (Harnack inequality) *Suppose that $p \in \mathcal{P}_d$, $U \subset \mathbb{R}^d$ is open and connected, and K is a compact subset of U. Then there exist $c = c(K, U, p) < \infty$ and positive integer $N = N(K, U, p)$ such that if $n \geq N$,*

$$U_n = \{x \in \mathbb{Z}^d : n^{-1} x \in U\}, \quad K_n = \{x \in \mathbb{Z}^d : n^{-1} x \in K\},$$

and $f : \overline{U_n} \to [0, \infty)$ is harmonic in U_n, then

$$f(x) \leq c\, f(y), \quad x, y \in K_n.$$

♣ This is the discrete analogue of the Harnack principle for positive harmonic functions in $\mathbb{R}^d$. Suppose that $K \subset U \subset \mathbb{R}^d$ where K is compact and U is open. Then there exists $c(K, U) < \infty$ such that if $f : U \to (0, \infty)$ is harmonic, then

$$f(x) \leq c(K, U)\, f(y), \quad x, y \in K.$$

Proof Without loss of generality we will assume that U is bounded. In (6.21) we showed that there exists $\delta > 0, c_0 < \infty$ such that

$$f(x) \leq c_0 f(y) \quad \text{if} \quad |x - y| \leq \delta \operatorname{dist}(x, \partial U_n). \tag{6.22}$$

Let us call two points z, w in U *adjacent* if $|z - w| < (\delta/4) \max\{\operatorname{dist}(z, \partial U), \operatorname{dist}(w, \partial U)\}$. Let ρ denote the graph distance associated to this adjacency, i.e. $\rho(z, w)$ is the minimum k such that there exists a sequence $z = z_0, z_1, \ldots, z_k = w$ of points in U such that z_j is adjacent to z_{j-1} for $j = 1, \ldots, k$. Fix $z \in U$, and let $V_k = \{w \in U : \rho(z, w) \leq k\}$, $V_{n,k} = \{x \in \mathbb{Z}^d : n^{-1} x \in V_k\}$. For $k \geq 1$, V_k is open, and the connectedness of U implies that $\cup V_k = U$. For n sufficiently large, if $x, y \in V_{n,k}$, there is a sequence of points $x = x_0, x_1, \ldots, x_k = y$ in $V_{n,k}$ such that $|x_j - x_{j-1}| < (\delta/2) \max\{\operatorname{dist}(x_j, \partial U), \operatorname{dist}(x_{j-1}, \partial U)\}$. Repeated application of (6.22) gives $f(x) \leq c_0^k f(y)$. Compactness of K implies that $K \subset V_k$ for some finite k, and hence $K_n \subset V_{n,k}$. □

6.4 Further estimates

In this section we will collect some more facts about random walks in $\mathcal{P}_d$ restricted to the set $\mathcal{C}_n$. The first three propositions are similar to Lemma 6.3.4.

Proposition 6.4.1 *If $p \in \mathcal{P}_2$, $m < n$, $T = \tau_{\mathcal{C}_n \setminus \mathcal{C}_m}$, then for $x \in \mathcal{C}_n \setminus \mathcal{C}_m$,*

$$\mathbb{P}^x\{S_T \in \partial \mathcal{C}_n\} = \frac{\log \mathcal{J}(x) - \log m + O(m^{-1})}{\log n - \log m}.$$

Proof Let $q = \mathbb{P}^x\{S_T \in \partial \mathcal{C}_n\}$. The optional sampling theorem applied to the bounded martingale $M_j = a(S_{j\wedge T})$ gives

$$a(x) = \mathbb{E}^x[a(S_T)] = (1-q)\,\mathbb{E}^x[a(S_T) \mid S_T \in \partial_i \mathcal{C}_m] + q\,\mathbb{E}^x[a(S_T) \mid S_T \in \partial \mathcal{C}_n].$$

From (6.11) and Proposition 6.3.1 we know that

$$a(x) = C_2 \log \mathcal{J}(x) + \gamma_2 + O(|x|^{-2}),$$
$$\mathbb{E}^x[a(S_T) \mid S_T \in \partial_i \mathcal{C}_m] = C_2 \log m + \gamma_2 + O(m^{-1}),$$
$$\mathbb{E}^x[a(S_T) \mid S_T \in \partial \mathcal{C}_n] = C_2 \log n + \gamma_2 + O(n^{-1}).$$

Solving for q gives the result. □

Proposition 6.4.2 *If $p \in \mathcal{P}_d, d \geq 3$, $T = \tau_{\mathbb{Z}^d \setminus \mathcal{C}_m}$, then for $x \in \mathbb{Z}^d \setminus \mathcal{C}_m$,*

$$\mathbb{P}^x\{T < \infty\} = \left(\frac{m}{\mathcal{J}(x)}\right)^{d-2} \left[1 + O(m^{-1})\right].$$

Proof Since $G(y)$ is a bounded harmonic function on $\tau_{\mathbb{Z}^d \setminus \mathcal{C}_m}$ with $G(\infty) = 0$, (6.5) gives

$$G(x) = \mathbb{E}^x[G(S_T); T < \infty] = \mathbb{P}^x\{T < \infty\}\,\mathbb{E}^x[G(S_T) \mid T < \infty].$$

But (6.10) gives

$$G(x) = C_d\, \mathcal{J}(x)^{2-d}\, [1 + O(|x|^{-2})],$$
$$\mathbb{E}^x[G(S_T) \mid T < \infty] = C_d\, m^{2-d}\, [1 + O(m^{-1})].$$

□

Proposition 6.4.3 *If $p \in \mathcal{P}_2$, $n > 0$, and $T = \tau_{\mathcal{C}_n \setminus \{0\}}$, then for $x \in \mathcal{C}_n$,*

$$\mathbb{P}^x\{S_T = 0\} = \left[1 - \frac{\log \mathcal{J}(x) + O(|x|^{-1})}{\log n}\right] \left[1 + O\left(\frac{1}{\log n}\right)\right].$$

Proof Recall that $\mathbb{P}^x\{S_T = 0\} = G_{\mathcal{C}_n}(x, 0)/G_{\mathcal{C}_n}(0, 0)$. The estimate then follows immediately from Propositions 6.3.2 and 6.3.5. The $O(|x|^{-1})$ term is superfluous except for x very close to $\partial \mathcal{C}_n$. □

Suppose that $m \leq n/2, x \in \mathcal{C}_m, z \in \partial \mathcal{C}_n$. By applying Theorem 6.3.8 $O(m)$ times we can see that (for n sufficiently large)

$$H_{\mathcal{C}_n}(x,z) = \mathbb{P}^x\{S_{\xi_n} = z\} = H_{\mathcal{C}_n}(0,z)\left[1 + O\left(\frac{m}{n}\right)\right]. \tag{6.23}$$

We will use this in the next two propositions to estimate some conditional probabilities.

Proposition 6.4.4 *Suppose that $p \in \mathcal{P}_d, d \geq 3$, $m < n/4$, and $\mathcal{C}_n \setminus \mathcal{C}_m \subset A \subset \mathcal{C}_n$. Suppose that $x \in \mathcal{C}_{2m}$ with $\mathbb{P}^x\{S_{\tau_A} \in \partial \mathcal{C}_n\} > 0$ and $z \in \partial \mathcal{C}_n$. Then for n sufficiently large,*

$$\mathbb{P}^x\{S_{\tau_A} = z \mid S_{\tau_A} \in \partial \mathcal{C}_n\} = H_{\mathcal{C}_n}(0,z)\left[1 + O\left(\frac{m}{n}\right)\right]. \tag{6.24}$$

Proof It is easy to check (using optional stopping) that it suffices to verify (6.24) for $x \in \partial \mathcal{C}_{2m}$. Note that (6.23) gives

$$\mathbb{P}^x\{S_{\xi_n} = z\} = H_{\mathcal{C}_n}(0,z)\left[1 + O\left(\frac{m}{n}\right)\right],$$

and since $\partial A \setminus \partial \mathcal{C}_n \subset \mathcal{C}_m$,

$$\mathbb{P}^x\{S_{\xi_n} = z \mid S_{\tau_A} \notin \partial \mathcal{C}_n\} = H_{\mathcal{C}_n}(0,z)\left[1 + O\left(\frac{m}{n}\right)\right].$$

This implies that

$$\mathbb{P}^x\{S_{\xi_n} = z; S_{\tau_A} \notin \partial \mathcal{C}_n\} = \mathbb{P}\{S_{\tau_A} \notin \partial \mathcal{C}_n\}\, H_{\mathcal{C}_n}(0,z)\left[1 + O\left(\frac{m}{n}\right)\right].$$

The last estimate, combined with (6.24), yields

$$\mathbb{P}^x\{S_{\xi_n} = z; S_{\tau_A} \in \partial \mathcal{C}_n\} = \mathbb{P}\{S_{\tau_A} \in \partial \mathcal{C}_n\}\, H_{\mathcal{C}_n}(0,z) + H_{\mathcal{C}_n}(0,z)\, O\left(\frac{m}{n}\right).$$

Using Proposition 6.4.2, we can see there is a c such that

$$\mathbb{P}^x\{S_{\tau_A} \in \partial \mathcal{C}_n\} \geq \mathbb{P}^x\{S_j \notin \mathcal{C}_m \text{ for all } j\} \geq c, \quad x \in \partial \mathcal{C}_{2m},$$

which allows us to write the preceding expression as

$$\mathbb{P}^x\{S_{\xi_n} = z; S_{\tau_A} \in \partial \mathcal{C}_n\} = \mathbb{P}\{S_{\tau_A} \in \partial \mathcal{C}_n\}\, H_{\mathcal{C}_n}(0,z)\left[1 + O\left(\frac{m}{n}\right)\right].$$

□

For $d = 2$ we get a similar result but with a slightly larger error term.

Proposition 6.4.5 *Suppose that $p \in \mathcal{P}_2$, $m < n/4$, and $\mathcal{C}_n \setminus \mathcal{C}_m \subset A \subset \mathcal{C}_n$. Suppose that $x \in \mathcal{C}_{2m}$ with $\mathbb{P}^x\{S_{\tau_A} \in \partial\mathcal{C}_n\} > 0$ and $z \in \partial\mathcal{C}_n$. Then, for n sufficiently large,*

$$\mathbb{P}^x\{S_{\tau_A} = z \mid S_{\tau_A} \in \partial\mathcal{C}_n\} = H_{\mathcal{C}_n}(0,z)\left[1 + O\left(\frac{m \log(n/m)}{n}\right)\right]. \tag{6.25}$$

Proof The proof is essentially the same, except for the last step, where Proposition 6.4.1 gives us

$$\mathbb{P}^x\{S_{\tau_A} \in \partial\mathcal{C}_n\} \geq \frac{c}{\log(n/m)}, \qquad x \in \mathcal{C}_{2m},$$

so that

$$H_{\mathcal{C}_n}(0,z)\, O\left(\frac{m}{n}\right)$$

can be written as

$$\mathbb{P}^x\{S_{\tau_A} \in \partial\mathcal{C}_n\}\, H_{\mathcal{C}_n}(0,z)\, O\left(\frac{m \log(n/m)}{n}\right).$$

□

The next proposition is a stronger version of Proposition 6.2.2. Here we show that the boundedness assumption of that proposition can be replaced with an assumption of sublinearity.

Proposition 6.4.6 *Suppose that $p \in \mathcal{P}_d, d \geq 3$ and $A \subset \mathbb{Z}^d$ with $\mathbb{Z}^d \setminus A$ finite. Suppose that $f : \mathbb{Z}^d \to \mathbb{R}$ is harmonic on A and satisfies $f(x) = o(|x|)$ as $x \to \infty$. Then there exists $b \in \mathbb{R}$ such that for all x,*

$$f(x) = \mathbb{E}^x[f(S_{\overline{\tau}_A}); \overline{\tau}_A < \infty] + b\,\mathbb{P}^x\{\overline{\tau}_A = \infty\}.$$

Proof Without loss of generality, we may assume that $0 \notin A$. Also, we may assume that $f \equiv 0$ on $\mathbb{Z}^d \setminus A$; otherwise, we can consider

$$\hat{f}(x) = f(x) - \mathbb{E}^x[f(S_{\overline{\tau}_A}); \overline{\tau}_A < \infty].$$

The assumptions imply that there is a sequence of real numbers ϵ_n decreasing to 0 such that $|f(x)| \leq \epsilon_n\, n$ for all $x \in \overline{\mathcal{C}_n}$ and hence

$$|f(x) - f(y)| \leq 2\epsilon_n\, n, \quad x, y \in \partial\mathcal{C}_n.$$

Since $\mathcal{L}f \equiv 0$ on A, (6.8) gives

$$0 = f(0) = \mathbb{E}[f(S_{\xi_n})] - \sum_{y \in \mathbb{Z}^d \setminus A} G_{\mathcal{C}_n}(0,y)\, \mathcal{L}f(y), \tag{6.26}$$

and since $\mathbb{Z}^d \setminus A$ is finite, this implies that

$$\lim_{n\to\infty} \mathbb{E}[f(S_{\xi_n})] = b := \sum_{y \in \mathbb{Z}^d \setminus A} G(0,y)\, \mathcal{L}f(y).$$

If $x \in A \cap \mathcal{C}_n$, the optional sampling theorem implies that

$$\begin{aligned} f(x) &= \mathbb{E}^x[f(S_{\tau_A \wedge \xi_n})] = \mathbb{E}^x[f(S_{\xi_n}); \tau_A > \xi_n] \\ &= \mathbb{P}^x\{\tau_A > \xi_n\}\, \mathbb{E}^x[f(S_{\xi_n}) \mid \tau_A > \xi_n]. \end{aligned}$$

For every $w \in \partial \mathcal{C}_n$, we can write

$$\begin{aligned} &\mathbb{E}^x[f(S_{\xi_n}) \mid \tau_A > \xi_n] - \mathbb{E}[f(S_{\xi_n})] \\ &\quad = \sum_{z \in \partial \mathcal{C}_n} f(z)\, [\mathbb{P}^x\{S_{\xi_n} = z \mid \tau_A > \xi_n\} - H_{\mathcal{C}_n}(0,z)] \\ &\quad = \sum_{z \in \partial \mathcal{C}_n} [f(z) - f(w)]\, [\mathbb{P}^x\{S_{\xi_n} = z \mid \tau_A > \xi_n\} - H_{\mathcal{C}_n}(0,z)]. \end{aligned}$$

For n large, apply $\sum_{w \in \partial \mathcal{C}_n}$ and divide by $|\partial \mathcal{C}_n|$ the above identity, and note that (6.24) now implies that

$$\left| \mathbb{E}^x[f(S_{\xi_n}) \mid \tau_A > \xi_n] - \mathbb{E}[f(S_{\xi_n})] \right| \le c\, \frac{|x|}{n} \sup_{y,z \in \partial \mathcal{C}_n} |f(z) - f(y)| \le c\, |x|\, \epsilon_n. \tag{6.27}$$

Therefore,

$$\begin{aligned} f(x) &= \lim_{n\to\infty} \mathbb{P}^x\{\tau_A > \xi_n\}\, \mathbb{E}^x[f(S_{\xi_n}) \mid \tau_A > \xi_n] \\ &= \mathbb{P}^x\{\tau_A = \infty\} \lim_{n\to\infty} \mathbb{E}[f(S_{\xi_n})] = b\, \mathbb{P}^x\{\tau_A = \infty\}. \end{aligned}$$

□

Proposition 6.4.7 *Suppose that $p \in \mathcal{P}_2$ and A is a finite subset of $\mathbb{Z}^2$ containing the origin. Let $T = T_A = \overline{\tau}_{\mathbb{Z}^2 \setminus A} = \min\{j \ge 0 : S_j \in A\}$. Then for each $x \in \mathbb{Z}^2$ the limit*

$$g_A(x) := \lim_{n\to\infty} C_2 (\log n)\, \mathbb{P}^x\{\xi_n < T\} \tag{6.28}$$

exists. Moreover, if $y \in A$,

$$g_A(x) = a(x-y) - \mathbb{E}^x[a(S_T - y)]. \tag{6.29}$$

Proof If $y \in A$ and $x \in \mathcal{C}_n \setminus A$, the optional sampling theorem applied to the bounded martingale $M_j = a(S_{j\wedge T\wedge \xi_n} - y)$ implies that

$$\begin{aligned} a(x-y) &= \mathbb{E}^x[a(S_{T\wedge\xi_n} - y)] \\ &= \mathbb{P}^x\{\xi_n < T\}\,\mathbb{E}^x[a(S_{\xi_n} - y) \mid \xi_n < T] + \mathbb{E}^x[a(S_T - y)] \\ &\quad - \mathbb{P}^x\{\xi_n < T\}\,\mathbb{E}^x[a(S_T - y) \mid \xi_n < T]. \end{aligned}$$

As $n \to \infty$,

$$\mathbb{E}^x[a(S_{\xi_n} - y) \mid \xi_n < T] \sim C_2 \log n.$$

Letting $n \to \infty$, we obtain the result. □

Remark As mentioned before, it follows that the right-hand side of (6.29) is the same for all $y \in A$. Also, since there exists δ such that $\mathcal{C}_{\delta n} \subset \mathcal{B}_n \subset \mathcal{C}_{n/\delta}$ we can replace (6.28) with

$$g_A(x) := \lim_{n\to\infty} C_2\,(\log n)\,\mathbb{P}^x\{\xi_n^* < T\}.$$

The astute reader will note that we already proved this proposition in Proposition 4.6.3.

Proposition 6.4.8 *Suppose that $p \in \mathcal{P}_2$ and A is a finite subset of $\mathbb{Z}^2$. Suppose that $f : \mathbb{Z}^2 \to \mathbb{R}$ is harmonic on $\mathbb{Z}^2 \setminus A$; vanishes on A; and satisfies $f(x) = o(|x|)$ as $|x| \to \infty$. Then $f = b\,g_A$ for some $b \in \mathbb{R}$.*

Proof Without loss of generality, assume that $0 \in A$ and let $T = T_A$ be as in the previous proposition. Using (6.8) and (6.12), we get

$$\mathbb{E}[f(S_{\xi_n})] = \sum_{y\in A} G_{\mathcal{C}_n}(0,y)\,\mathcal{L}f(y) = C_2 \log n \sum_{y\in A} \mathcal{L}f(y) + O(1).$$

(Here and below, the error terms may depend on A.) As in the argument deducing (6.27), we use (6.25) to see that

$$\begin{aligned} &|\mathbb{E}^x[f(S_{T\wedge\xi_n}) \mid \xi_n < T] - \mathbb{E}[f(S_{\xi_n})]| \\ &\quad \le c\,\frac{|x|\,\log n}{n} \sup_{y,z\in\partial\mathcal{C}_n} |f(y) - f(z)| \le c\,|x|\,\epsilon_n\,\log n, \end{aligned}$$

and combining the last two estimates we get

$$\begin{aligned} f(x) = \mathbb{E}^x[f(S_{T\wedge\xi_n})] &= \mathbb{P}^x\{\xi_n < T\}\,\mathbb{E}^x[f(S_{T_A\wedge\xi_n}) \mid \xi_n < T] \\ &= \mathbb{P}^x\{\xi_n < T\}\,\mathbb{E}[f(S_{\xi_n})] + |x|\,o(1) \\ &= b\,g_A(x) + o(1), \end{aligned}$$

where $b = \sum_{y\in A} \mathcal{L}f(y)$. □

6.5 Capacity, transient case

If A is a finite subset of $\mathbb{Z}^d$, we let

$$T_A = \tau_{\mathbb{Z}^d\setminus A}, \quad \overline{T}_A = \overline{\tau}_{\mathbb{Z}^d\setminus A},$$

$$\mathrm{rad}(A) = \sup\{|x| : x \in A\}.$$

If $p \in \mathcal{P}_d, d \geq 3$, define

$$\mathrm{Es}_A(x) = \mathbb{P}^x\{T_A = \infty\}, \qquad g_A(x) = \mathbb{P}^x\{\overline{T}_A = \infty\}.$$

Note that $\mathrm{Es}_A(x) = 0$ if $x \in A \setminus \partial_i A$. Furthermore, due to Proposition 6.4.6, g_A is the unique function on $\mathbb{Z}^d$ that is zero on A; harmonic on $\mathbb{Z}^d \setminus A$; and satisfies $g_A(x) \sim 1$ as $|x| \to \infty$. In particular, if $x \in A$,

$$\mathcal{L}g_A(x) = \sum_y p(y)\,g_A(x+y) = \mathrm{Es}_A(x).$$

Definition If $d \geq 3$, the *capacity* of a finite set A is given by

$$\mathrm{cap}(A) = \sum_{x\in A} \mathrm{Es}_A(x) = \sum_{z\in\partial_i A} \mathrm{Es}_A(z) = \sum_{x\in A} \mathcal{L}g_A(x) = \sum_{z\in\partial_i A} \mathcal{L}g_A(z).$$

♣ The motivation for the above definition is given by the following property (stated as the next proposition): as $z \to \infty$, the probability that a random walk starting at z ever hits A is comparable to $|z|^{2-d}\,\mathrm{cap}(A)$.

Proposition 6.5.1 *If $p \in \mathcal{P}_d, d \geq 3$ and $A \subset \mathbb{Z}^d$ is finite, then*

$$\mathbb{P}^x\{T_A < \infty\} = \frac{C_d\,\mathrm{cap}(A)}{\mathcal{J}(x)^{d-2}}\left[1 + O\left(\frac{\mathrm{rad}(A)}{|x|}\right)\right], \qquad |x| \geq 2\,\mathrm{rad}(A).$$

Proof There is a δ such that $\mathcal{B}_n \subset \mathcal{C}_{n/\delta}$ for all n. We will first prove the result for $x \notin \mathcal{C}_{2\text{rad}(A)/\delta}$. By the last-exit decomposition, Proposition 4.6.4,

$$\mathbb{P}^x\{\overline{T}_A < \infty\} = \sum_{y \in A} G(x, y)\, \text{Es}_A(y).$$

For $y \in A$, $\mathcal{J}(x - y) = \mathcal{J}(x) + O(|y|)$. Therefore,

$$G(x, y) = C_d\, \mathcal{J}(x)^{2-d} + O\left(\frac{|y|}{|x|^{d-1}}\right) = \frac{C_d}{\mathcal{J}(x)^{d-2}}\left[1 + O\left(\frac{\text{rad}(A)}{|x|}\right)\right].$$

This gives the result for $x \notin \mathcal{C}_{2\text{rad}(A)/\delta}$. We can extend this to $|x| \geq 2\text{rad}(A)$ by using the Harnack inequality (Theorem 6.3.9) on the set

$$\{z : 2\,\text{rad}(A) \leq |z|; \mathcal{J}(z) \leq (3/\delta)\,\text{rad}(A)\}.$$

Note that for x in this set, $\text{rad}(A)/|x|$ is of order 1, so it suffices to show that there is a c such that, for any two points x, z in this set,

$$\mathbb{P}^x\{\overline{T}_A < \infty\} \leq c\, \mathbb{P}^z\{\overline{T}_A < \infty\}.$$

□

Proposition 6.5.2 *If* $p \in \mathcal{P}_d, d \geq 3$,

$$\text{cap}(\mathcal{C}_n) = C_d^{-1}\, n^{d-2} + O(n^{d-1}).$$

Proof By Proposition 4.6.4,

$$1 = \mathbb{P}\{\overline{T}_{\mathcal{C}_n} < \infty\} = \sum_{y \in \partial_i \mathcal{C}_n} G(0, y)\, \text{Es}_{\mathcal{C}_n}(y),$$

But for $y \in \partial_i \mathcal{C}_n$, Proposition 6.3.1 gives

$$G(0, y) = C_d\, n^{2-d}\, [1 + O(n^{-1})].$$

Hence,

$$1 = C_d\, n^{2-d}\, \text{cap}(\mathcal{C}_n)\left[1 + O(n^{-1})\right].$$

□

Let

$$T_{A,n} = T_A \wedge \xi_n = \inf\{j \geq 1 : S_j \in A \text{ or } S_j \notin \mathcal{C}_n\}.$$

If $x \in A \subset \mathcal{C}_n$,

$$\mathbb{P}^x\{T_A > \xi_n\} = \sum_{y \in \partial \mathcal{C}_n} \mathbb{P}^x\{S_{T_{A,n}} = y\} = \sum_{y \in \partial \mathcal{C}_n} \mathbb{P}^y\{S_{T_{A,n}} = x\}.$$

The last equation uses symmetry of the walk. As a consequence,

$$\sum_{x \in A} \mathbb{P}^x\{T_A > \xi_n\} = \sum_{x \in A} \sum_{y \in \partial \mathcal{C}_n} \mathbb{P}^y\{S_{T_{A,n}} = x\} = \sum_{y \in \partial \mathcal{C}_n} \mathbb{P}^y\{T_A < \xi_n\}. \tag{6.30}$$

Therefore,

$$\text{cap}(A) = \sum_{x \in A} \text{Es}_A(x) = \lim_{n \to \infty} \sum_{x \in A} \mathbb{P}^x\{T_A > \xi_n\} = \lim_{n \to \infty} \sum_{y \in \partial \mathcal{C}_n} \mathbb{P}^y\{T_A < \xi_n\}. \tag{6.31}$$

♣ The identities (6.30)–(6.31) relate, for a given finite set A, the probability that a random walker started uniformly in A "escapes" A and the probability that a random walker started uniformly on the boundary of a large ellipse (far away from A) ever hits A. Formally, every path from A to infinity can also be considered as a path from infinity to A by reversal. This correspondence is manifested again in Proposition 6.5.4.

Proposition 6.5.3 *If $p \in \mathcal{P}_d, d \geq 3$, and A, B are finite subsets of $\mathbb{Z}^d$, then*

$$\text{cap}(A \cup B) \leq \text{cap}(A) + \text{cap}(B) - \text{cap}(A \cap B).$$

Proof Choose n such that $A \cup B \subset \mathcal{C}_n$. Then, for $y \in \partial \mathcal{C}_n$,

$$\begin{aligned} \mathbb{P}^y\{T_{A \cup B} < \xi_n\} &= \mathbb{P}^x\{T_A < \xi_n \text{ or } T_B < \xi_n\} \\ &= \mathbb{P}^y\{T_A < \xi_n\} + \mathbb{P}^y\{T_B < \xi_n\} - \mathbb{P}^y\{T_A < \xi_n, T_B < \xi_n\} \\ &\leq \mathbb{P}^y\{T_A < \xi_n\} + \mathbb{P}^y\{T_B < \xi_n\} - \mathbb{P}^y\{T_{A \cap B} < \xi_n\}. \end{aligned}$$

The proposition then follows from (6.31). □

Definition If $p \in \mathcal{P}_d, d \geq 3$, and $A \subset \mathbb{Z}^d$ is finite, the *harmonic measure of A (from infinity)* is defined by

$$\text{hm}_A(x) = \frac{\text{Es}_A(x)}{\text{cap}(A)}, \quad x \in A.$$

Note that hm_A is a probability measure supported on $\partial_i A$. As the next proposition shows, it can be considered as the hitting measure of A by a random walk "started at infinity conditioned to hit A".

Proposition 6.5.4 *If $p \in \mathcal{P}_d, d \geq 3$, and $A \subset \mathbb{Z}^d$ is finite, then for $x \in A$,*

$$\mathrm{hm}_A(x) = \lim_{|y| \to \infty} \mathbb{P}^y\{S_{T_A} = x \mid T_A < \infty\}.$$

In fact, if $A \subset \mathcal{C}_{n/2}$ and $y \notin \mathcal{C}_n$, then

$$\mathbb{P}^y\{S_{T_A} = x \mid T_A < \infty\} = \mathrm{hm}_A(x)\left[1 + O\left(\frac{\mathrm{rad}(A)}{|y|}\right)\right]. \tag{6.32}$$

Proof If $A \subset \mathcal{C}_n$ and $y \notin \mathcal{C}_n$, the last-exit decomposition (Proposition 4.6.4) gives

$$\mathbb{P}^y\{S_{T_A} = x\} = \sum_{z \in \partial \mathcal{C}_n} G_{\mathbb{Z}^d \setminus A}(y, z)\, \mathbb{P}^z\{S_{T_{A,n}} = x\},$$

where, as before, $T_{A,n} = T_A \wedge \xi_n$. By symmetry and (6.24),

$$\begin{aligned}\mathbb{P}^z\{S_{T_{A,n}} = x\} = \mathbb{P}^x\{S_{T_{A,n}} = z\} &= \mathbb{P}^x\{\xi_n < T_A\}\, H_{\mathcal{C}_n}(0, z)\left[1 + O\left(\frac{\mathrm{rad}(A)}{n}\right)\right] \\ &= \mathrm{Es}_A(x)\, H_{\mathcal{C}_n}(0, z)\left[1 + O\left(\frac{\mathrm{rad}(A)}{n}\right)\right].\end{aligned}$$

The last equality uses

$$\mathrm{Es}_A(x) = \mathbb{P}^x\{T_A = \infty\} = \mathbb{P}^x\{T_A > \xi_n\}\left[1 + O\left(\frac{\mathrm{rad}(A)^{d-2}}{n^{d-2}}\right)\right],$$

which follows from Proposition 6.4.2. Therefore,

$$\mathbb{P}^y\{S_{T_A} = x\} = \mathrm{Es}_A(x)\left[1 + O\left(\frac{\mathrm{rad}(A)}{n}\right)\right] \sum_{z \in \partial \mathcal{C}_n} G_{\mathbb{Z}^d \setminus A}(y, z)\, H_{\mathcal{C}_n}(0, z),$$

and by summing over x,

$$\mathbb{P}^y\{T_A < \infty\} = \mathrm{cap}(A)\left[1 + O\left(\frac{\mathrm{rad}(A)}{n}\right)\right] \sum_{z \in \partial \mathcal{C}_n} G_{\mathbb{Z}^d \setminus A}(y, z)\, H_{\mathcal{C}_n}(0, z).$$

We obtain (6.32) by dividing the last two expressions. □

Proposition 6.5.5 *If $p \in \mathcal{P}_d, d \geq 3$, and $A \subset \mathbb{Z}^d$ is finite, then*

$$\operatorname{cap}(A) = \sup \sum_{x \in A} f(x), \tag{6.33}$$

where the supremum is over all functions $f \geq 0$ supported on A such that

$$Gf(y) := \sum_{x \in \mathbb{Z}^d} G(y,x) f(x) = \sum_{x \in A} G(y,x) f(x) \leq 1$$

for all $y \in \mathbb{Z}^d$.

Proof Let $\hat{f}(x) = \mathrm{Es}_A(x)$. Note that Proposition 4.6.4 implies that for $y \in \mathbb{Z}^d$,

$$1 \geq \mathbb{P}^y\{\overline{T}_A < \infty\} = \sum_{x \in A} G(y,x)\, \mathrm{Es}_A(x).$$

Hence, $G\hat{f} \leq 1$ and the supremum in (6.33) is at least as large as cap(A). Note also that $G\hat{f}$ is the unique bounded function on $\mathbb{Z}^d$ that is harmonic on $\mathbb{Z}^d \setminus A$; equals 1 on A; and approaches 0 at infinity. Suppose that $f \geq 0, f = 0$ on $\mathbb{Z}^d \setminus A$, with $Gf(y) \leq 1$ for all $y \in \mathbb{Z}^d$. Then Gf is the unique bounded function on $\mathbb{Z}^d$ that is harmonic on $\mathbb{Z}^d \setminus A$; equals $Gf \leq 1$ on A; and approaches zero at infinity. By the maximum principle, $Gf(y) \leq G\hat{f}(y)$ for all y. In particular, $G(\hat{f} - f)$ is harmonic on $\mathbb{Z}^d \setminus A$, is nonnegative on $\mathbb{Z}^d$, and approaches zero at infinity. We need to show that

$$\sum_{x \in A} f(x) \leq \sum_{x \in A} \hat{f}(x).$$

If $x, y \in A$, let

$$K_A(x,y) = \mathbb{P}^x\{S_{T_A} = y\}.$$

Note that $K_A(x,y) = K_A(y,x)$ and

$$\sum_{y \in A} K_A(x,y) = 1 - \mathrm{Es}_A(x).$$

If h is a bounded function on $\mathbb{Z}^d$ that is harmonic on $\mathbb{Z}^d \setminus A$ and has $h(\infty) = 0$, then $h(z) = \mathbb{E}[h(S_{\overline{T}_A}); \overline{T}_A < \infty], z \in \mathbb{Z}^d$. Using this, one can easily check that for $x \in A$,

$$\mathcal{L}h(x) = \left[\sum_{y \in A} K_A(x,y)\, h(y)\right] - h(x).$$

Also, if $h \geq 0$,

$$\sum_{x \in A} \sum_{y \in A} K_A(x,y)\, h(y) = \sum_{y \in A} h(y) \sum_{x \in A} K_A(y,x)$$
$$= \sum_{y \in A} h(y)\, [1 - \mathrm{Es}_A(y)] \leq \sum_{y \in A} h(y),$$

which implies that

$$\sum_{x \in \mathbb{Z}^d} \mathcal{L}h(x) = \sum_{x \in A} \mathcal{L}h(x) \leq 0.$$

Then, using (4.25),

$$\sum_{x \in A} f(x) = -\sum_{x \in A} \mathcal{L}[Gf](x) \leq -\sum_{x \in A} \mathcal{L}[Gf](x) - \sum_{x \in A} \mathcal{L}[G(\hat{f} - f)](x)$$
$$= \sum_{x \in A} \mathrm{Es}_A(x).$$

□

Our definition of capacity depends on the random walk p. The next proposition shows that capacities for different p's in the same dimension are comparable.

Proposition 6.5.6 *Suppose that $p, q \in \mathcal{P}_d, d \geq 3$ and let $\mathrm{cap}_p, \mathrm{cap}_q$ denote the corresponding capacities. Then, there is a $\delta = \delta(p,q) > 0$ such that for all finite $A \subset \mathbb{Z}^d$,*

$$\delta \,\mathrm{cap}_p(A) \leq \mathrm{cap}_q(A) \leq \delta^{-1} \,\mathrm{cap}_p(A).$$

Proof It follows from Theorem 4.3.1 that there exists δ such that

$$\delta\, G_p(x,y) \leq G_q(x,y) \leq \delta^{-1}\, G_p(x,y),$$

for all x, y. The proposition then follows from Proposition 6.5.5. □

Definition If $p \in \mathcal{P}_d, d \geq 3$, and $A \subset \mathbb{Z}^d$, then A is *transient* if

$$\mathbb{P}\{S_n \in A \text{ i.o.}\} = 0.$$

Otherwise, the set is called *recurrent*.

Lemma 6.5.7 *If $p \in \mathcal{P}_d, d \geq 3$, then a subset A of $\mathbb{Z}^d$ is recurrent if and only if for every $x \in \mathbb{Z}^d$,*

$$\mathbb{P}^x\{S_n \in A \text{ i.o.}\} = 1.$$

Proof The if direction of the statement is trivial. To show the only if direction, let $F(y) = \mathbb{P}^y\{S_n \in A \text{ i.o.}\}$, and note that F is a bounded harmonic function on $\mathbb{Z}^d$, so it must be constant by Proposition 6.1.2. Now, if $F(y) \geq \epsilon > 0, y \in \mathbb{Z}^d$, then for each x there is an N_x such that

$$\mathbb{P}^x\{S_n \in A \text{ for some } n \leq N_x\} \geq \epsilon/2.$$

By iterating this we can see for all x,

$$\mathbb{P}^x\{S_n \in A \text{ for some } n < \infty\} = 1,$$

and the lemma follows easily. □

♣ Alternatively, $\{S_n \in A \text{ i.o.}\}$ is an exchangeable event with respect to the independent, identically distributed steps of the random walk, and therefore $\mathbb{P}^x(S_n \in A \text{ i.o.}) \in \{0, 1\}$.

Clearly, all finite sets are transient; in fact, finite unions of transient sets are transient. If A is a subset such that

$$\sum_{x \in A} G(x) < \infty, \tag{6.34}$$

then A is transient. To see this, let S_n be a random walk starting at the origin and let V denote the number of visits to A,

$$V_A = \sum_{j=0}^{\infty} 1\{S_n \in A\}.$$

Then (6.34) implies that $\mathbb{E}[V_A] < \infty$ which implies that $\mathbb{P}\{V_A < \infty\} = 1$. In Exercise 6.3, it is shown that the converse is not true, i.e. there exist transient sets A with $\mathbb{E}[V_A] = \infty$.

Lemma 6.5.8 *Suppose that $p \in \mathcal{P}_d, d \geq 3$, and $A \subset \mathbb{Z}^d$. Then A is transient if and only if*

$$\sum_{k=1}^{\infty} \mathbb{P}\{T^k < \infty\} < \infty, \tag{6.35}$$

where $T^k = T_{A_k}$ and $A_k = A \cap (\mathcal{C}_{2^k} \setminus \mathcal{C}_{2^{k-1}})$.

Proof Let E_k be the event $\{T^k < \infty\}$. Since the random walk is transient, A is transient if and only if $\mathbb{P}\{E_k \text{ i.o.}\} = 0$. Hence the Borel–Cantelli lemma implies that any A satisfying (6.35) is transient.

Suppose that

$$\sum_{k=1}^{\infty} \mathbb{P}\{T^k < \infty\} = \infty.$$

Then either the sum over even k or the sum over odd k is infinite. We will assume the former; the argument if the latter holds is almost identical. Let $B_{k,+} = A_k \cap \{(z^1, \ldots, z^d) : z^1 \geq 0\}$ and $B_{k,-} = A_k \cap \{(z^1, \ldots, z^d) : z^1 \leq 0\}$. Since $\mathbb{P}\{T^{2k} < \infty\} \leq \mathbb{P}\{T_{B_{2k,+}} < \infty\} + \mathbb{P}\{T_{B_{2k,-}} < \infty\}$, we know that either

$$\sum_{k=1}^{\infty} \mathbb{P}\{T_{B_{2k,+}} < \infty\} = \infty, \tag{6.36}$$

or the same equality with $B_{2k,-}$ replacing $B_{2k,+}$. We will assume that (6.36) holds and write $\sigma_k = T_{B_{2k,+}}$. An application of the Harnack inequality (we leave the details as Exercise 6.11) shows that there is a c such that for all $j \neq k$,

$$\mathbb{P}\{\sigma_j < \infty \mid \sigma_j \wedge \sigma_k = \sigma_k < \infty\} \leq c\, \mathbb{P}\{\sigma_j < \infty\}.$$

This implies that

$$\mathbb{P}\{\sigma_j < \infty, \sigma_k < \infty\} \leq 2c\, \mathbb{P}\{\sigma_j < \infty\}\, \mathbb{P}\{\sigma_k < \infty\}.$$

Using this and a special form of the Borel–Cantelli lemma (Corollary A.6.2) we can see that

$$\mathbb{P}\{\sigma_j < \infty \text{ i.o.}\} > 0,$$

which implies that A is not transient. □

Corollary 6.5.9 (Wiener's test) *Suppose that $p \in \mathcal{P}_d, d \geq 3$, and $A \subset \mathbb{Z}^d$, then A is transient if and only if*

$$\sum_{k=1}^{\infty} 2^{(2-d)k} \operatorname{cap}(A_k) < \infty \tag{6.37}$$

where $A_k = A \cap (\mathcal{C}_{2^k} \setminus \mathcal{C}_{2^{k-1}})$. In particular, if A is transient for some $p \in \mathcal{P}_d$, then it is transient for all $p \in \mathcal{P}_d$.

Proof Due to Proposition 6.5.1, we have that $\mathbb{P}\{T^k < \infty\} \asymp 2^{(2-d)k} \operatorname{cap}(A_k)$. □

Theorem 6.5.10 *Suppose that $d \geq 3$, $p \in \mathcal{P}_d$, and S_n is a p-walk. Let A be the set of points visited by the random walk,*

$$A = S[0, \infty) = \{S_n : n = 0, 1, \ldots\}.$$

If $d = 3, 4$, then with probability one A is a recurrent set. If $d \geq 5$, then with probability one A is a transient set.

Proof Since a set is transient if and only if all its translates are transient, we see that for each n, A is recurrent if and only if the set

$$\{S_m - S_n : m = n, n+1, \ldots\}$$

is recurrent. Hence the event $\{A \text{ is recurrent}\}$ is a tail event, and the Kolmogorov 0-1 law now implies that it has the probability zero or one.

Let Y denote the random variable that equals the expected number of visits to A by an independent random walker $\tilde{S}_n$ starting at the origin. In other words,

$$Y = \sum_{x \in A} G(x) = \sum_{x \in \mathbb{Z}^d} 1\{x \in A\}\, G(x).$$

Then,

$$\mathbb{E}(Y) = \sum_{x \in \mathbb{Z}^d} \mathbb{P}\{x \in A\}\, G(x) = G(0)^{-1} \sum_{x \in \mathbb{Z}^d} G(x)^2.$$

Since $G(x) \asymp |x|^{2-d}$, we have $G(x)^2 \asymp |x|^{4-2d}$. By examining the sum, we see that $\mathbb{E}(Y) = \infty$ for $d = 3, 4$ and $\mathbb{E}(Y) < \infty$ for $d \geq 5$. If $d \geq 5$, this gives $Y < \infty$ with probability one which implies that A is transient with probability one.

We now focus on $d = 4$ (it is easy to see that if the result holds for $d = 4$, then it also holds for $d = 3$). It suffices to show that $\mathbb{P}\{A \text{ is recurrent}\} > 0$. Let S^1, S^2 be independent random walks with increment distribution p starting at the origin, and let

$$\sigma_k^j = \min\{n : S_n^j \notin \mathcal{C}_{2^k}\}.$$

Let

$$\begin{aligned} V_k^j &= [\mathcal{C}_{2^k} \setminus \mathcal{C}_{2^{k-1}}] \cap S^j[0, \sigma_{k+1}^j) \\ &= \{x \in \mathcal{C}_{2^k} \setminus \mathcal{C}_{2^{k-1}} : S_n^j = x \text{ for some } n \leq \sigma_{k+1}^j\}. \end{aligned}$$

Let E_k be the event $\{V_k^1 \cap V_k^2 \neq \emptyset\}$. We will show that $\mathbb{P}\{E_k \text{ i.o.}\} > 0$ which will imply that with positive probability, $\{S_n^1 : n = 0, 1, \ldots\}$ is recurrent. Using Corollary A.6.2, one can see that it suffices to show that

$$\sum_{k=1}^{\infty} \mathbb{P}(E_{3k}) = \infty, \tag{6.38}$$

and that there exists a constant $c < \infty$ such that for $m < k$,

$$\mathbb{P}(E_{3m} \cap E_{3k}) \leq c\, \mathbb{P}(E_{3m})\, \mathbb{P}(E_{3k}). \tag{6.39}$$

The event E_{3m} depends only on the values of S_n^j with $\sigma_{3m-1}^j \leq n \leq \sigma_{3m+1}^j$. Hence, the Harnack inequality implies that $\mathbb{P}(E_{3k} \mid E_{3m}) \leq c\, \mathbb{P}(E_{3k})$ so (6.39) holds. To prove (6.38), let $J^j(k, x)$ denote the indicator function of the event that $S_n^j = x$ for some $n \leq \sigma_k^j$. Then,

$$Z_k := \#(V_k^1 \cap V_k^2) = \sum_{x \in \mathcal{C}_{2^k} \setminus \mathcal{C}_{2^{k-1}}} J^1(k, x)\, J^2(k, x).$$

There exist c_1, c_2 such that if $x, y \in \mathcal{C}_{2^k} \setminus \mathcal{C}_{2^{k-1}}$ (recall $d - 2 = 2$),

$$\mathbb{E}[J^j(k, x)] \geq c_1\, (2^k)^{-2},$$

$$\mathbb{E}[J^j(k, x)\, J^j(k, y)] \leq c_2\, (2^k)^{-2}\, [1 + |x - y|]^{-2}.$$

(The latter inequality is obtained by noting that the probability that a random walker hits both x and y given that it hits at least one of them is bounded above by the probability that a random walker starting at the origin visits $y - x$.) Therefore,

$$\mathbb{E}[Z_k] = \sum_{x \in \mathcal{C}_{2^k} \setminus \mathcal{C}_{2^{k-1}}} \mathbb{E}[J^1(k,x)]\, \mathbb{E}[J^2(k,x)] \geq c \sum_{x \in \mathcal{C}_{2^k} \setminus \mathcal{C}_{2^{k-1}}} (2^k)^{-4} \geq c,$$

$$\mathbb{E}[Z_k^2] = \sum_{x,y \in \mathcal{C}_{23k} \setminus \mathcal{C}_{2^{k-1}}} \mathbb{E}[J^1(k,x)\, J^1(k,y)]\, \mathbb{E}[J^2(k,x)\, J^2(k,y)]$$

$$\leq c \sum_{x,y \in \mathcal{C}_{2^k} \setminus \mathcal{C}_{2^{k-1}}} (2^k)^{-4} \frac{1}{[1 + |x-y|^2]^2} \leq ck,$$

where for the last inequality, note that there are $O(2^{4k})$ points in $\mathcal{C}_{2^k} \setminus \mathcal{C}_{2^{k-1}}$, and that for $x \in \mathcal{C}_{2^k} \setminus \mathcal{C}_{2^{k-1}}$ there are $O(\ell^3)$ points $y \in \mathcal{C}_{2^k} \setminus \mathcal{C}_{2^{k-1}}$ at distance ℓ from x. The second moment estimate, Lemma A.6.1 now implies that $\mathbb{P}\{Z_k > 0\} \geq c/k$, hence (6.38) holds. □

♣ The CLT implies that the number of points in $\mathcal{B}_n$ visited by a random walk is of order n^2. Roughly speaking, we can say that a random walk path is a "two-dimensional" set. Asking whether or not this is recurrent is asking whether or not two random two-dimensional sets intersect. Using the example of planes in $\mathbb{R}^d$, one can guess that the critical dimension is four.

6.6 Capacity in two dimensions

The theory of capacity in two dimensions is somewhat similar to that for $d \geq 3$, but there are significant differences due to the fact that the random walk is recurrent. We start by recalling a few facts from Propositions 6.4.7 and 6.4.8. If $p \in \mathcal{P}_2$ and $0 \in A \subset \mathbb{Z}^2$ is finite, let

$$g_A(x) = a(x) - \mathbb{E}^x[a(S_{\overline{T}_A})] = \lim_{n \to \infty} C_2 (\log n)\, \mathbb{P}^x\{\xi_n < \overline{T}_A\}. \tag{6.40}$$

The function g_A is the unique function on $\mathbb{Z}^2$ that vanishes on A, is harmonic on $\mathbb{Z}^2 \setminus A$, and satisfies $g_A(x) \sim C_2 \log \mathcal{J}(x) \sim C_2 \log |x|$ as $x \to \infty$. If $y \in A$, we can also write

$$g_A(x) = a(x-y) - \mathbb{E}^x[a(S_{\overline{T}_A} - y)].$$

To simplify notation, we will mostly assume that $0 \in A$, and then $a(x) - g_A(x)$ is the unique bounded function on $\mathbb{Z}^2$ that is harmonic on $\mathbb{Z}^2 \setminus A$ and has boundary value a on A. We define the *harmonic measure of A (from infinity)* by

$$\mathrm{hm}_A(x) = \lim_{|y|\to\infty} \mathbb{P}^y\{S_{T_A} = x\}. \tag{6.41}$$

Since $\mathbb{P}^y\{T_A < \infty\} = 1$, this is the same as $\mathbb{P}^y\{S_{T_A} = x \mid T_A < \infty\}$ and hence agrees with the definition of harmonic measure for $d \geq 3$. It is not clear a priori that the limit exists; this fact is established in the next proposition.

Proposition 6.6.1 *Suppose that $p \in \mathcal{P}_2$ and $0 \in A \subset \mathbb{Z}^2$ is finite. Then the limit in (6.41) exists and equals $\mathcal{L}g_A(x)$.*

Proof Fix A and let $r_A = \mathrm{rad}(A)$. Let n be sufficiently large so that $A \subset \mathcal{C}_{n/4}$. Using (6.25) on the set $\mathbb{Z}^2 \setminus A$, we see that if $x \in \partial_i A, y \in \partial \mathcal{C}_n$,

$$\mathbb{P}^y\{S_{T_A \wedge \xi_n} = x\} = \mathbb{P}^x\{S_{T_A \wedge \xi_n} = y\} = \mathbb{P}^x\{\xi_n < T_A\}\, H_{\mathcal{C}_n}(0, y) \times \left[1 + O\left(\frac{r_A \log n}{n}\right)\right].$$

If $z \in \mathbb{Z}^2 \setminus \mathcal{C}_n$, the last-exit decomposition (Proposition 4.6.4) gives

$$\mathbb{P}^z\{S_{T_A} = x\} = \sum_{y \in \partial \mathcal{C}_n} G_{\mathbb{Z}^2 \setminus A}(z, y)\, \mathbb{P}^y\{S_{T_A \wedge \xi_n} = x\}.$$

Therefore,

$$\mathbb{P}^z\{S_{T_A} = x\} = \mathbb{P}^x\{\xi_n < T_A\}\, J(n, z) \left[1 + O\left(\frac{r_A \log n}{n}\right)\right], \tag{6.42}$$

where

$$J(n, z) = \sum_{y \in \partial \mathcal{C}_n} H_{\mathcal{C}_n}(0, y)\, G_{\mathbb{Z}^2 \setminus A}(z, y).$$

If $x \in A$, the definition of $\mathcal{L}$, the optional sampling theorem, and the asymptotic expansion of g_A respectively imply that

$$\begin{aligned} \mathcal{L}g_A(x) = \mathbb{E}^x[g_A(S_1)] &= \mathbb{E}^x[g_A(S_{T_A \wedge \xi_n})] \\ &= \mathbb{E}^x[g_A(S_{\xi_n}); \xi_n < T_A] \\ &= \mathbb{P}^x\{\xi_n < T_A\}\ [C_2 \log n + O_A(1)]. \end{aligned} \tag{6.43}$$

In particular,

$$\mathcal{L}g_A(x) = \lim_{n\to\infty} C_2\, (\log n)\, \mathbb{P}^x\{\xi_n < T_A\}, \quad x \in A. \tag{6.44}$$

(This is the $d = 2$ analogue of the relation $\mathcal{L}g_A(x) = \mathrm{Es}_A(x)$ for $d \geq 3$.)

Note that (as in (6.30))

$$\sum_{x\in\partial_i A} \mathbb{P}^x\{\xi_n < T_A\} = \sum_{x\in\partial_i A}\sum_{y\in\partial\mathcal{C}_n} \mathbb{P}^x\{S_{\xi_n\wedge T_A} = y\}$$
$$= \sum_{y\in\partial\mathcal{C}_n}\sum_{x\in\partial_i A} \mathbb{P}^y\{S_{\xi_n\wedge T_A} = x\} = \sum_{y\in\partial\mathcal{C}_n} \mathbb{P}^y\{T_A < \xi_n\}.$$

Proposition 6.4.3 shows that if $x \in A$, then the probability that a random walk starting at x reaches $\partial\mathcal{C}_n$ before visiting the origin is bounded above by $c \log r_A/\log n$. Therefore,

$$\mathbb{P}^y\{T_A < \xi_n\} = \mathbb{P}^y\{T_{\{0\}} < \xi_n\}\left[1 + O\left(\frac{\log r_A}{\log n}\right)\right].$$

As a consequence,

$$\sum_{x\in\partial_i A} \mathbb{P}^x\{\xi_n < T_A\} = \sum_{y\in\partial\mathcal{C}_n} \mathbb{P}^y\{S_{\xi_n\wedge T_{\{0\}}} = 0\}\left[1 + O\left(\frac{\log r_A}{\log n}\right)\right]$$
$$= \mathbb{P}\{\xi_n < T_{\{0\}}\}\left[1 + O\left(\frac{\log r_A}{\log n}\right)\right]$$
$$= [C_2 \log n]^{-1}\left[1 + O\left(\frac{\log r_A}{\log n}\right)\right].$$

Combining this with (6.44) gives

$$\sum_{x\in A} \mathcal{L}g_A(x) = \sum_{x\in\partial_i A} \mathcal{L}g_A(x) = 1. \tag{6.45}$$

Here, we see a major difference between the recurrent and transient cases. If $d \geq 3$, the sum above equals $\mathrm{cap}(A)$ and increases in A, while it is constant in A if $d = 2$. (In particular, it would not be a very useful definition for a capacity!)

Using (6.42) together with $\sum_{x\in A} \mathbb{P}^z\{S_{T_A} = x\} = 1$, we see that

$$J(n,z)\sum_{x\in A} \mathbb{P}^x\{\xi_n < T_A\} = 1 + O\left(\frac{r_A \log n}{n}\right),$$

which by (6.43)–(6.45) implies that

$$J(n,z) = C_2 \log n\left[1 + O\left(\frac{r_A \log n}{n}\right)\right],$$

uniformly in $z \in \mathbb{Z}^2 \setminus \mathcal{C}_n$, and the claim follows by (6.42). □

We define the *capacity* of A by

$$\text{cap}(A) := \lim_{y\to\infty} [a(y) - g_A(y)] = \sum_{x\in A} \text{hm}_A(x)\, a(x-z),$$

where $z \in A$. The last proposition establishes the limit if $z = 0 \in A$, and for other z use (6.29) and $\lim_{y\to\infty} a(x) - a(x-z) = 0$. We have the expansion

$$g_A(x) = C_2 \log \mathcal{J}(x) + \gamma_2 - \text{cap}(A) + o_A(1), \qquad |x| \to \infty.$$

It is easy to check from the definition that the capacity is translation invariant, that is, $\text{cap}(A+y) = \text{cap}(A), y \in \mathbb{Z}^d$. Note that singleton sets have capacity zero.

Proposition 6.6.2 *Suppose that $p \in \mathcal{P}_2$.*

(a) If $0 \in A \subset B \subset \mathbb{Z}^d$ are finite, then $g_A(x) \geq g_B(x)$ for all x. In particular, $\text{cap}(A) \leq \text{cap}(B)$.

(b) If $A, B \subset \mathbb{Z}^d$ are finite subsets containing the origin, then for all x

$$g_{A\cup B}(x) \geq g_A(x) + g_B(x) - g_{A\cap B}(x). \tag{6.46}$$

In particular,

$$\text{cap}(A \cup B) \leq \text{cap}(A) + \text{cap}(B) - \text{cap}(A \cap B).$$

Proof The inequality $g_A(x) \geq g_B(x)$ follows immediately from (6.40). The inequality (6.46) follows from (6.40) and the observation (recall also the argument for Proposition 6.5.3)

$$\begin{aligned}\mathbb{P}^x\{\overline{T}_{A\cup B} < \xi_n\} &= \mathbb{P}^x\{\overline{T}_A < \xi_n \text{ or } \overline{T}_B < \xi_n\}\\ &= \mathbb{P}^x\{\overline{T}_A < \xi_n\} + \mathbb{P}^x\{\overline{T}_B < \xi_n\} - \mathbb{P}^x\{\overline{T}_A < \xi_n, \overline{T}_B < \xi_n\}\\ &\leq \mathbb{P}^x\{\overline{T}_A < \xi_n\} + \mathbb{P}^x\{\overline{T}_B < \xi_n\} - \mathbb{P}^x\{\overline{T}_{A\cap B} < \xi_n\},\end{aligned}$$

which implies that

$$\mathbb{P}^x\{\overline{T}_{A\cup B} > \xi_n\} \geq \mathbb{P}^x\{\overline{T}_A > \xi_n\} + \mathbb{P}^x\{\overline{T}_B > \xi_n\} - \mathbb{P}^x\{\overline{T}_{A\cap B} > \xi_n\}.$$

□

We next derive an analogue of Proposition 6.5.5. If A is a finite set, let a_A denote the $\#(A) \times \#(A)$ symmetric matrix with entries $a(x, y)$. Let a_A also denote the operator

$$a_A f(x) = \sum_{y \in A} a(x,y) f(y)$$

which is defined for all functions $f : A \to \mathbb{R}$ and all $x \in \mathbb{Z}^2$. Note that $x \mapsto a_A f(x)$ is harmonic on $\mathbb{Z}^2 \setminus A$.

Proposition 6.6.3 *Suppose that $p \in \mathcal{P}_2$ and $0 \in A \subset \mathbb{Z}^2$ is finite. Then*

$$\operatorname{cap}(A) = \left[\sup \sum_{y \in A} f(y) \right]^{-1},$$

where the supremum is over all nonnegative functions f on A satisfying $a_A f(x) \leq 1$ for all $x \in A$.

If $A = \{0\}$ is a singleton set, the proposition is trivial, since $a_A f(0) = 0$ for all f and hence the supremum is infinity. A natural first guess for other A (which turns out to be correct) is that the supremum is obtained by a function f satisfying $a_A f(x) = 1$ for all $x \in A$. If $\{a_A(x,y)\}_{x,y \in A}$ is invertible, there is a unique such function that can be written as $f = a_A^{-1} \mathbf{1}$ (where $\mathbf{1}$ denotes the vector of all 1s). The main ingredient in the proof of Proposition 6.6.3 is the next lemma that shows that this inverse is well defined, assuming that A has at least two points.

Lemma 6.6.4 *Suppose that $p \in \mathcal{P}_2$ and $0 \in A \subset \mathbb{Z}^2$ is finite with at least two points. Then a_A^{-1} exists and*

$$a_A^{-1}(x,y) = \mathbb{P}^x\{S_{T_A} = y\} - \delta(y - x) + \frac{\mathcal{L} g_A(x) \, \mathcal{L} g_A(y)}{\operatorname{cap}(A)}, \qquad x, y \in A.$$

Proof We will first show that for all $x \in \mathbb{Z}^2$,

$$\sum_{z \in A} a(x,z) \, \mathcal{L} g_A(z) = \operatorname{cap}(A) + g_A(x). \tag{6.47}$$

To prove this, we will need the following fact (see Exercise 6.7):

$$\lim_{n \to \infty} [G_{\mathcal{C}_n}(0,0) - G_{\mathcal{C}_n}(x,y)] = a(x,y). \tag{6.48}$$

Consider the function

$$h(x) = \sum_{z \in A} a(x,z) \, \mathcal{L} g_A(z).$$

We first claim that h is constant on A. By a last-exit decomposition (Proposition 4.6.4), if $x, y \in A$,

$$1 = \mathbb{P}^x\{\overline{T}_A < \xi_n\} = \sum_{z\in A} G_{\mathcal{C}_n}(x,z)\,\mathbb{P}^z\{\xi_n < T_A\} = \sum_{z\in A} G_{\mathcal{C}_n}(y,z)\,\mathbb{P}^z\{\xi_n < T_A\}.$$

Hence,

$$(C_2 \log n) \sum_{z\in A} [G_{\mathcal{C}_n}(0,0) - G_{\mathcal{C}_n}(x,z)]\mathbb{P}^z\{\xi_n < T_A\}$$
$$= (C_2 \log n) \sum_{z\in A} [G_{\mathcal{C}_n}(0,0) - G_{\mathcal{C}_n}(y,z)]\mathbb{P}^z\{\xi_n < T_A\}.$$

Letting $n \to \infty$, and recalling that $C_2 (\log n)\,\mathbb{P}^z\{\xi_n < T_A\} \to \mathcal{L}g_A(z)$, we conclude that $h(x) = h(y)$. Theorem 4.4.4 and (6.45) imply that

$$\lim_{x\to\infty} [a(x) - h(x)] = 0.$$

Hence, $a(x) - h(x)$ is a bounded function that is harmonic in $\mathbb{Z}^2 \setminus A$ and takes the value $a - h_A$ on A, where h_A denotes the constant value of h on A. Now, Theorem 6.2.1 implies that $a(x) - h(x) = a(x) - g_A(x) - h_A$. Therefore,

$$h_A = \lim_{x\to\infty} [a(x) - g_A(x)] = \mathrm{cap}(A).$$

This establishes (6.47).

An application of the optional sampling theorem gives for $z \in A$

$$G_{\mathcal{C}_n}(x,z) = \delta(z-x) + \mathbb{E}^x[G_{\mathcal{C}_n}(S_1,z)]$$
$$= \delta(z-x) + \sum_{y\in A} \mathbb{P}^x\{S_{T_A\wedge\xi_n} = y\}\, G_{\mathcal{C}_n}(y,z).$$

Hence,

$$G_{\mathcal{C}_n}(0,0) - G_{\mathcal{C}_n}(x,z) = -\delta(z-x) + G_{\mathcal{C}_n}(0,0)\,\mathbb{P}^x\{\xi_n < T_A\}$$
$$+ \sum_{y\in A} \mathbb{P}^x\{S_{T_A\wedge\xi_n} = y\}\,[G_{\mathcal{C}_n}(0,0) - G_{\mathcal{C}_n}(y,z)].$$

Letting $n \to \infty$ and using (6.12) and (6.48), as well as Proposition 6.6.1, this gives

$$\delta(z-x) = -a(x,z) + \mathcal{L}g_A(x) + \sum_{y\in A} \mathbb{P}^x\{S_{T_A} = y\}\, a(y,z).$$

If $x, z \in A$, we can use (6.47) to write the previous identity as

$$\delta(z-x) = \sum_{y\in A}\left[\mathbb{P}^x\{S_{T_A} = y\} - \delta(y-x) + \frac{\mathcal{L}g_A(x)\,\mathcal{L}g_A(y)}{\mathrm{cap}(A)}\right] a(y,z),$$

provided that $\mathrm{cap}(A) > 0$. □

Proof of Proposition 6.6.3. Let $\hat{f}(x) = \mathcal{L}g_A(x)/\mathrm{cap}(A)$. Applying (6.47) to $x \in A$ gives

$$\sum_{y\in A} a(x,y)\hat{f}(y) = 1, \qquad x \in A.$$

Suppose that f satisfies the conditions in the statement of the proposition, and let $h = a_A\hat{f} - a_A f$ which is nonnegative in A. Then, using Lemma 6.6.4,

$$\begin{aligned}\sum_{x\in A}[\hat{f}(x) - f(x)] &= \sum_{x\in A}\left[\sum_{y\in A} a_A^{-1}(x,y)\,h(y)\right]\\ &\geq \sum_{x\in A}\sum_{y\in A}\mathbb{P}^x\{S_{\tau_A} = y\}\,h(y) - \sum_{x\in A} h(x)\\ &= \sum_{y\in A} h(y)\sum_{x\in A}\mathbb{P}^y\{S_{\tau_A} = x\} - \left[\sum_{y\in A} h(y)\right] = 0.\end{aligned}$$

Proposition 6.6.5 *If $p \in \mathcal{P}_2$,*

$$\mathrm{cap}(\mathcal{C}_n) = C_2\log n + \gamma_2 + O(n^{-1}).$$

Proof Recall the asymptotic expansion for $g_{\mathcal{C}_n}$. By definition of capacity, we have,

$$g_{\mathcal{C}_n}(x) = C_2\,\log\mathcal{J}(x) + \gamma_2 - \mathrm{cap}(\mathcal{C}_n) + o(1), \qquad x\to\infty.$$

But for $x \notin \mathcal{C}_n$,

$$\begin{aligned}g_{\mathcal{C}_n}(x) = a(x) - \mathbb{E}^x[a(S_{T_{\mathcal{C}_n}})] &= C_2\log\mathcal{J}(x)\\ &\quad + \gamma_2 + O(|x|^{-2}) - [C_2\log n + \gamma_2 + O(n^{-1})].\end{aligned}$$

□

Lemma 6.6.6 *If $p \in \mathcal{P}_2$, and $A \subset B \subset \mathbb{Z}^2$ are finite, then*

$$\mathrm{cap}(A) = \mathrm{cap}(B) - \sum_{y \in B} \mathrm{hm}_B(y)\, g_A(y).$$

Proof $g_A - g_B$ is a bounded function that is harmonic on $\mathbb{Z}^2 \setminus B$ with boundary value g_A on B. Therefore,

$$\begin{aligned}\mathrm{cap}(B) - \mathrm{cap}(A) &= \lim_{x \to \infty} [g_A(x) - g_B(x)] \\ &= \lim_{x \to \infty} \mathbb{E}^x[g_A(S_{\overline{T}_B}) - g_B(S_{\overline{T}_B})] = \sum_{y \in B} \mathrm{hm}_B(y)\, g_A(y).\end{aligned}$$

□

♣ Proposition 6.6.5 tells us that the capacity of an ellipse of diameter n is $C_2 \log n + O(1)$. The next lemma shows that this is also true for any connected set of diameter n. In particular, the capacities of the ball of radius n and a line of radius n are asymptotic as $n \to \infty$. This is not true for capacities in $d \geq 3$.

Lemma 6.6.7 *If $p \in \mathcal{P}_2$, there exist c_1, c_2 such that the following holds. If A is a finite subset of $\mathbb{Z}^2$ with* $\mathrm{rad}(A) < n$ *satisfying*

$$\#\{x \in A : k - 1 \leq |x| < k\} \geq 1, \qquad k = 1, \ldots, n,$$

then

(a) if $x \in \partial \mathcal{C}_{2n}$,

$$\mathbb{P}^x\{T_A < \xi_{4n}\} \geq c_1,$$

(b) $|\mathrm{cap}(A) - C_2 \log n| \leq c_2$,

(c) if $x \in \partial \mathcal{C}_{2n}$, $m \geq 4n$, and $A_n = A \cap \mathcal{C}_n$, then

$$c_1 \leq \mathbb{P}^x \left\{T_{A_n} > \xi_m\right\} \log(m/n) \leq c_2. \tag{6.49}$$

Proof (a) Let δ be such that $\mathcal{B}_{\delta n} \subset \mathcal{C}_n$, and let B denote a subset of A contained in $\mathcal{B}_{\delta n}$ such that

$$\#\{x \in B : k - 1 \leq |x| < k\} = 1$$

for each positive integer $k < \delta n$. We will prove the estimate for B which will clearly imply the estimate for A. Let $V = V_{n,B}$ denote the number of visits to B before leaving $\mathcal{C}_{4n}$,

$$V = \sum_{j=0}^{\xi_{4n}-1} 1\{S_j \in B\} = \sum_{j=0}^{\infty} \sum_{z \in B} 1\{S_j = z; j < \xi_{4n}\}.$$

The strong Markov property implies that if $x \in \partial \mathcal{C}_{2n}$,

$$\mathbb{E}^x[V] = \mathbb{P}^x\{T_B < \xi_{4n}\} \, \mathbb{E}^x[V \mid T_B < \xi_{4n}] \le \mathbb{P}^x\{T_B < \xi_{4n}\} \max_{z \in B} \mathbb{E}^z[V].$$

Hence, we need only find a c_1 such that $\mathbb{E}^x[V] \ge c_1 \, \mathbb{E}^z[V]$ for all $x \in \partial \mathcal{C}_{2n}, z \in B$. Note that $\#(B) = \delta n + O(1)$. By Exercise 6.13, we can see that $G_{\mathcal{C}_{4n}}(x, z) \ge c$ for $x, z \in \mathcal{C}_{2n}$. Therefore, $\mathbb{E}^x[V] \ge c\, n$. If $z \in B$, there are at most $2k$ points w in $B \setminus \{z\}$ satisfying $|z - w| \le k + 1$, $k = 1, \ldots, \delta n$. Using Proposition 6.3.5, we see that

$$G_{\mathcal{C}_{4n}}(z, w) \le C_2 \, [\log n - \log |z - w| + O(1)].$$

Therefore,

$$\mathbb{E}^z[V] = \sum_{w \in B} G_{\mathcal{C}_{4n}}(z, w) \le \sum_{k=1}^{\delta n} 2\, C_2 \, [\log n - \log k + O(1)] \le c\, n.$$

The last inequality uses the estimate

$$\sum_{k=1}^{n} \log k = O(\log n) + \int_1^n \log x \, dx = n \, \log n - n + O(\log n).$$

(b) There exists a δ such that $\mathcal{B}_n \subset \mathcal{C}_{n/\delta}$ for all n and hence

$$\operatorname{cap}(A) \le \operatorname{cap}(\mathcal{B}_n) \le \operatorname{cap}(\mathcal{C}_{n/\delta}) \le C_2 \, \log n + O(1).$$

Hence, we only need to give a lower bound on $\operatorname{cap}(A)$. By the previous lemma it suffices to find a uniform upper bound for g_A on $\partial \mathcal{C}_{4n}$. For $m > 4n$, let

$$r_m = r_{m,n,A} = \max_{y \in \mathcal{C}_{2n}} \mathbb{P}^y\{\xi_m < T_A\},$$

$$r_m^* = r_{m,n,A} = \max_{y \in \mathcal{C}_{4n}} \mathbb{P}^y\{\xi_m < T_A\}.$$

Using part (a) and the strong Markov property, we see that there is a $\rho < 1$ such that $r_m \le \rho\, r_m^*$. Also, if $y \in \mathcal{C}_{4n}$

$$\mathbb{P}^y\{\xi_m < T_A\} = \mathbb{P}^y\{\xi_m < T_{\mathcal{C}_{2n}}\} + \mathbb{P}^y\{\xi_m > T_{\mathcal{C}_{2n}}\}\, \mathbb{P}^y\{\xi_m < T_A \mid \xi_m > T_{\mathcal{C}_{2n}}\}$$
$$\le \mathbb{P}^y\{\xi_m < T_{\mathcal{C}_{2n}}\} + \rho\, r_m^*.$$

Proposition 6.4.1 tells us that there is a c_3 such that for $y \in \mathcal{C}_{4n}$,

$$\mathbb{P}^y\{\xi_m < T_{\mathcal{C}_{2n}}\} \le \frac{c_3}{\log m - \log n + O(1)}.$$

Therefore,

$$g_A(y) = \lim_{m\to\infty} C_2\,(\log m)\, \mathbb{P}^y\{\xi_m < T_A\} \le \frac{C_2\, c_3}{1-\rho}.$$

(c) The lower bound for (6.49) follows from Proposition 6.4.1 and the observation

$$\mathbb{P}^x\{T_{A_n} > \xi_m\} \ge \mathbb{P}^x\{T_{\mathcal{C}_n} > \xi_m\}.$$

For the upper bound let

$$u = u_n = \max_{x\in\partial\mathcal{C}_{2n}} \mathbb{P}^x\{T_{A_n} > \xi_m\}.$$

Consider a random walk starting at $y \in \partial\mathcal{C}_{2n}$ and consider $T_{\mathcal{C}_n} \wedge \xi_m$. Clearly,

$$\mathbb{P}^y\{T_{A_n} > \xi_m\} = \mathbb{P}^y\{\xi_m < T_{\mathcal{C}_n}\} + \mathbb{P}^y\{\xi_m > T_{\mathcal{C}_n}; \xi_m < T_{A_n}\}.$$

By Proposition 6.4.1, for all $y \in \partial\mathcal{C}_{2n}$

$$\mathbb{P}^y\left\{\xi_m < T_{\mathcal{C}_n}\right\} \le \frac{c}{\log(m/n)}.$$

Let $\sigma = \sigma_n = \min\{j \ge T_{\mathcal{C}_n} : S_j \in \partial\mathcal{C}_{2n}\}$. Then, by the Markov property,

$$\mathbb{P}^y\{\xi_m > T_{\mathcal{C}_n}, \xi_m \le T_{A_n}\} \le u\, \mathbb{P}^y\{S[0,\sigma] \cap A_n = \emptyset\}.$$

Part (a) shows that there is a $\rho < 1$ such that $\mathbb{P}^y\{S[0,\sigma] \cap A_n = \emptyset\} \le \rho$ and hence, we get

$$\mathbb{P}^y\{T_{A_n} > \xi_m\} \le \frac{c}{\log(m/n)} + \rho\, u.$$

Since this holds, for all $y \in \partial \mathcal{C}_{2n}$, this implies that

$$u \leq \frac{c}{\log(m/n)} + \rho\, u,$$

which gives us the upper bound. □

♣ A major example of a set satisfying the condition of the theorem is a connected (with respect to simple random walk) subset of $\mathbb{Z}^2$ with radius between $n-1$ and n. In the case of simple random walk, there is another proof of part (a) based on the observation that the simple random walk starting anywhere on $\partial \mathcal{C}_{2n}$ makes a closed loop about the origin contained in $\mathcal{C}_n$ with a probability uniformly bounded away from 0. One can justify this rigorously by using an approximation by Brownian motion. If the random walk makes a closed loop, then it must intersect any connected set. Unfortunately, it is not easy to modify this argument for random walks that take nonnearest neighbor steps.

6.7 Neumann problem

We will consider the following "Neumann problem." Suppose that $p \in \mathcal{P}_d$ and $A \subset \mathbb{Z}^d$ with nonempty boundary ∂A. If $f : \overline{A} \to \mathbb{R}$ is a function, we define its *normal derivative* at $y \in \partial A$ by

$$Df(y) = \sum_{x \in A} p(y,x)\, [f(x) - f(y)].$$

Given $D^* : \partial A \to A$, the Neumann problem is to find a function $f : \overline{A} \to \mathbb{R}$ such that

$$\mathcal{L}f(x) = 0, \qquad x \in A, \tag{6.50}$$

$$Df(y) = D^*(y), \qquad y \in \partial A. \tag{6.51}$$

♣ The term normal derivative is motivated by the case of simple random walk and a point $y \in \partial A$ such that there is a unique $x \in A$ with $|y - x| = 1$. Then $Df(y) = [f(x) - f(y)]/2d$, which is a discrete analogue of the normal derivative.

A solution to (6.50)–(6.51) will not always exist. The next lemma, which is a form of Green's theorem, shows that if A is finite, a necessary condition for existence is

$$\sum_{y \in \partial A} D^*(y) = 0. \tag{6.52}$$

Lemma 6.7.1 *Suppose that $p \in \mathcal{P}_d$, A is a finite subset of $\mathbb{Z}^d$ and $f : \overline{A} \to \mathbb{R}$ is a function. Then*

$$\sum_{x \in A} \mathcal{L}f(x) = -\sum_{y \in \partial A} Df(y).$$

Proof

$$\begin{aligned}\sum_{x \in A} \mathcal{L}f(x) &= \sum_{x \in A} \sum_{y \in \overline{A}} p(x,y)\,[f(y) - f(x)] \\ &= \sum_{x \in A} \sum_{y \in A} p(x,y)\,[f(y) - f(x)] + \sum_{x \in A} \sum_{y \in \partial A} p(x,y)\,[f(y) - f(x)]\end{aligned}$$

However,

$$\sum_{x \in A} \sum_{y \in A} p(x,y)\,[f(y) - f(x)] = 0,$$

since $p(x,y)\,[f(y) - f(x)] + p(y,x)\,[f(x) - f(y)] = 0$ for all $x, y \in A$. Therefore,

$$\sum_{x \in A} \mathcal{L}f(x) = \sum_{y \in \partial A} \sum_{x \in A} p(x,y)\,[f(y) - f(x)] = -\sum_{y \in \partial A} Df(y).$$

□

Given A, the *excursion Poisson kernel* is the function

$$H_{\partial A} : \partial A \times \partial A \longrightarrow [0,1],$$

defined by

$$H_{\partial A}(y,z) = \mathbb{P}^y\left\{S_1 \in A, S_{\tau_A} = z\right\} = \sum_{x \in A} p(y,x)\,H_A(x,z),$$

where $H_A : \overline{A} \times \partial A \to [0,1]$ is the Poisson kernel. If $z \in \partial A$ and $H(x) = H_A(x,z)$, then

$$\begin{aligned}DH(y) &= H_{\partial A}(y,z), \qquad y \in \partial A \setminus \{z\}, \\ DH(z) &= H_{\partial A}(z,z) - \mathbb{P}^z\{S_1 \in A\}.\end{aligned}$$

More generally, if $f : \overline{A} \to \mathbb{R}$ is harmonic in A, then $f(y) = \sum_{z \in \partial A} f(z) H_A(y,z)$ so that

$$Df(y) = \sum_{z \in \partial A} H_{\partial A}(y,z)\,[f(z) - f(y)]. \tag{6.53}$$

Note that if $y \in \partial A$, then

$$\sum_{z \in \partial A} H_{\partial A}(y,z) = \mathbb{P}^y\{S_1 \in A\} \le 1.$$

It is sometimes useful to consider the Markov transition probabilities $\hat{H}_{\partial A}$ where $\hat{H}_{\partial A}(y,z) = H_{\partial A}(y,z)$ for $y \neq z$, and $\hat{H}_{\partial A}(y,y)$ is chosen so that

$$\sum_{z \in \partial A} \hat{H}_{\partial A}(y,z) = 1.$$

Note that again (compare with (6.53))

$$Df(y) = \sum_{z \in \partial A} \hat{H}_{\partial A}(y,z)\,[f(z) - f(y)],$$

which we can write in matrix form

$$Df = [\hat{H}_{\partial A} - I] f.$$

If A is finite, then the $\#(\partial A) \times \#(\partial A)$ matrix $\hat{H}_{\partial A} - I$ is sometimes called the *Dirichlet-to-Neumann map* because it takes the boundary values f (Dirichlet conditions) of a harmonic function to the derivatives Df (Neumann conditions). The matrix is not invertible since constant functions f are mapped to zero derivatives. We also know that the image of the map is contained in the subspace of functions D^* satisfying (6.52). The next proposition shows that the rank of the matrix is $\#(\partial A) - 1$.

It will be useful to define random walk "reflected off ∂A". There are several natural ways to do this. We define this to be the Markov chain with state space $\overline{A}$ and transition probabilities q where $q(x,y) = p(x,y)$ if $x \in A$ or $y \in A$; $q(x,y) = 0$ if $x, y \in \partial A$ are distinct; and $q(y,y)$ is defined for $y \in \partial A$ so that $\sum_{z \in \overline{A}} q(y,z) = 1$. In words, this chain moves like random walk with transition probability p while in A, and whenever its current position y is in ∂A, the only moves allowed are those into $A \cup \{y\}$. While the original walk could step out of $A \cup \{y\}$ with some probability $\tilde{p}(y) = \tilde{p}(y, A, p)$, the modified walk stays at y with probability $p(y,y) + \tilde{p}(y)$.

Proposition 6.7.2 *Suppose that $p \in \mathcal{P}_d$, A is a finite, connected subset of $\mathbb{Z}^d$, and $D^* : \partial A \to \mathbb{R}$ is a function satisfying (6.52). Then there is a function $f : \overline{A} \to \mathbb{R}$ satisfying (6.50) and (6.51). The function f is unique up to an* *additive constant. One such function is given by*

$$f(x) = -\lim_{n\to\infty} \mathbb{E}^x \left[\sum_{j=0}^{n} D^*(Y_j)\, 1\{Y_j \in \partial A\} \right], \tag{6.54}$$

where Y_j is a Markov chain with transition probabilities q as defined in the previous paragraph.

Proof It suffices to show that f as defined in (6.54) is well defined and satisfies (6.50) and (6.51). Indeed, if this is true, then $f + c$ also satisfies it. Since the image of the matrix $\hat{H}_{\partial A} - I$ contains the set of functions satisfying (6.52) and this is a subspace of dimension $\#(\partial A) - 1$, we get the uniqueness.

Note that q is an irreducible, symmetric Markov chain and hence has the uniform measure as the invariant measure $\pi(y) = 1/m$ where $m = \#(\overline{A})$. Because the chain also has points with $q(y, y) > 0$, it is aperiodic. Also,

$$\begin{aligned}\mathbb{E}^x \left[\sum_{j=0}^{n} D^*(Y_j)\, 1\{Y_j \in \partial A\} \right] &= \sum_{j=0}^{n} \sum_{z\in\partial A} q_j(x,z)\, D^*(z) \\ &= \sum_{j=0}^{n} \sum_{z\in\partial A} \left[q_j(x,z) - \frac{1}{m} \right] D^*(z).\end{aligned}$$

By standard results about Markov chains (see Section A.4), we know that

$$\left| q_j(x,z) - \frac{1}{m} \right| \le c\, e^{-\alpha j},$$

for some positive constants c, α. Hence the sum is convergent. It is then straightforward to check that it satisfies (6.50) and (6.51). □

6.8 Beurling estimate

The Beurling estimate is an important tool for estimating hitting (avoiding) probabilities of sets in two dimensions. The Beurling estimate is a discrete analogue of what is known as the Beurling projection theorem for Brownian motion in $\mathbb{R}^2$.

Recall that a set $A \subset \mathbb{Z}^d$ is connected (for simple random walk) if any two points in A can be connected by a nearest neighbor path of points in A.

Theorem 6.8.1 (Beurling estimate) *If $p \in \mathcal{P}_2$, there exists a constant c such that if A is an infinite connected subset of $\mathbb{Z}^d$ containing the origin and S is simple random walk, then*

$$\mathbb{P}\{\xi_n < T_A\} \leq \frac{c}{n^{1/2}}, \qquad d = 2. \tag{6.55}$$

We prove the result for simple random walk, and then we describe the extension to more general walks.

Definition Let $\mathcal{A}_d$ denote the collection of infinite subsets of $\mathbb{Z}^d$ with the property that for each positive integer j,

$$\#\{z \in A : (j-1) \leq |z| < j\} = 1.$$

One important example of a set in $\mathcal{A}_d$ is the half-infinite line

$$L = \{j\mathbf{e}_1 : j = 0, 1, \ldots\}.$$

We state two immediate facts about $\mathcal{A}_d$.

- If A' is an infinite connected subset of $\mathbb{Z}^d$ containing the origin, then there exists a (not necessarily connected) $A \in \mathcal{A}_d$ with $A \subset A'$.
- If $z \in A \in \mathcal{A}_d$, then for every real $r > 0$.

$$\#\{w \in A : |z - w| \leq r\} \leq \#\{w \in A : |z| - r \leq |w| \leq |z| + r\} \leq 2r + 1. \tag{6.56}$$

Theorem 6.8.1 for simple random walk is implied by the following stronger result.

Theorem 6.8.2 *For simple random walk in $\mathbb{Z}^2$ there is a c such that if $A \in \mathcal{A}_d$, then*

$$\mathbb{P}\{\xi_n < T_A\} \leq \frac{c}{n^{1/2}}. \tag{6.57}$$

Proof We fix n and let $V = V_n = \{y_1, \ldots, y_n\}$ where y_j denotes the unique point in A with $j \leq |y_j| < j + 1$. We let $K = K_n = \{x_1, \ldots, x_n\}$ where $x_j = j\mathbf{e}_1$. Let $G_n = G_{\mathcal{B}_n}, \mathcal{B} = \mathcal{B}_{n^3}, G = G_{\mathcal{B}}, \xi = \xi_{n^3}$. Let

$$v(z) = \mathbb{P}^z\{\xi < T_{V_n}\}, \quad q(z) = \mathbb{P}^z\{\xi < T_{K_n}\}.$$

By (6.49), there exist c_1, c_2 such that for $z \in \partial \mathcal{B}_{2n}$,

$$\frac{c_1}{\log n} \leq v(z) \leq \frac{c_2}{\log n}, \quad \frac{c_1}{\log n} \leq q(z) \leq \frac{c_2}{\log n}.$$

We will establish

$$v(0) \leq \frac{c}{n^{1/2} \log n}$$

and then the Markov property will imply that (6.57) holds. Indeed, note that

$$v(0) \geq \mathbb{P}(\xi_{2n} < T_{V_{2n}}) \mathbb{P}(\xi < \xi_n | \xi_{2n} < T_{V_{2n}}).$$

By (5.17) and (6.49), we know that there is a c such that for $j = 1, \ldots, n$,

$$q(x_j) \leq c\, n^{-1/2} \left[j^{-1/2} + (n - j + 1)^{-1/2} \right] [\log n]^{-1}. \tag{6.58}$$

In particular, $q(0) \leq c/(n^{1/2} \log n)$ and hence it suffices to prove that

$$v(0) - q(0) \leq \frac{c}{n^{1/2} \log n}. \tag{6.59}$$

If $|x|, |y| \leq n$, then

$$G_{n^3 - 2n}(0, y - x) \leq G_{n^3}(x, y) \leq G_{n^3 + 2n}(0, y - x),$$

and hence (4.28) and Theorem 4.4.4 imply

$$G(x, y) = \frac{2}{\pi} \log n^3 + \gamma_2 - a(x, y) + O\left(\frac{1}{n^2}\right), \quad |x|, |y| \leq n. \tag{6.60}$$

Using Proposition 4.6.4, we write

$$\begin{aligned} v(0) - q(0) &= \mathbb{P}\{\xi < T_V\} - \mathbb{P}\{\xi < T_K\} \\ &= \mathbb{P}\{\xi > T_K\} - \mathbb{P}\{\xi > T_V\} \\ &= \sum_{j=1}^{n} G(0, x_j)\, q(x_j) - \sum_{j=1}^{n} G(0, y_j)\, v(y_j) \\ &= \sum_{j=1}^{n} [G(0, x_j) - G(0, y_j)]\, q(x_j) + \sum_{j=1}^{n} G(0, y_j)\, [q(x_j) - v(y_j)]. \end{aligned}$$

Using (6.58) and (6.60), we get

$$(\log n)\left|\sum_{j=1}^{n}[G(0,x_j)-G(0,y_j)]\,q(x_j)\right| \le O(n^{-1})$$
$$+\,c\sum_{j=1}^{n}|a(x_j)-a(y_j)|\,(j^{-1/2}+(n-j)^{-1/2})\,n^{-1/2}.$$

Since $|x_j|=j, |y_j|=j+O(1)$, (4.4.4) implies that

$$|a(x_j)-a(y_j)| \le \frac{c}{j},$$

and hence

$$(\log n)\left|\sum_{j=1}^{n}[G(0,x_j)-G(0,y_j)]\,q(x_j)\right| \le O(n^{-1})+c\sum_{j=1}^{n}\frac{1}{j^{3/2}\,n^{1/2}} \le \frac{c}{n^{1/2}}.$$

For the last estimate we note that

$$\sum_{j=1}^{n}\frac{1}{j(n-j)^{1/2}} \le \sum_{j=1}^{n}\frac{1}{j^{3/2}}.$$

In fact, if $a,b\in\mathbb{R}^n$ are two vectors such that a has nondecreasing components (that is, $a^1\le a^2\le\ldots\le a^n$) then $a\cdot b\le a\cdot b^*$ where $b^*=(b^{\pi(1)},\ldots,b^{\pi(n)})$ and π is any permutation that makes $b^{\pi(1)}\le b^{\pi(2)}\le\ldots\le b^{\pi(n)}$.

Therefore, to establish (6.59), it suffices to show that

$$\sum_{j=1}^{n}G(0,y_j)\,[q(x_j)-v(y_j)] \le \frac{c}{n^{1/2}\,\log n}. \tag{6.61}$$

Note that we are not taking absolute values on the left-hand side. Consider the function

$$F(z)=\sum_{j=1}^{n}G(z,y_j)\,[q(x_j)-v(y_j)],$$

and note that F is harmonic on $\mathcal{B}\setminus V$. Since $F\equiv 0$ on $\partial\mathcal{B}$, either $F\le 0$ everywhere (in which case (6.61) is trivial) or it takes its maximum on V.

Therefore, it suffices to find a c such that for all $k = 1, \dots, n$,

$$\sum_{j=1}^{n} G(y_k, y_j)\,[q(x_j) - v(y_j)] \le \frac{c}{n^{1/2}\log n}.$$

By using Proposition 4.6.4 once again, we get

$$\sum_{j=1}^{n} G(y_k, y_j)\, v(y_j) = \mathbb{P}^{y_k}\{\overline{T}_V \le \xi\} = 1 = \mathbb{P}^{x_k}\{\overline{T}_K \le \xi\} = \sum_{j=1}^{n} G(x_k, x_j)\, q(x_j).$$

Plugging in, we get

$$\sum_{j=1}^{n} G(y_k, y_j)\,[q(x_j) - v(y_j)] = \sum_{j=1}^{n} [G(y_k, y_j) - G(x_k, x_j)]\, q(x_j).$$

We will now bound the right-hand side. Note that $|x_k - x_j| = |k - j|$ and $|y_k - y_j| \ge |k - j| - 1$. Hence, using (6.60),

$$G(y_k, y_j) - G(x_k, x_j) \le \frac{c}{|k - j| + 1}$$

and therefore for each $k = 1, \dots, n$

$$\sum_{j=1}^{n} [G(y_k, y_j) - G(x_k, x_j)]\, q(x_j) \le c \sum_{j=1}^{n} \frac{1}{(|k - j| + 1)\, j^{1/2} \log n} \le \frac{c}{n^{1/2}\log n}.$$

□

One can now generalize this result.

Definition If $p \in \mathcal{P}_2$ and k is a positive integer, let $\mathcal{A}^* = \mathcal{A}^*_{2,k,p}$ denote the collection of infinite subsets of $\mathbb{Z}^2$ with the property that for each positive integer j,

$$\#\{z \in A : (j-1)k \le \mathcal{J}(z) < jk\} \ge 1,$$

and let $\mathcal{A}$ denote the collection of subsets with

$$\#\{z \in A : (j-1)k \le \mathcal{J}(z) < jk\} = 1.$$

If $A \in \mathcal{A}^*$ then A contains a subset in $\mathcal{A}$.

Theorem 6.8.3 *If $p \in \mathcal{P}_2$ and k is a positive integer, there is a c such that if $A \in \mathcal{A}^*$, then*

$$\mathbb{P}\{\xi_n < T_A\} \leq \frac{c}{n^{1/2}}.$$

The proof is done similarly to that of the last theorem. We let $K = \{x_1, \ldots, x_n\}$ where $x_j = jl\mathbf{e}_1$ and l is chosen sufficiently large so that $\mathcal{J}(l\mathbf{e}_1) > k$, and set $V = \{y_1, \ldots, y_n\}$ where $y_j \in A$ with $j\mathcal{J}^*(l\mathbf{e}_1) \leq |y_j| < (j+1)\mathcal{J}^*(l\mathbf{e}_1)$. See Exercise 5.2.

6.9 Eigenvalue of a set

Suppose that $p \in \mathcal{P}_d$ and $A \subset \mathbb{Z}^d$ is finite and connected (with respect to p) with $\#(A) = m$. The *(first) eigenvalue* of A is defined to be the number $\alpha_A = e^{-\lambda_A}$ such that for each $x \in A$, as $n \to \infty$,

$$\mathbb{P}^x\{\tau_A > n\} \asymp \alpha_A^n = e^{-\lambda_A n}.$$

Let P^A denote the $m \times m$ matrix $[p(x,y)]_{x,y \in A}$ and, as before, let $\mathcal{L}^A = P^A - I$. Note that $(P^A)^n$ is the matrix $[p_n^A(x,y)]$ where $p_n^A(x,y) = \mathbb{P}^x\{S_n = y; n < \tau_A\}$. We will say that $p \in \mathcal{P}_d$ is *aperiodic restricted to A* if there exists an n such that $(P^A)^n$ has all entries strictly positive; otherwise, we say that p is *bipartite restricted to A*. In order for p to be aperiodic restricted to A, p must be aperiodic. However, it is possible for p to be aperiodic but for p to be bipartite restricted to A (Exercise 6.16). The next two propositions show that α_A is the largest eigenvalue for the matrix P^A, or, equivalently, $1 - \alpha_A$ is the smallest eigenvalue for the matrix $\mathcal{L}^A$.

Proposition 6.9.1 *If $p \in \mathcal{P}_d$, $A \subset \mathbb{Z}^d$ is finite and connected, and p restricted to A is aperiodic, then there exist numbers $0 < \beta = \beta_A < \alpha = \alpha_A < 1$ such that if $x, y \in A$,*

$$p_n^A(x,y) = \alpha^n\, g_A(x)\, g_A(y) + O_A(\beta^n). \tag{6.62}$$

Here $g_A : A \to \mathbb{R}$ is the unique positive function satisfying

$$P^A g_A(x) = \alpha_A\, g_A(x), \qquad x \in A,$$

$$\sum_{x \in A} g_A(x)^2 = 1.$$

In particular,

$$\mathbb{P}^x\{\tau_A > n\} = \tilde{g}_A(x)\,\alpha^n + O_A(\beta^n),$$

where

$$\tilde{g}_A(x) = g_A(x) \sum_{y\in A} g_A(y).$$

We write O_A to indicate that the implicit constant in the error term depends on A.

Proof This is a general fact about irreducible Markov chains; see Proposition A.4.3. In the notation of that proposition $v = w = g$. Note that

$$\mathbb{P}^x\{\tau_A > n\} = \sum_{y\in A} p_n^A(x, y).$$

□

Proposition 6.9.2 *If $p \in \mathcal{P}_d$, $A \subset \mathbb{Z}^d$ is finite and connected, and p is bipartite restricted to A, then there exist numbers $0 < \beta = \beta_A < \alpha = \alpha_A < 1$ such that if $x, y \in A$ for all n sufficiently large,*

$$p_n^A(x, y) + p_{n+1}^A(x, y) = 2\,\alpha^n\, g_A(x)\, g_A(y) + O_A(\beta^n).$$

Here $g_A : A \to \mathbb{R}$ is the unique positive function satisfying

$$\sum_{x\in A} g_A(x)^2 = 1, \qquad P^A g_A(x) = \alpha\, g_A(x),\ x \in A.$$

Proof This can be proved similarly using Markov chains. We omit the proof.

□

Proposition 6.9.3 *Suppose that $p \in \mathcal{P}_d$; $\epsilon \in (0, 1)$, and $p_\epsilon = \epsilon\,\delta_0 + (1-\epsilon)\,p$ is the corresponding lazy walker. Suppose that A is a finite, connected subset of $\mathbb{Z}^d$ and let $\alpha, \alpha_\epsilon, g, g_\epsilon$ be the eigenvalues and eigenfunctions for A using p, p_ϵ, respectively. Then $1 - \alpha_\epsilon = (1-\epsilon)\,(1-\alpha)$ and $g_\epsilon = g$.*

Proof Let P^A, P_ϵ^A be the corresponding matrices. Then $P_\epsilon^A = (1-\epsilon)\,P^A + \epsilon\, I$ and hence

$$P_\epsilon^A\, g_A = [(1-\epsilon)\,\alpha + \epsilon]\, g_A.$$

□

♣ A standard problem is to estimate λ_A or α_A as A gets large and $\alpha_A \to 1, \lambda_A \to 0$. In these cases it usually suffices to consider the eigenvalue of the lazy walker with $\epsilon = 1/2$. Indeed, let $\tilde{\lambda}_A$ be the eigenvalue for the lazy walker. Since,

$$\lambda_A = 1 - \alpha_A + O((1-\alpha_A)^2), \quad \alpha_A \to 1-.$$

we get

$$\tilde{\lambda}_A = \frac{1}{2}\lambda_A + O(\lambda_A^2), \quad \lambda_A \to 0.$$

Proposition 6.9.1 gives no bounds for the β. The optimal β is the maximum of the absolute values of the eigenvalues other than α. In general, it is hard to estimate β, and it is possible for β to be very close to α. We will show that in the case of the nice set $\mathcal{C}_n$ there is an upper bound for β independent of n. We fix $p \in \mathcal{P}_d$ with $p(x,x) > 0$ and let $e^{-\lambda_m} = \alpha_{\mathcal{C}_m}, g_m = g_{\mathcal{C}_m}$, and $p_n^m(x,y) = p_n^{\mathcal{C}_m}(x,y)$. For $x \in \mathcal{C}_m$ we let

$$\rho_m(x) = \frac{\mathrm{dist}(x, \partial\mathcal{C}_m) + 1}{m},$$

and we set $\rho_m \equiv 0$ on $\mathbb{Z}^d \setminus \mathcal{C}_m$.

Proposition 6.9.4 *There exist c_1, c_2 such that for all m sufficiently large and all $x, y \in \mathcal{C}_m$,*

$$c_1\, \rho_m(x)\, \rho_m(y) \le m^d\, p_{m^2}^m(x,y) \le c_2\, \rho_m(x)\, \rho_m(y). \tag{6.63}$$

Also, there exist c_3, c_4 such that for every $n \ge m^2$, and all $x, y \in \mathcal{C}_m$,

$$c_3\, \rho_m(y)\, m^{-d} \le \mathbb{P}^x\{S_n = y \mid \tau_{\mathcal{C}_m} > n\} \le c_4\, \rho_m(y)\, m^{-d}.$$

♣ This proposition is an example of a *parabolic boundary Harnack principle*. At any time larger than $\mathrm{rad}^2(\mathcal{C}_m)$, the position of the random walker, given that it has stayed in $\mathcal{C}_m$ up to the current time, is independent of the initial state up to a multiplicative constant.

Proof For notational ease, we will restrict ourselves to the case where m is even. (If m is odd, essentially the same proof works, except that $m^2/4$ must be replaced with $\lfloor m^2/4 \rfloor$, etc.) We write $\rho = \rho_m$. Note that

$$p_{m^2}^m(x,y) = \sum_{z,w} p_{m^2/4}^m(x,z)\, p_{m^2/2}^m(z,w)\, p_{m^2/4}^m(w,y). \tag{6.64}$$

The LCLT implies that there is a c such that for all z, w, $p^m_{m^2/2}(z,w) \leq p_{m^2/2}(z,w) \leq c\, m^{-d}$. Therefore,

$$p^m_{m^2}(x,y) \leq m^{-d}\, \mathbb{P}^x\{\tau_{\mathcal{C}_m} > m^2/4\}\, \mathbb{P}^y\{\tau_{\mathcal{C}_m} > m^2/4\}.$$

Gambler's ruin (see Proposition 5.1.6) implies that $\mathbb{P}^x\{\tau_{\mathcal{C}_m} > m^2/4\} \leq c\, \rho_m(x)$. This gives the upper bound for (6.63).

For the lower bound, we first note that there is an $\epsilon > 0$ such that

$$p^m_{\lfloor \epsilon m^2 \rfloor}(z,w) \geq \epsilon\, m^{-d}, \qquad |z|, |w| \leq \epsilon\, m.$$

Indeed, the Markov property implies that

$$p^m_{\lfloor \epsilon m^2 \rfloor}(z,w) \geq p_{\lfloor \epsilon m^2 \rfloor}(z,w) - \max\{p_k(\tilde{z},w) : k \leq \lfloor \epsilon m^2 \rfloor, \tilde{z} \in \mathbb{Z}^d \setminus \mathcal{C}_m\}, \tag{6.65}$$

and the LCLT establishes the estimate. Using this estimate and the invariance principle, one can see that for every $\epsilon > 0$, there is a c such that for $z, w \in \mathcal{C}_{(1-\epsilon)m}$,

$$p_{m^2/2}(z,w) \geq c\, m^{-d}.$$

Indeed, in order to estimate $p_{m^2/2}(z,w)$, we split the path into three pieces: the first $m^2/8$ steps, the middle $m^2/4$ steps, and the final $m^2/8$ steps (here, we are assuming that $m^2/8$ is an integer for notational ease). We estimate both the probability that the walk starting at z has not left $\mathcal{C}_m$ and is in the ball of radius ϵm at time $m^2/8$ and the corresponding probability for the walk in reverse time starting at w using the invariance principle. There is a positive probability for this, where the probability depends on ϵ. For the middle piece we use (6.65), and then we "connect" the paths to obtain the lower bound on the probability.

Using (6.64), we can then see that it suffices to find $\epsilon > 0$ and $c > 0$ such that

$$\sum_{z \in \mathcal{C}_{(1-\epsilon)m}} p^m_{m^2/4}(x,z) \geq c\, \rho(x). \tag{6.66}$$

Let $T = \tau_{\mathcal{C}_m} \setminus \tau_{\mathcal{C}_{m/2}}$ as in Lemma 6.3.4 and let $T_m = T \wedge (m^2/4)$. Using that lemma and Theorem 5.1.7, we can see that

$$\mathbb{P}^x\{S_{T_m} \in \mathcal{C}_m\} \leq c_1\, \rho(x).$$

Propositions 6.4.1 and 6.4.2 can be used to see that

$$\mathbb{P}^x\{S_T \in \mathcal{C}_{m/2}\} \geq c_2\, \rho(x).$$

We can write

$$\mathbb{P}^x\{S_T \in \mathcal{C}_{m/2}\} = \sum_z \mathbb{P}^x\{S_{T_m} = z\}\, \mathbb{P}^x\{S_T \in \mathcal{C}_{m/2} \mid S_{T_m} = z\}.$$

The conditional expectation can be estimated again by Lemma 6.3.4; in particular, we can find an ϵ such that

$$\mathbb{P}^z\{S_T \in \mathcal{C}_{m/2}\} \leq \frac{c_2}{2c_1}, \qquad z \notin \mathcal{C}_{(1-\epsilon)m}.$$

This implies that

$$\sum_{z \in \mathcal{C}_{(1-\epsilon)m}} \mathbb{P}^x\{S_{T_m} = z\} \geq \sum_{z \in \mathcal{C}_{(1-\epsilon)m}} \mathbb{P}^x\{S_{T_m} = z\}\, \mathbb{P}^x\{S_T \in \mathcal{C}_{m/2} \mid S_{T_m} = z\}$$
$$\geq \frac{c_2}{2}\, \rho(x).$$

A final appeal to the CLT shows that if $\epsilon \leq 1/4$,

$$\sum_{z \in \mathcal{C}_{(1-\epsilon)m}} p^m_{m^2/4}(x, z) \geq c \sum_{z \in \mathcal{C}_{(1-\epsilon)m}} \mathbb{P}^x\{S_{T_m} = z\}.$$

The last assertion follows for $n = m^2$ by noting that

$$\mathbb{P}^x\{S_{m^2} = y \mid \tau_{\mathcal{C}_m} > m^2\} = \frac{p^m_{m^2}(x, y)}{\sum_z p^m_{m^2}(x, z)}$$

and

$$\sum_z p^m_{m^2}(x, z) \asymp \rho_m(x)\, m^{-d} \sum_z \rho_m(z) \asymp \rho_m(x).$$

For $n > m^2$, we can argue similarly by conditioning on the walk at time $n - m^2$. □

Corollary 6.9.5 *There exist c_1, c_2 such that*

$$c_1 \leq m^2 \lambda_m \leq c_2.$$

Proof See Exercise 6.10. □

Corollary 6.9.6 *There exist c_1, c_2 such that for all m and all $x \in \mathcal{C}_{m/2}$,*

$$c_1 \, e^{-\lambda_m n} \le \mathbb{P}^x\{\xi_m > n\} \le c_2 e^{-\lambda_m n}.$$

Proof Using the previous corollary, it suffices to prove the estimates for $n = km^2, k \in \{1, 2, \ldots\}$. Let $\beta_k(x) = \beta_k(x, m) = \mathbb{P}^x\{\xi_m > km^2\}$ and let $\beta_k = \max_{x \in \mathcal{C}_m} \beta_k(x)$. Using the previous proposition, we see that there is a c_1 such that

$$\beta_k \ge \beta_k(x) \ge c_1 \, \beta_k, \quad x \in \mathcal{C}_{m/2}.$$

Due to the same estimates,

$$\mathbb{P}^x\{S_{\xi_m} \in \mathcal{C}_{m/2} \mid \xi_m > km^2\} \ge c_2.$$

Therefore, there is a c_3 such that

$$c_3 \beta_j \, \beta_k \le \beta_{j+k} \le \beta_j \, \beta_k,$$

which implies (see Corollary A.7.2)

$$e^{-\lambda_m m^2 k} \le \beta_k \le c_3^{-1} \, e^{-\lambda_m m^2 k},$$

and hence for $x \in \mathcal{C}_{m/2}$,

$$c_1 \, e^{-\lambda_m m^2 k} \le \beta_k(x) \le c_3^{-1} \, e^{-\lambda_m m^2 k}.$$

□

Exercises

Exercise 6.1 Show that Proposition 6.1.2 holds for $p \in \mathcal{P}^*$.

Exercise 6.2 (i) Show that if $p \in \mathcal{P}_d$ and $x \in \mathcal{C}_n$,

$$\mathbb{E}^x[\xi_n] = \sum_{y \in \mathcal{C}_n} G_{\mathcal{C}_n}(x, y) = n^2 - \mathcal{J}(x) + O(n).$$

(Hint: see Exercise 1.5.)

(ii) Show that if $p \in \mathcal{P}'_d$ and $x \in \mathcal{C}_n$,

$$\mathbb{E}^x[\xi_n] = \sum_{y \in \mathcal{C}_n} G_{\mathcal{C}_n}(x, y) = n^2 - \mathcal{J}(x) + o(n^2).$$

Exercise 6.3 In this exercise we construct a transient subset A of $\mathbb{Z}^3$ with

$$\sum_{y \in A} G(0, y) = \infty. \tag{6.67}$$

Here, G denotes the Green's function for simple random walk. Our set will be of the form

$$A = \bigcup_{k=1}^{\infty} A_k, \qquad A_k = \{z \in \mathbb{Z}^3 : |z - 2^k\, \mathbf{e}_1| \leq \epsilon_k\, 2^k\}.$$

for some $\epsilon_k \to 0$.

(i) Show that (6.67) holds if and only if $\sum_{k=1}^{\infty} \epsilon_k^3\, 2^{2k} = \infty$.
(ii) Show that A is transient if and only if $\sum_{k=1}^{\infty} \epsilon_k < \infty$.
(iii) Find a transient A satisfying (6.67).

Exercise 6.4 Show that there is a $c < \infty$ such that the following holds. Suppose that S_n is simple random walk in $\mathbb{Z}^2$ and let $V = V_{n,N}$ be the event that the path $S[0, \xi_N]$ does not disconnect the origin from $\partial \mathcal{B}_n$. Then, if $x \in \mathcal{B}_{2n}$,

$$\mathbb{P}^x(V) \leq \frac{c}{\log(N/n)}.$$

(Hint: there is a $\rho > 0$ such that the probability that a walk starting at $\partial \mathcal{B}_{n/2}$ disconnects the origin before reaching $\partial \mathcal{B}_n$ is at least ρ; see Exercise 3.4.)

Exercise 6.5 Suppose that $p \in \mathcal{P}_d, d \geq 3$. Show that there exists a sequence $K_n \to \infty$ such that if $A \subset \mathbb{Z}^d$ is a finite set with at least n points, then $\mathrm{cap}(A) \geq K_n$.

Exercise 6.6 Suppose that $p \in \mathcal{P}_d$ and $r < 1$. Show that there exists $c = c_r < \infty$ such that the following holds.

(i) If $|\mathbf{e}| = 1$ and $x \in \mathcal{C}_{rn}$,

$$\sum_{y \in \mathcal{C}_n} \left| G_{\mathcal{C}_n}(x + \mathbf{e}, y) - G_{\mathcal{C}_n}(x, y) \right| \leq c\, n,$$

(ii) Suppose that f, g, F are as in Corollary 6.2.4 with $A = \mathcal{C}_n$. Then, if $x \in \mathcal{C}_{rn}$,

$$|\nabla_j f(x)| \leq \frac{c}{n}\left[\|F\|_\infty - n^2\,\|g\|_\infty\right].$$

Exercise 6.7 Show that if $p \in \mathcal{P}_2$ and $r > 0$,

$$\lim_{n\to\infty}[G_{\mathcal{C}_{n+r}}(0,0) - G_{\mathcal{C}_n}(0,0)] = 0.$$

Use this and (6.16) to conclude that for all x, y,

$$\lim_{n\to\infty}[G_{\mathcal{C}_n}(0,0) - G_{\mathcal{C}_n}(x,y)] = a(x,y).$$

Exercise 6.8 Suppose that $p \in \mathcal{P}_d$ and $A \subset \mathbb{Z}^d$ is finite. Define

$$\mathcal{Q}_A(f,g) = \sum_{x,y\in\overline{A}} p(x,y)\,[f(y) - f(x)]\,[g(y) - g(x)],$$

and $\mathcal{Q}_A(f) = \mathcal{Q}_A(f,f)$. Let $F : \partial A \to \mathbb{R}$ be given. Show that the infimum of $\mathcal{Q}_A(f)$ restricted to functions $f : \overline{A} \to \mathbb{R}$ with $f \equiv F$ on ∂A is obtained by the unique harmonic function with boundary value F.

Exercise 6.9 Write the two-dimensional integer lattice in complex form, $\mathbb{Z}^2 = \mathbb{Z} + i\mathbb{Z}$ and let A be the upper half plane $A = \{j + ik \in \mathbb{Z}^2 : k > 0\}$. Show that for simple random walk

$$G_A(x,y) = a(\overline{x},y) - a(x,y), \qquad x,y \in A,$$

$$H_A(x,j) = \frac{1}{4}\,[a(x,j-i) - a(x,j+i)] + \delta(x-j), \qquad x \in \overline{A}, j \in \mathbb{Z},$$

where $\overline{j+ik} = j - ik$ denotes complex conjugate. Find

$$\lim_{k\to\infty} k\,H_A(ik,j).$$

Exercise 6.10 Prove Corollary 6.9.5.

Exercise 6.11 Provide the details of the Harnack inequality argument in Lemma 6.5.8 and Theorem 6.5.10.

Exercise 6.12 Suppose that $p \in \mathcal{P}_d$.

(i) Show that there is a $c < \infty$ such that if $x \in A \subset \mathcal{C}_n$ and $z \in \mathcal{C}_{2n}$,

$$\mathbb{P}^x\{S_{\xi_{2n}} = z \mid \xi_{2n} < T_A\} \leq c\,n^{1-d}\,\frac{\mathbb{P}^x\{\xi_n < T_A\}}{\mathbb{P}^x\{\xi_{2n} < T_A\}}.$$

(ii) Let A be the line $\{j\mathbf{e}_1 : j \in \mathbb{Z}\}$. Show that there is an $\epsilon > 0$ such that for all n sufficiently large,

$$\mathbb{P}\{\text{dist}(S_{\xi_n}, A) \geq \epsilon n \mid \xi_n < T_A\} \geq \epsilon.$$

(Hint: you can use the gambler's ruin estimate to estimate $\mathbb{P}^x\{\xi_{n/2} < T_A\}/\mathbb{P}^x\{\xi_n < T_A\}$.)

Exercise 6.13 Show that for each $p \in \mathcal{P}_2$ and each $r \in (0, 1)$, there is a c such that for all n sufficiently large,

$$G_{\mathcal{C}_n}(x, y) \geq c, \quad x, y \in \mathcal{C}_{rn},$$
$$G_{\mathcal{B}_n}(x, y) \geq c, \quad x, y \in \mathcal{B}_{rn}.$$

Exercise 6.14 Suppose that $p \in \mathcal{P}_2$ and let $A = \{x_1, x_2\}$ be a two-point set.

(i) Prove that $\text{hm}_A(x_1) = 1/2$.
(ii) Show that there is a $c < \infty$ such that if $A \subset \mathcal{C}_n$, then for $y \in \mathbb{Z}^2 \setminus \mathcal{C}_{2n}$,

$$\left| \mathbb{P}^y\{S_{T_A} = x_1\} - \frac{1}{2} \right| \leq \frac{c}{\log n}.$$

(Hint: suppose that $\mathbb{P}^y\{S_{T_A} = x_j\} \geq 1/2$ and let V be the set of z such that $\mathbb{P}^z\{S_{T_A} = x_j\} \leq 1/2$. Let $\sigma = \min\{j : S_j \in V\}$. Then, it suffices to prove that $\mathbb{P}^y\{T_A < \sigma\} \leq c/\log n$.)
(iii) Show that there is a $c < \infty$ such that if $A = \mathbb{Z}^2 \setminus \{x\}$ with $x \neq 0$, then

$$\left| G_A(0, 0) - \frac{4}{\pi} \log |x| \right| \leq c.$$

Exercise 6.15 Suppose that $p \in \mathcal{P}_2$. Show that there exist $c_1, c_2 > 0$ such that the following holds.

(i) If n is sufficiently large, A is a set as in Lemma 6.6.7, and $A_n = A \cap \{|z| \geq n/2\}$, then for $x \in \partial\mathcal{B}_{n/2}$,

$$\mathbb{P}^x\{T_A < \xi_n^*\} \geq c.$$

(ii) If $x \in \partial\mathcal{B}_{n/2}$,

$$G_{\mathbb{Z}^2 \setminus A}(x, 0) \leq c.$$

(iii) If A' is a set with $\mathcal{B}_{n/2} \subset A' \subset \mathbb{Z}^2 \setminus A_n$,

$$\left| G_{A'}(0,0) - \frac{2}{\pi} \log n \right| \leq c.$$

Exercise 6.16 Give an example of an aperiodic $p \in \mathcal{P}_d$ and a finite connected (with respect to p) set A for which p is bipartite restricted to A.

Exercise 6.17 Suppose that S_n is simple random walk in $\mathbb{Z}^d$ so that $\xi_n = \xi_n^*$. If $|x| < n$, let

$$u(x,n) = \mathbb{E}^x \left[|S_{\xi_n}| - n \right]$$

and note that $0 \leq u(x,n) \leq 1$.

(i) Show that

$$n^2 - |x|^2 + 2n\, u(x,n) \leq \mathbb{E}^x[\xi_n] \leq n^2 - |x|^2 + (2n+1)\, u(x,n).$$

(ii) Show that if $d = 2$,

$$\frac{\pi}{2} G_{\mathcal{B}_n}(0,x) = \log n - \log |x| + \frac{u(x,n)}{n} + O(|x|^{-2}).$$

(iii) Show that if $d \geq 3$,

$$C_d^{-1}\, G_{\mathcal{B}_n}(0,x) = \frac{1}{|x|^{d-2}} - \frac{1}{n^{d-2}} + \frac{(d-2)\, u(x,n)}{n^{d-1}} + O(|x|^{-d}).$$

Exercise 6.18 Suppose that S_n is simple random walk in $\mathbb{Z}^d$ with $d \geq 3$. For this exercise assume that we know that

$$G(x) \sim \frac{C_d}{|x|^{d-2}}, \quad |x| \to \infty$$

for some constant C_d but no further information on the asymptotics. The purpose of this exercise is to find C_d. Let V_d be the volume of the unit ball in $\mathbb{R}^d$ and $\omega_d = d\, V_d$ the surface area of the boundary of the unit ball.

(i) Show that as $n \to \infty$,

$$\sum_{x \in \mathcal{B}_n} G(0,x) \sim \frac{C_d\, \omega_d\, n^2}{2} = \frac{C_d\, d\, V_d\, n^2}{2}.$$

(ii) Show that as $n \to \infty$,

$$\sum_{x \in \mathcal{B}_n} [G(0,x) - G_{\mathcal{B}_n}(0,x)] \sim C_d \, V_d \, n^2.$$

(iii) Show that as $n \to \infty$,

$$\sum_{x \in \mathcal{B}_n} G_{\mathcal{B}_n}(0,x) \sim n^2.$$

(iv) Conclude that

$$C_d \, V_d \left(\frac{d}{2} - 1 \right) = 1.$$

7
Dyadic coupling

7.1 Introduction

In this chapter we will study the *dyadic* or *KMT coupling*, which is a coupling of Brownian motion and random walk for which the paths are significantly closer to each other than in the Skorokhod embedding. Recall that if (S_n, B_n) are coupled by the Skorokhod embedding, then typically one expects $|S_n - B_n|$ to be of order $n^{1/4}$. In the dyadic coupling, $|S_n - B_n|$ will be of order $\log n$. We mainly restrict our consideration to one dimension, although we discuss some higher dimensional versions in Section 7.6.

Suppose that $p \in \mathcal{P}_1'$ and

$$S_n = X_1 + \cdots + X_n$$

is a p-walk. Suppose that there exists $b > 0$ such that

$$\mathbb{E}[X_1^2] = \sigma^2, \quad \mathbb{E}[e^{b|X_1|}] < \infty. \tag{7.1}$$

Then, by Theorem 2.3.11, there exist N, c, ϵ such that if we define $\delta(n,x)$ by

$$p_n(x) := \mathbb{P}\{S_n = x\} = \frac{1}{\sqrt{2\pi\sigma^2 n}} e^{-\frac{x^2}{2\sigma^2 n}} \exp\{\delta(n,x)\},$$

then for all $n \geq N$ and $|x| \leq \epsilon n$,

$$|\delta(n,x)| \leq c \left[\frac{1}{\sqrt{n}} + \frac{|x|^3}{n^2}\right]. \tag{7.2}$$

Theorem 7.1.1 *Suppose that $p \in \mathcal{P}_d'$ satisfies (7.1) and (7.2). Then one can define on the same probability space $(\Omega, \mathcal{F}, \mathbb{P})$, a Brownian motion B_t with* *variance parameter σ^2 and a random walk with increment distribution p such that the following holds. For each $\alpha < \infty$, there is a c_α such that*

$$\mathbb{P}\left\{\max_{1\le j\le n} |S_j - B_j| \ge c_\alpha \log n\right\} \le c_\alpha\, n^{-\alpha}. \tag{7.3}$$

Remark From the theorem it is easy to conclude the corresponding result for bipartite or continuous-time walks with $p \in \mathcal{P}_1$. In particular, the result holds for discrete-time and continuous-time simple random walk.

♣ We will describe the dyadic coupling formally in Section 7.4, but we will give a basic idea here. Suppose that $n = 2^m$. One starts by defining S_{2^m} as closely to B_{2^m} as possible. Using the LCLT, we can do this in a way so that with very high probability $|S_{2^m} - B_{2^m}|$ is of order 1. We then define $S_{2^{m-1}}$ using the values of $B_{2^m}, B_{2^{m-1}}$, and again get an error of order 1. We keep subdividing intervals using binary splitting, and every time we construct the value of S at the middle point of a new interval. If at each subdivision we get an error of order 1, the total error should be at most of order m, the number of subdivisions needed. (Typically it might be less because of cancellation.)

♣ The assumption $\mathbb{E}[e^{b|X_1|}] < \infty$ for some $b > 0$ is necessary for (7.3) to hold at $j = 1$. Suppose that $p \in \mathcal{P}_1'$ such that for each n there is a coupling with

$$\mathbb{P}\{|S_1 - B_1| \ge \hat{c}\,\log n\} \le \hat{c}\, n^{-1}.$$

It is not difficult to show that as $n \to \infty$, $\mathbb{P}\{|B_1| \ge \hat{c}\,\log n\} = o(n^{-1})$, and hence

$$\mathbb{P}\{|S_1| \ge 2\hat{c}\,\log n\} \le \mathbb{P}\{|S_1 - B_1| \ge \hat{c}\,\log n\} + \mathbb{P}\{|B_1| \ge \hat{c}\,\log n\} \le 2\,\hat{c}\, n^{-1}$$

for n sufficiently large. If we let $x = 2\hat{c}\log n$, this becomes

$$\mathbb{P}\{|X_1| \ge x\} \le 2\hat{c}\, e^{-x/(2\hat{c})},$$

for all x sufficiently large which implies that $\mathbb{E}[e^{b|X_1|}] < \infty$ for $b < (2\hat{c})^{-1}$.

Some preliminary estimates and definitions are given in Sections 7.2 and 7.3, the coupling is defined in Section 7.4, and we show that it satisfies (7.3) in Section 7.5. The proof is essentially the same for all values of σ^2. For ease of notation we will assume that $\sigma^2 = 1$. It also suffices to prove the result for $n = 2^m$ and we will assume this in Sections 7.4 and 7.5.

For the remainder of this chapter, we fix b, ϵ, c_0, N and assume that p is an increment distribution satisfying

$$\mathbb{E}[e^{b|X_1|}] < \infty, \tag{7.4}$$

and

$$p_n(x) = \frac{1}{\sqrt{2\pi n}} e^{-\frac{x^2}{2n}} \exp\{\delta(n,x)\},$$

where

$$|\delta(n,x)| \le c_0 \left[\frac{1}{\sqrt{n}} + \frac{|x|^3}{n^2}\right], \quad n \ge N, \ |x| \le \epsilon n. \tag{7.5}$$

7.2 Some estimates

In this section we collect a few lemmas about random walk that will be used in establishing (7.3). The reader may wish to skip this section at first reading and come back to the estimates as they are needed.

Lemma 7.2.1 *Suppose that S_n is a random walk with increment distribution p satisfying (7.4) and (7.5). Define $\delta_n^*(n,x,y)$ by*

$$\mathbb{P}\{S_n = x \mid S_{2n} = y\} = \frac{1}{\sqrt{\pi n}} \exp\left\{-\frac{(x-(y/2))^2}{n}\right\} \exp\{\delta_*(n,x,y)\}.$$

Then, if $n \ge N$, $|x|, |y| \le \epsilon n/2$,

$$|\delta_*(n,x,y)| \le 9\, c_0 \left[\frac{1}{\sqrt{n}} + \frac{|x|^3}{n^2} + \frac{|y|^3}{n^2}\right].$$

♣ Without the conditioning, S_n is approximately normal with mean zero and variance n. Conditioned on the event $S_{2n} = y$, S_n is approximately normal with mean $y/2$ and variance $n/2$. Note that specifying the value at time $2n$ reduces the variance of S_n.

Proof Note that

$$\mathbb{P}\{S_n = x \mid S_{2n} = y\} = \frac{\mathbb{P}\{S_n = x, S_{2n} - S_n = y - x\}}{\mathbb{P}\{S_{2n} = y\}} = \frac{p_n(x)\, p_n(y-x)}{p_{2n}(y)}.$$

Since $|x|, |y|, |x-y| \le \epsilon n$, we can apply (7.5). Note that

$$|\delta_*(n,x,y)| \le |\delta(n,x)| + |\delta(n,y-x)| + |\delta(2n,y)|.$$

We use the simple estimate $|y-x|^3 \le 8(|x|^3 + |y|^3)$. □

Lemma 7.2.2 *If* $x_1, x_2, \ldots, x_n \in \mathbb{R}$, *then*

$$\sum_{j=1}^{n} \frac{(x_1 + \cdots + x_j)^2}{2^j} \le 2\,\frac{\sqrt{2}+1}{\sqrt{2}-1}\left[\sum_{j=1}^{n} \frac{x_j^2}{2^j}\right]. \tag{7.6}$$

Proof Due to the homogeneity of (7.6) we may assume that $\sum 2^{-j} x_j^2 = 1$. Let $y_j = 2^{-j/2} x_j, \mathbf{y} = (y_1, \ldots, y_n)$. Then

$$\begin{aligned}
\sum_{i=1}^{n} \frac{(x_1 + \cdots + x_i)^2}{2^i} &= \sum_{i=1}^{n} \sum_{1 \le j,k \le i} \frac{x_j x_k}{2^i} \\
&= \sum_{j=1}^{n} \sum_{k=1}^{n} x_j x_k \sum_{i=j \vee k}^{n} 2^{-i} \\
&\le 2 \sum_{j=1}^{n} \sum_{k=1}^{n} 2^{-(j \vee k)}\, x_j x_k \\
&= 2 \sum_{j=1}^{n} \sum_{k=1}^{n} 2^{-|k-j|/2}\, y_j y_k \\
&= 2\langle A\mathbf{y}, \mathbf{y}\rangle \le 2\lambda\, \|\mathbf{y}\|^2 = 2\lambda,
\end{aligned}$$

where $A = A_n$ is the $n \times n$ symmetric matrix with entries $a(j,k) = 2^{-|k-j|/2}$ and $\lambda = \lambda_n$ denotes the largest eigenvalue of A. Since λ is bounded by the maximum of the row sums,

$$\lambda \le 1 + 2\sum_{j=1}^{\infty} 2^{-j/2} = \frac{\sqrt{2}+1}{\sqrt{2}-1}.$$

□

♣ We will use the fact that the left-hand side of (7.6) is bounded by a constant times the term in brackets on the right-hand side. The exact constant is not important.

Lemma 7.2.3 *Suppose that S_n is a random walk with increment distribution satisfying (7.4) and (7.5). Then, for every α there exists a $c = c(\alpha)$ such that*

$$\mathbb{P}\left\{ \sum_{\log_2 n < j \le n} \frac{S_{2^j}^2}{2^j} \ge c\, n \right\} \le c\, e^{-\alpha n}. \tag{7.7}$$

♣ Consider the random variables $U_j = S_{2^j}^2/2^j$ and note that $\mathbb{E}[U_j] = 1$. Suppose that $U_1, U_2, \ldots$ were independent. If it were also true that there exist t, c such that $\mathbb{E}[e^{tU_j}] \le c$ for all j, then (7.7) would be a standard large deviation estimate similar to Theorem A.2.5. To handle the lack of independence, we consider the independent random variables $[S_{2^j} - S_{2^{j-1}}]^2/2^j$ and use (7.6). If the increment distribution is bounded, then we also get $\mathbb{E}[e^{tU_j}] \le c$ for some t; see Exercise 2.6. However, if the range is infinite this expectation may be infinite for all $t > 0$; see Exercise 7.3. To overcome this difficulty, we use a striaghtforward truncation argument.

Proof We fix $\alpha > 0$ and allow constants in this proof to depend on α. Using (7.4), we see that there is a β such that

$$\mathbb{P}\{|S_n| \ge n\} \le e^{-\beta n}.$$

Hence, we can find c_1 such that

$$\sum_{\log_2 n < j \le n} [\,\mathbb{P}\{|S_{2^j}| \ge c_1 2^j\} + \mathbb{P}\{|S_{2^j} - S_{2^{j-1}}| \ge c_1 2^j\}\,] = O(e^{-\alpha n}).$$

Fix this c_1, and let $j_0 = \lfloor \log_2 n + 1 \rfloor$ be the smallest integer greater than $\log_2 n$. Let $Y_j = 0$ for $j < j_0$, $Y_{j_0} = S_{2^{j_0}}$, and for $j > j_0$, let $Y_j = S_{2^j} - S_{2^{j-1}}$. Then, except for an event of probability $O(e^{-\alpha n})$, $|Y_j| \le c_1 2^j$ for $j \ge j_0$ and hence

$$\mathbb{P}\left\{ \sum_{j=j_0}^{n} \frac{Y_j^2}{2^j} \ne \sum_{j=j_0}^{n} \frac{Y_j^2}{2^j}\, 1\{|Y_j| \le c_1 2^j\} \right\} \le O(e^{-\alpha n}).$$

Note that

$$\sum_{\log_2 n < j \le n} \frac{S_{2^j}^2}{2^j} = \sum_{j=j_0}^{n} \frac{S_{2^j}^2}{2^j} = \sum_{j=1}^{n} \frac{(Y_1 + \cdots + Y_j)^2}{2^j} \le c \sum_{j=j_0}^{n} \frac{Y_j^2}{2^j}.$$

The last step uses (7.6). Therefore it suffices to prove that

$$\mathbb{P}\left\{\sum_{j=1}^{n} \frac{Y_j^2}{2^j} 1\{|Y_j| \le c_1 2^j\} \ge cn\right\} \le e^{-\alpha n}.$$

The estimates (7.4) and (7.5) imply that there is a $t > 0$ such that for each n,

$$\mathbb{E}\left[\exp\left\{\frac{t S_n^2}{n}\right\}; |S_n| \le c_1 n\right] \le e$$

(see Exercise 7.2). Therefore,

$$\mathbb{E}\left[\exp\left\{t \sum_{j=1}^{n} \frac{Y_j^2}{2^j} 1\{|Y_j| \le c_1 2^j\}\right\}\right] \le e^n,$$

which implies that

$$\mathbb{P}\left\{\sum_{j=1}^{n} \frac{Y_j^2}{2^j} 1\{|Y_j| \le c_1 2^j\} \ge t^{-1}(\alpha + 1)\, n\right\} \le e^{-\alpha n}.$$

□

7.3 Quantile coupling

In this section we consider the simpler problem of coupling S_n and B_n for a fixed n. The following is a general definition of quantile coupling. We will only use quantile coupling in a particular case where F is supported on $\mathbb{Z}$ or on $(1/2)\mathbb{Z}$.

Definition Suppose that F is the distribution function of a discrete random variable supported on the locally finite set

$$\cdots < a_{-1} < a_0 < a_1 < \cdots$$

and Z is a random variable with a continuous, strictly increasing distribution function G. Let r_k be defined by $G(r_k) = F(a_k)$, i.e. if $F(a_k) > F(a_k-)$,

$$G(r_k) - G(r_{k-1}) = F(a_k) - F(a_k-).$$

Let f be the step function

$$f(z) = a_k \text{ if } r_{k-1} < z \le r_k,$$

and let X be the random variable $f(Z)$. We call X the *quantile coupling* of F with Z, and f the *quantile coupling function* of F and G.

Note that the event $\{X = a_k\}$ is the same as the event $\{r_{k-1} < Z \leq r_k\}$. Hence,

$$\mathbb{P}\{X = a_k\} = \mathbb{P}\{r_{k-1} < Z \leq r_k\} = G(r_k) - G(r_{k-1}) = F(a_k) - F(a_k-),$$

and X has distribution function F. Also, if

$$G(a_k - t) \leq F(a_{k-1}) < F(a_k) \leq G(a_k + t), \tag{7.8}$$

then it is immediate from the above definitions that $\{X = a_k\} \subset \{|X - Z| = |a_k - Z| \leq t\}$. Hence, if we wish to prove that $|X - Z| \leq t$ on the event $\{X = a_k\}$, it suffices to establish (7.8).

As an intermediate step in the construction of the dyadic coupling, we study the quantile coupling of the random walk distribution with a normal random variable that has the same mean and variance. Let Φ denote the standard normal distribution function, and let Φ_β (where $\beta > 0$) denote the distribution function of a mean zero normal random variable with variance β.

Proposition 7.3.1 *For every ϵ, b, c_0, N there exist c, δ such that if S_n is a random walk with increment distribution p satisfying (7.4) and (7.5), the following holds for $n \geq N$. Let F_n denote the distribution function of S_n, and suppose that Z has distribution function Φ_n. Let (X, Z) be the quantile coupling of F_n with Z. Then,*

$$|X - Z| \leq c\left[1 + \frac{X^2}{n}\right], \qquad |X| \leq \delta n.$$

Proposition 7.3.2 *For every ϵ, b, c_0, N there exist c, δ such that the following holds for $n \geq N$. Suppose that S_n is a random walk with increment distribution p satisfying (7.4) and (7.5). Suppose that $|y| \leq \delta n$ with $\mathbb{P}\{S_{2n} = y\} > 0$. Let $F_{n,y}$ denote the conditional distribution function of $S_n - (S_{2n}/2)$ given that $S_{2n} = y$, and suppose that Z has a distribution function $\Phi_{n/2}$. Let (X, Z) be the quantile coupling of $F_{n,y}$ with Z. Then,*

$$|X - Z| \leq c\left[1 + \frac{X^2}{n} + \frac{y^2}{n}\right], \qquad |X|, |y| \leq \delta n.$$

Using (7.8), we see that in order to prove the above propositions, it suffices to show the following estimate for the corresponding distribution functions.

Lemma 7.3.3 *For every ϵ, b, c_0, N there exist c, δ such that if S_n is a random walk with increment distribution p satisfying (7.4) and (7.5), the following holds for $n \geq N$. Let $F_n, F_{n,y}$ be as in the propositions above. Then for $y \in \mathbb{Z}$, $|x|, |y| \leq \delta n$,*

$$\Phi_n\left(x - c\left[1 + \frac{x^2}{n}\right]\right) \leq F_n(x-1)$$
$$\leq F_n(x) \leq \Phi_n\left(x + c\left[1 + \frac{x^2}{n}\right]\right), \tag{7.9}$$
$$\Phi_{n/2}\left(x - c\left[1 + \frac{x^2}{n} + \frac{y^2}{n}\right]\right) \leq F_{n,y}(x-1)$$
$$\leq F_{n,y}(x) \leq \Phi_{n/2}\left(x + c\left[1 + \frac{x^2}{n} + \frac{y^2}{n}\right]\right),$$

Proof It suffices to establish the inequalities in the case where x is a nonnegative integer. Implicit constants in this proof are allowed to depend on ϵ, b, c_0 and we assume $n \geq N$. If F is a distribution function, we write $\overline{F} = 1 - F$. Since for $t > 0$,

$$\frac{(t+1)^2}{2n} = \frac{t^2}{n} + O\left(\frac{1}{\sqrt{n}} + \frac{t^3}{n^2}\right),$$

(consider $t \leq \sqrt{n}$ and $t \geq \sqrt{n}$), we can see that (7.4) and (7.5) imply that we can write

$$p_n(x) = \int_x^{x+1} \frac{1}{\sqrt{2\pi n}} e^{-t^2/(2n)} \exp\left\{O\left(\frac{1}{\sqrt{n}} + \frac{t^3}{n^2}\right)\right\} dt, \qquad |x| \leq \epsilon n.$$

Hence, using (A.12), for some a and all $|x| \leq \epsilon n$,

$$\overline{F}_n(x) = \mathbb{P}\{S_n \geq \epsilon n\} + \mathbb{P}\{x \leq S_n < \epsilon n\}$$
$$= O(e^{-an}) + \int_x^{\epsilon n} \frac{1}{\sqrt{2\pi n}} e^{-t^2/(2n)} \exp\left\{O\left(\frac{1}{\sqrt{n}} + \frac{t^3}{n^2}\right)\right\} dt.$$

From this we can conclude that for $|x| \leq \epsilon n$,

$$\overline{F}_n(x) = \overline{\Phi}_n(x) \exp\left\{O\left(\frac{1}{\sqrt{n}} + \frac{x^3}{n^2}\right)\right\}, \tag{7.10}$$

and from this we can conclude (7.9). The second inequality is done similarly by using Lemma 7.2.1 to derive

$$\overline{F}_{n,y}(x) = \overline{\Phi}_{n/2}(x) \exp\left\{O\left(\frac{1}{\sqrt{n}} + \frac{x^3}{n^2} + \frac{y^3}{n^2}\right)\right\},$$

for $|x|, |y| \leq \delta n$. Details are left for Exercise 7.4.

□

♣ To derive Proposition 7.3.1 and Proposition 7.3.2 we use only estimates on the distribution functions F_n, $F_{n,y}$ and not pointwise estimates (LCLT). However, the pointwise estimate (7.5) is used in the proof of Lemma 7.2.1 which is used in turn to estimate $F_{n,y}$.

7.4 The dyadic coupling

In this section we define the dyadic coupling. Fix $n = 2^m$ and assume that we are given a standard Brownian motion defined on some probability space. We will define the random variables $S_1, S_2, \ldots, S_{2^m}$ as functions of the random variables $B_1, B_2, \ldots, B_{2^m}$ so that $S_1, \ldots, S_{2^m}$ has the distribution of a random walk with increment distribution p.

In Chapter 3, we constructed a Brownian motion from a collection of independent normal random variables by a dyadic construction. Here we reverse the process, starting with the Brownian motion, B_t, and obtaining the independent normals. We will only use the random variables $B_1, B_2, \ldots, B_{2^m}$. Define $\Gamma_{k,j}$ by

$$\Gamma_{k,j} = B_{k2^{m-j}} - B_{(k-1)2^{m-j}}, \qquad j = 0, 1, \ldots, m; \;\; k = 1, 2, 3, \ldots, 2^j.$$

For each j, $\{\Gamma_{k,j} : k = 1, 2, 3, \ldots, 2^j\}$ are independent normal random variables with mean zero and variance 2^{m-j}. Let $Z_{1,0} = B_{2^m}$ and define

$$Z_{2k+1,j}, \quad j = 1, \ldots, m, \;\; k = 0, 1, \ldots, 2^{j-1} - 1,$$

recursively by

$$\Gamma_{2k+1,j} = \frac{1}{2}\,\Gamma_{k+1,j-1} + Z_{2k+1,j}, \tag{7.11}$$

so that also

$$\Gamma_{2k+2,j} = \frac{1}{2}\,\Gamma_{k+1,j-1} - Z_{2k+1,j}.$$

One can check (see Corollary A.3.1) that the random variables $\{Z_{2k+1,j} : j = 0, \ldots, 2^m,\ k = 0, 1, \ldots, 2^{m-1} - 1\}$ are independent, mean zero, normal random variables with $\mathbb{E}[Z_{1,0}^2] = 2^m$ and that $\mathbb{E}[Z_{2k+1,j}^2] = 2^{m-j-1}$ for $j \geq 1$. We can rewrite (7.11) as

$$B_{(2k+1)2^{m-j}} = \frac{1}{2}\left[B_{k2^{m-j+1}} + B_{(k+1)2^{m-j+1}}\right] + Z_{2k+1,j}. \tag{7.12}$$

Let $f_m(\cdot)$ denote the quantile coupling function for the distribution functions of S_{2^m} and B_{2^m}. If $y \in \mathbb{Z}$, let $f_j(\cdot, y)$ denote the quantile coupling function for the conditional distribution of

$$S_{2^j} - \frac{1}{2}S_{2^{j+1}}$$

given that $S_{2^{j+1}} = y$ and a normal random variable with mean zero and variance 2^{j-1}. This is well defined as long as $\mathbb{P}\{S_{2^{j+1}} = y\} > 0$. Note that the range of $f_j(\cdot, y)$ is contained in $(1/2)\mathbb{Z}$. This conditional distribution is symmetric about the origin (see Exercise 7.1), so $f_j(-z, y) = -f_j(z, y)$. We can now define the dyadic coupling.

- Let $S_{2^m} = f_m(B_{2^m})$.
- Suppose that the values of $S_{l2^{m-j+1}}, l = 1, \ldots, 2^{j-1}$ are known. Let

$$\Delta_{k,i} = S_{k2^{m-i}} - S_{(k-1)2^{m-i}}.$$

Then we let

$$S_{(2k-1)2^{m-j}} = \frac{1}{2}\left[S_{(k-1)2^{m-j+1}} + S_{k2^{m-j+1}}\right] + f_{m-j}(Z_{2k-1,j}, \Delta_{k,j-1}),$$

so that

$$\Delta_{2k-1,j} = \frac{1}{2}\Delta_{k,j-1} + f_{m-j}(Z_{2k-1,j}, \Delta_{k,j-1}),$$

$$\Delta_{2k,j} = \frac{1}{2}\Delta_{k,j-1} - f_{m-j}(Z_{2k-1,j}, \Delta_{k,j-1}).$$

It follows immediately from the definition that $(S_1, S_2, \ldots, S_{2^m})$ has the distribution of the random walk with increment p. Also, Exercise 7.1 shows that $\Delta_{2k-1,j}$ and $\Delta_{2k,j}$ have the same conditional distribution, given that $\Delta_{k,j-1}$.

It is convenient to rephrase this definition in terms of random variables indexed by dyadic intervals. Let $I_{k,j}$ denote the interval

$$I_{k,j} = [(k-1)2^{m-j}, k2^{m-j}], \quad j = 0, \ldots, m; \quad k = 1, \ldots, 2^j.$$

We write $Z(I)$ for the normal random variable associated to the midpoint of I,

$$Z(I_{k,j}) = Z_{2k-1,j+1}$$

Then the $Z(I)$ are independent mean zero normal random variables indexed by the dyadic intervals with variance $|I|/4$ where $|\cdot|$ denotes length. We also write

$$\Gamma(I_{k,j}) = \Gamma_{k,j}, \qquad \Delta(I_{k,j}) = \Delta_{k,j}.$$

Then the definition can be given as follows.

- Let $\Gamma(I_{1,0}) = B_{2^m}, \ \Delta(I_{1,0}) = f_m(B_{2^m})$.
- Suppose that I is a dyadic interval of length 2^{m-j+1} that is the union of consecutive dyadic intervals I^1, I^2 of length 2^{m-j}. Then

$$\Gamma(I^1) = \frac{1}{2}\Gamma(I) + Z(I), \qquad \Gamma(I^2) = \frac{1}{2}\Gamma(I) - Z(I) \tag{7.13}$$

$$\Delta(I^1) = \frac{1}{2}\Delta(I) + f_j(Z(I), \Delta(I)), \qquad \Delta(I^2) = \frac{1}{2}\Delta(I) - f_j(Z(I), \Delta(I)). \tag{7.14}$$

- Note that if $j \geq 1$ and $k \in \{1, \ldots, 2^j\}$, then

$$B_{k2^{m-j}} = \sum_{i \leq k} \Gamma([(i-1)2^{m-j}, i2^{m-j}]),$$

$$S_{k2^{m-j}} = \sum_{i \leq k} \Delta([(i-1)2^{m-j}, i2^{m-j}]). \tag{7.15}$$

We next note a few important properties of the coupling.

- If $I = I^1 \cup I^2$ as above, then $\Gamma(I^1), \Gamma(I^2), \ \Delta(I^1), \Delta(I^2)$ are deterministic functions of $\Gamma(I), \Delta(I), Z(I)$. The conditional distributions of $(\Gamma(I^1), \Delta(I^1))$ and $(\Gamma(I^2), \Delta(I^2))$ given $(\Gamma(I), \Delta(I))$ are the same.
- By iterating this, we get the following. For each interval $I_{k,j}$ consider the joint distribution random variables

$$(\Gamma(I_{l,i}), \Delta(I_{l,i})), \quad i = 0, \ldots, j,$$

where $l = l(i,k,j)$ is chosen so that $I_{k,j} \subset I_{l,i}$. Then, this distribution is the same for all $k = 1, 2, \ldots, 2^j$. In particular, if

$$R_{k,j} = \sum_{i=0}^{j} |\Gamma(I_{l,i}) - \Delta(I_{l,i})|,$$

then the random variables $R_{1,j}, \ldots, R_{2^j,j}$ are identically distributed. (They are not independent.)

- For $k = 1$,

$$\Gamma(I_{1,j}) - \Delta(I_{1,j}) = \frac{1}{2}\left[\Gamma(I_{1,j-1}) - \Delta(I_{1,j-1})\right] + \left[Z_{1,j} - f_j(Z_{1,j}, S_{2^{m-j+1}})\right]$$

By iterating this, we get

$$R_{1,j} \leq |S_{2^m} - B_{2^m}| + 2\sum_{l=1}^{j} |f_{m-l}(Z_{1,l}, S_{2^{m-l+1}}) - Z_{1,l}|. \tag{7.16}$$

- Define $\Theta(I_{1,0}) = |B_{2^m} - S_{2^m}| = |\Gamma(I_{1,0}) - \Delta(I_{1,0})|$. Suppse that $j \geq 1$ and $I_{k,j}$ is an interval with "parent" interval I'. Define $\Theta(I_{k,j})$ to be the maximum of $|B_t - S_t|$ where the maximum is over three values of t: the left endpoint, midpoint, and right endpoint of I. We claim that

$$\Theta(I_{k,j}) \leq \Theta(I') + |\Gamma(I_{k,j}) - \Delta(I_{k,j})|.$$

Since the endpoints of $I_{k,j}$ are either endpoints or midpoints of I', it suffices to show that

$$|B_t - S_t| \leq \max\left\{|B_{s_-} - S_{s_-}|, |B_{s_+} - S_{s_+}||\right\} + |\Gamma(I_{k,j}) - \Delta(I_{k,j})|,$$

where t, s_-, s_+ denote the midpoint, left endpoint, and right endpoint of $I_{k,j}$, respectively. But using (7.13), (7.4), and (7.15), we see that

$$B_t - S_t = \frac{1}{2}\left[(B_{s_-} - S_{s_-}) + (B_{s_+} - S_{s_+})\right] + |\Gamma(I_{k,j}) - \Delta(I_{k,j})|,$$

and hence the claim follows from the simple inequality $|x + y| \leq 2\max\{|x|, |y|\}$. Hence, by induction, we see that

$$\Theta(I_{k,j}) \leq R_{k,j}. \tag{7.17}$$

7.5 Proof of Theorem 7.1.1

Recall that $n = 2^m$. It suffices to show that for each α there is a c_α such that for each integer j,

$$\mathbb{P}\{|S_i - B_i| \geq c_\alpha \log n\} \leq c_\alpha\, n^{-\alpha}. \tag{7.18}$$

Indeed, if the above holds, then

$$\mathbb{P}\left\{\max_{1\le i\le n} |S_i - B_i| \ge c_\alpha \log n\right\} \le \sum_{i=1}^{n} \mathbb{P}\{|S_i - B_i| \ge c_\alpha \log n\} \le c_\alpha\, n^{-\alpha+1}.$$

We claim, in fact, that it suffices to find a sequence $0 = i_0 < i_1 < \cdots < i_l = n$ such that $|i_k - i_{k-1}| \le c_\alpha \log n$ and such that (7.18) holds for these indices. Indeed, if we prove this and $|j - i_k| \le c_\alpha \log n$, then exponential estimates show that there is a c'_α such that

$$\mathbb{P}\{|S_j - S_{i_k}| \ge c'_\alpha \log n\} + \mathbb{P}\{|B_j - B_{i_k}| \ge c'_\alpha \log n\} \le c'_\alpha\, n^{-\alpha},$$

and hence the triangle inequality gives (7.18) (with a different constant).

For the remainder of this section we fix α and allow constants to depend on α. By the reasoning of the previous paragraph and (7.17), it suffices to find a c such that for $\log_2 m + c \le j \le m$, and $k = 1, \ldots, 2^{m-j}$,

$$\mathbb{P}\left\{R_{k,j} \ge c_\alpha\, m\right\} \le c_\alpha\, e^{-\alpha m},$$

and as pointed out in the previous section, it suffices to consider the case $k = 1$, and show

$$\mathbb{P}\left\{R_{1,j} \ge c_\alpha\, m\right\} \le c_\alpha\, e^{-\alpha m}, \quad \text{for } j = \log_2 m + c, \ldots, m. \tag{7.19}$$

Let δ be the minimum of the two values given in Propositions 7.3.1 and 7.3.2, and recall that there is a $\beta = \beta(\delta)$ such that

$$\mathbb{P}\{|S_{2^j}| \ge \delta 2^j\} \le \exp\{-\beta 2^j\}$$

In particular, we can find a c_3 such that

$$\sum_{\log_2 m + c_3 \le j \le m} \mathbb{P}\{|S_{2^j}| \ge \delta 2^j\} \le O(e^{-\alpha m}).$$

Proposition 7.3.1 tells us that on the event $\{|S_{2^m}| \le \delta 2^m\}$,

$$|S_{2^m} - B_{2^m}| \le c\left[1 + \frac{S_{2^m}^2}{2^m}\right].$$

Similarly, Proposition 7.3.2 tells us that on the event $\{\max\{|S_{2^{m-l}}|, |S_{2^{m-l+1}}|\} \leq \delta 2^{m-l}\}$, we have

$$|Z_{1,l} - f_{m-l}(Z_{1,l}, S_{2^{m-l+1}})| \leq c \left[1 + \frac{S^2_{2^{m-l+1}} + S^2_{2^{m-l}}}{2^{m-l}}\right].$$

Hence, by (7.16), we see that on the same event, simultaneously for all $j \in [\log_2 m + c_3, m]$,

$$|S_{2^{m-j}} - B_{2^{m-j}}| \leq R_{1,j} + |S_{2^m} - B_{2^m}| \leq c \left[m + \sum_{\log_2 m - c_3 \leq i \leq m} \frac{S^2_{2^i}}{2^i}\right].$$

We now use (7.7) (due to the extra term $-c_3$ in the lower limit of the sum, one may have to apply (7.7) twice) to conclude (7.19) for $j \geq \log_2 m + c_3$.

7.6 Higher dimensions

Without trying to extend the result of the previous section to the general (bounded exponential moment) walks in higher dimensions, we indicate two immediate consequences.

Theorem 7.6.1 *One can define on the same probability space $(\Omega, \mathcal{F}, \mathbb{P})$, a Brownian motion B_t in $\mathbb{R}^2$ with covariance matrix $(1/2)\, I$ and a simple random walk in $\mathbb{Z}^2$ such that the following holds. For each $\alpha < \infty$, there is a c_α such that*

$$\mathbb{P}\left\{\max_{1 \leq j \leq n} |S_j - B_j| \geq c_\alpha \log n\right\} \leq c_\alpha\, n^{-\alpha}.$$

Proof We use the trick from Exercise 1.7. Let $(S_{n,1}, B_{n,1}), (S_{n,2}, B_{n,2})$ be independent dyadic couplings of one-dimensional simple random walk and Brownian motion. Let

$$S_n = \left(\frac{S_{n,1} + S_{n,2}}{2}, \frac{S_{n,1} - S_{n,2}}{2}\right),$$

$$B_n = \left(\frac{B_{n,1} + B_{n,2}}{2}, \frac{B_{n,1} - B_{n,2}}{2}\right).$$

□

Theorem 7.6.2 *If $p \in \mathcal{P}_d$, one can define on the same probability space $(\Omega, \mathcal{F}, \mathbb{P})$, a Brownian motion B_t in $\mathbb{R}^d$ with covariance matrix Γ and a* *continuous-time random walk $\tilde{S}_t$ with increment distribution p such that the following holds. For each $\alpha < \infty$, there is a c_α such that*

$$\mathbb{P}\left\{\max_{1\le j\le n} |\tilde{S}_j - B_j| \ge c_\alpha \log n\right\} \le c_\alpha\, n^{-\alpha}. \tag{7.20}$$

Proof Recall from (1.3) that we can write any such $\tilde{S}_t$ as

$$\tilde{S}_t = \tilde{S}^1_{q_1 t}\, x_1 + \cdots + \tilde{S}^l_{q_l t}\, x_l,$$

where $q_1, \ldots, q_l > 0$, $x_1, \ldots, x_l \in \mathbb{Z}^d$, and $\tilde{S}^1, \ldots, \tilde{S}^l$ are independent one-dimensional simple continuous-time random walks. Choose l independent couplings as in Theorem 7.1.1,

$$(S^1_t, B^1_t), (S^2_t, B^2_t), \ldots, (S^l_t, B^l_t),$$

where $B^1, \ldots, B^l$ are standard Brownian motions. Let

$$B_t = B^1_{q_1 t}\, x_1 + \cdots + B^l_{q_l t}\, x_l.$$

This satisfies (7.20). □

7.7 Coupling the exit distributions

Proposition 7.7.1 *Suppose that $p \in \mathcal{P}_d$. Then one can define on the same probability space a (discrete-time) random walk S_n with increment distribution p, a continuous-time random walk $\tilde{S}_t$ with increment distribution p, and a Brownian motion B_t with covariance matrix Γ such that for each $n, r > 0$,*

$$\mathbb{P}\left\{|S_{\xi_n} - B_{\xi'_n}| \ge r \log n\right\} = \mathbb{P}\left\{|\tilde{S}_{\tilde{\xi}_n} - B_{\xi'_n}| \ge r \log n\right\} \le \frac{c}{r},$$

where

$$\xi_n = \min\{j : \mathcal{J}(S_j) \ge n\}, \quad \tilde{\xi}_n = \min\{t : \mathcal{J}(\tilde{S}_t) \ge n\},$$
$$\xi'_n = \min\{t : \mathcal{J}(B_t) = n\}.$$

♣ We advise caution when using the dyadic coupling to prove results about random walk. If (S_n, B_t) are coupled as in the dyadic coupling, then S_n and B_t are Markov processes, but the joint process (S_n, B_n) is not Markov.

Proof It suffices to prove the result for $\tilde{S}_t, B_t$, for then we can define S_j to be the discrete-time "skeleton" walk obtained by sampling $\tilde{S}_t$ at times of its jumps. We may also assume $r \le n$; indeed, since $|\tilde{S}_{\tilde{\xi}_n}| + |B_{\xi'_n}| \le O(n)$, for all n sufficiently large

$$\mathbb{P}\left\{|\tilde{S}_{\tilde{\xi}_n} - B_{\xi'_n}| \ge n \log n\right\} = 0.$$

By Theorem 7.6.2 we can define $\tilde{S}, B$ on the same probability space such that except for an event of probability $O(n^{-4})$,

$$|\tilde{S}_t - B_t| \le c_1 \log n, \quad 0 \le t \le n^3.$$

We claim that $\mathbb{P}\{\tilde{\xi}_n > n^3\}$ decay exponentially in n. Indeed, the CLT shows that there is a $c > 0$ such that for n sufficiently large and $|x| < n$, $\mathbb{P}^x\{\tilde{\xi}_n \le n^2\} \ge c$. Iterating this gives $\mathbb{P}^x\{\tilde{\xi}_n > n^3\} \le (1-c)^n$. Similarly, $\mathbb{P}\{\xi'_n > n^3\}$ decays exponentially. Therefore, except on an event of probability $O(n^{-4})$,

$$|\tilde{S}_t - B_t| \le c_1 \log n, \quad 0 \le t \le \max\{\tilde{\xi}_n, \xi'_n\}. \tag{7.21}$$

Note that the estimate (7.21) is not sufficient to directly yield the claim, since it is possible that one of the two paths (say $\tilde{S}$) first exits $\mathcal{C}_n$ at some point y, then moves far away from y (while staying close to $\partial\mathcal{C}_n$) and that only then the other path exits $\mathcal{C}_n$, while all along the two paths stay close. The rest of the argument shows that such an event has small probability. Let

$$\tilde{\sigma}_n(c_1) = \min\{t : \text{dist}(\tilde{S}_t, \mathbb{Z}^d \setminus \mathcal{C}_n) \le c_1 \log n\} \text{ and}$$
$$\sigma'_n(c_1) = \min\{t : \text{dist}(B_t, \mathbb{Z}^d \setminus \mathcal{C}_n) \le c_1 \log n\},$$

and define

$$\rho_n := \sigma_n(c_1) \wedge \sigma'_n(c_1).$$

Since $\rho_n \le \max\{\tilde{\xi}_n, \xi'_n\}$, we conclude as in (7.21) that with an overwhelming (larger than $1 - O(n^{-4})$) probability,

$$|\tilde{S}_t - B_t| \le c_1 \log n, \quad 0 \le t \le \rho_n,$$

and, in particular, that

$$|\tilde{S}_{\rho_n} - B_{\rho_n}| \le c_1 \log n. \tag{7.22}$$

On the event in (7.22) we have

$$\max\{\mathrm{dist}(\tilde{S}_{\rho_n}, \mathbb{Z}^d \setminus \mathcal{C}_n), \mathrm{dist}(B_{\rho_n}, \mathbb{Z}^d \setminus \mathcal{C}_n)\} \le 2c_1 \log n,$$

by triangle inequality, so, in particular,

$$\max\{\tilde{\sigma}_n(2c_1), \sigma_n'(2c_1)\} \le \rho_n. \tag{7.23}$$

Using the gambler's ruin estimate (see Exercise 7.5) and strong Markov property for each process separately (recall, they are not jointly Markov)

$$\mathbb{P}\{|\tilde{S}_{\tilde{\sigma}_n(2c_1)} - \tilde{S}_j| \le r \ \log n \text{ for all } j \in [\tilde{\sigma}_n(2c_1), \tilde{\xi}_n]\} \ge 1 - \frac{c_2}{r}, \tag{7.24}$$

and also

$$\mathbb{P}\{|B_{\sigma_n'(2c_1)} - B_t| \le r \ \log n \text{ for all } t \in [\sigma_n'(2c_1), \xi_n']\} \ge 1 - \frac{c_2}{r}. \tag{7.25}$$

Applying the triangle inequality to

$$\tilde{S}_{\tilde{\xi}_n} - B_{\xi_n'} = (\tilde{S}_{\tilde{\xi}_n} - \tilde{S}_{\rho_n}) + (\tilde{S}_{\rho_n} - B_{\rho_n}) + (B_{\rho_n} - B_{\xi_n'}),$$

on the intersection of the four events from (7.22)–(7.25), yields $\tilde{S}_{\tilde{\xi}_n} - B_{\xi_n'} \le (2r + c_1) \log n$, and the complement has probability bounded by $O(1/r)$. □

Definition A finite subset A of $\mathbb{Z}^d$ is *simply connected* if both A and $\mathbb{Z}^d \setminus A$ are connected. If $x \in \mathbb{Z}^d$, let $\mathcal{S}_x$ denote the closed cube in $\mathbb{R}^d$ of side length one, centered at x, with sides parallel to the coordinate axes. If $A \subset \mathbb{Z}^d$, let D_A be the domain defined as the interior of $\cup_{x \in A} \mathcal{S}_x$. The *inradius* of A is defined by

$$\mathrm{inrad}(A) = \min\{|y| : y \in \mathbb{Z}^d \setminus A\}.$$

Proposition 7.7.2 *Suppose that $p \in \mathcal{P}_2$. Then one can define on the same probability space a (discrete-time) random walk S_n with increment distribution p and a Brownian motion B_t with covariance matrix Γ such that the following holds. If A is a finite, simply connected set containing the origin and*

$$\rho_A = \inf\{t : B_t \notin D_A\},$$

then if $r > 0$,

$$\mathbb{P}\{|S_{\tau_A} - B_{\rho_A}| \ge r \ \log[\mathrm{inrad}(A)]\} \le \frac{c}{\sqrt{r}}.$$

Proof This is similar to the last proposition except that the gambler's ruin estimate is replaced with the Beurling estimate. □

Exercises

Exercise 7.1 Suppose that $S_n = X_1 + \cdots + X_n$ where $X_1, X_2, \ldots$ are independent, identically distributed random variables. Suppose that $\mathbb{P}\{S_{2n} = 2y\} > 0$ for some $y \in \mathbb{R}$. Show that the conditional distribution of

$$S_n - y$$

conditioned on $\{S_{2n} = 2y\}$ is symmetric about the origin.

Exercise 7.2 Suppose that S_n is a random walk in $\mathbb{Z}$ whose increment distribution satisfies (7.4) and (7.5) and let $C < \infty$. Show that there exists a $t = t(b, \epsilon, c_0, C) > 0$ such that for all n,

$$\mathbb{E}\left[\exp\left\{\frac{t\,S_n^2}{n}\right\}; |S_n| \le Cn\right] \le e.$$

Exercise 7.3 Suppose that $\tilde{S}_t$ is continuous-time simple random walk in $\mathbb{Z}$.

(i) Show that there is a $c < \infty$ such that for all positive integers n,

$$\mathbb{P}\{\tilde{S}_n = n^2\} \ge c^{-1} \exp\{-cn^2 \log n\}.$$

(Hint: consider the event that the walk makes exactly n^2 moves by time n, each of them in the positive direction.)

(ii) Show that if $t > 0$,

$$\mathbb{E}\left[\exp\left\{\frac{t\,\tilde{S}_n^2}{n}\right\}\right] = \infty.$$

Exercise 7.4 Let Φ be the standard normal distribution function, and let $\overline{\Phi} = 1 - \Phi$.

(i) Show that as $x \to \infty$,

$$\bar{\Phi}(x) \sim \frac{e^{-x^2/2}}{\sqrt{2\pi}} \int_0^\infty e^{-xt}\, dt = \frac{1}{x\sqrt{2\pi}}\, e^{-x^2/2}.$$

(ii) Prove (7.10).

(iii) Show that for all $0 \le t \le x$,

$$\overline{\Phi}(x+t) \le e^{-tx}\, e^{-t^2/2}\, \overline{\Phi}(x) \le e^{tx}\, e^{-t^2/2}\, \overline{\Phi}(x) \le \overline{\Phi}(x-t).$$

(iv) For positive integer n, let $\Phi_n(x) = \Phi(x/\sqrt{n})$ denote the distribution function of a mean zero normal random variable with variance n, and $\overline{\Phi}_n = 1 - \Phi_n$. Show that for every $b > 0$, there exist $\delta > 0$ and $0 < c < \infty$ such that if $0 \le x \le \delta n$,

$$\overline{\Phi}_n\left(x + c\left[1 + \frac{x^2}{n}\right]\right) \exp\left\{2b\left[\frac{1}{\sqrt{n}} + \frac{x^3}{n^2}\right]\right\}$$
$$\le \overline{\Phi}_n(x) \exp\left\{b\left[\frac{1}{\sqrt{n}} + \frac{x^3}{n^2}\right]\right\}$$
$$\le \overline{\Phi}_n\left(x - c\left[1 + \frac{x^2}{n}\right]\right).$$

(v) Prove (7.9).

Exercise 7.5 In this exercise we prove the following version of the gambler's ruin estimate. Suppose that $p \in \mathcal{P}_d, d \ge 2$. Then, there exists c such that the following is true. If $\theta \in \mathbb{R}^d$ with $|\theta| = 1$ and $r \ge 0$,

$$\mathbb{P}\left\{S_j \cdot \theta \ge -r, \quad 0 \le j \le \xi_n^*\right\} \le \frac{c(r+1)}{n}. \tag{7.26}$$

Here ξ_n^* is as defined in Section 6.3.

(i) Let

$$q(x,n,\theta) = \mathbb{P}^x\{S_j \cdot \theta > 0, \quad 1 \le j \le \xi_n^*\}.$$

Show that there is a $c_1 > 0$ such that for all n sufficiently large and all $\theta \in \mathbb{R}^d$ with $|\theta| = 1$, the cardinality of the set of $x \in \mathbb{Z}^d$ with $|x| \le n/2$ and

$$q(x,n,\theta) \ge c_1\, q(0,2n,\theta)$$

is at least $c_1\, n^{d-1}$.

(ii) Use a last-exit decomposition to conclude

$$\sum_{x \in \mathcal{C}_n} G_{\mathcal{B}_n}(0,x)\, q(x,n,\theta) \le 1,$$

and use this to conclude the result for $r = 0$.

(iii) Use Lemma 5.1.6 and the invariance principle to show that there is a $c_2 > 0$ such that for all $|\theta| = 1$,

$$q(0, n, \theta) \geq \frac{c_2}{n}.$$

(iv) Prove (7.26) for all $r \geq 0$.

8
Additional topics on simple random walk

In this chapter we only consider simple random walk on $\mathbb{Z}^d$. In particular, S will always denote a simple random walk in $\mathbb{Z}^d$. If $d \geq 3$, G denotes the corresponding Green's function, and we simplify the notation by setting

$$G(z) = -a(z), \quad d = 1, 2,$$

where a is the potential kernel. Note that then the equation $\mathcal{L}G(z) = -\delta(z)$ holds for all $d \geq 1$.

8.1 Poisson kernel

Recall that if $A \subset \mathbb{Z}^d$ and $\tau_A = \min\{j \geq 0 : S_j \notin A\}, \overline{\tau}_A = \min\{j \geq 1 : S_j \notin A\}$, then the Poisson kernel is defined for $x \in A, y \in \partial A$ by

$$H_A(x, y) = \mathbb{P}^x\{S_{\tau_A} = y\}.$$

For simple random walk, we would expect the Poisson kernel to be very close to that of Brownian motion. If $D \subset \mathbb{R}^d$ is a domain with sufficiently smooth boundary, we let $h_D(x, y)$ denote the Poisson kernel for Brownian motion. This means that, for each $x \in D$, $h_D(x, \cdot)$ is the density with respect to surface measure on ∂D of the distribution of the point at which the Brownian motion visits ∂D for the first time. For sets A that are rectangles with sides perpendicular to the coordinate axes (with finite or infinite length), explicit expressions can be obtained for the Poisson kernel and one can show convergence to the Brownian quantities with relatively small error terms. We give some of these formulas in this section.

8.1.1 Half space

If $d \geq 2$, we define the discrete upper half space $\mathcal{H} = \mathcal{H}_d$ by

$$\mathcal{H} = \{(x, y) \in \mathbb{Z}^{d-1} \times \mathbb{Z} : y > 0\},$$

with boundary $\partial\mathcal{H} = \mathbb{Z}^{d-1} \times \{0\}$ and "closure" $\overline{\mathcal{H}} = \mathcal{H} \cup \partial\mathcal{H}$. Let $T = \overline{\tau}_{\mathcal{H}}$, and let $H_{\mathcal{H}}$ denote the Poisson kernel, which for convenience we will write as a function $H_{\mathcal{H}} : \overline{\mathcal{H}} \times \mathbb{Z}^{d-1} \to [0, 1]$,

$$H_{\mathcal{H}}(z, x) = H_{\mathcal{H}}(z, (x, 0)) = \mathbb{P}^z\{S_T = (x, 0)\}.$$

If $z = (x, y) \in \mathbb{Z}^{d-1} \times \mathbb{Z}$, we write $\overline{z}$ for its "conjugate", $\overline{z} = (x, -y)$. If $z \in \mathcal{H}$, then $\overline{z} \in -\mathcal{H}$. If $z \in \partial\mathcal{H}$, then $\overline{z} = z$. Recall the Green's function for a set defined in Section 4.6.

Proposition 8.1.1 *For simple random walk in $\mathbb{Z}^d$, $d \geq 2$, if $z, w \in \mathcal{H}$,*

$$G_{\mathcal{H}}(z, w) = G(z - w) - G(z - \overline{w}),$$

$$H_{\mathcal{H}}(z, 0) = \frac{1}{2d} \left[G(z - \mathbf{e}_d) - G(z + \mathbf{e}_d)\right]. \tag{8.1}$$

Proof To establish the first relation, note that for $w \in \mathcal{H}$, the function $f(z) = G(z - w) - G(z - \overline{w}) = G(z - w) - G(\overline{z} - w)$ is bounded on $\mathcal{H}$, $\mathcal{L}f(z) = -\delta_w(z)$, and $f \equiv 0$ on $\partial\mathcal{H}$. Hence $f(z) = G_{\mathcal{H}}(z, w)$ by the characterization of Proposition 6.2.3. For the second relation, we use a last-exit decomposition (focusing on the last visit to $\mathbf{e}_d$ before leaving $\mathcal{H}$) to see that

$$H_{\mathcal{H}}(z, 0) = \frac{1}{2d} G_{\mathcal{H}}(z, \mathbf{e}_d).$$

□

The Poisson kernel for Brownian motion in the upper half space

$$\mathbb{H} = \mathbb{H}_d = \{(x, y) \in \mathbb{R}^{d-1} \times (0, \infty)\}$$

is given by

$$h_{\mathbb{H}}((x, y), 0) = h_{\mathbb{H}}((x + z, y), z) = \frac{2y}{\omega_d \, |(x, y)|^d},$$

where $\omega_d = 2\pi^{d/2}/\Gamma(d/2)$ is the surface area of the $(d-1)$-dimensional sphere of radius 1 in $\mathbb{R}^d$. The next theorem shows that this is also the asymptotic value for the Poisson kernel for the random walk in $\mathcal{H} = \mathcal{H}_d$, and that the error term is small.

Theorem 8.1.2 *If $d \geq 2$ and $z = (x, y) \in \mathbb{Z}^{d-1} \times \{1, 2, \ldots\}$, then*

$$H_{\mathcal{H}}(z, 0) = \frac{2y}{\omega_d |z|^d} \left[1 + O\left(\frac{|y|}{|z|^2} \right) \right] + O\left(\frac{1}{|z|^{d+1}} \right). \tag{8.2}$$

Proof We use (8.1). If we did not need to worry about the error terms, we would naively estimate

$$\frac{1}{2d} \left[G(z - \mathbf{e}_d) - G(z + \mathbf{e}_d) \right]$$

by

$$\frac{C_d}{2d} \left[|z - \mathbf{e}_d|^{2-d} - |z + \mathbf{e}_d|^{2-d} \right], \quad d \geq 3, \tag{8.3}$$

$$\frac{C_2}{4} \log \frac{|z - \mathbf{e}_d|}{|z + \mathbf{e}_d|}, \quad d = 2. \tag{8.4}$$

Using the Taylor series expansion, one can check that the quantities in (8.3) and (8.4) equal

$$\frac{2y}{\omega_d |z|^d} + O\left(\frac{|y|^2}{|z|^{d+2}} \right).$$

However, the error term in the expansion of the Green's function or potential kernel is $O(|z|^{-d})$, so we need to do more work to show that the error term in (8.2) is of the order $O(|y|^2/|z|^{d+2}) + O(|z|^{-(d+1)})$.

Assume without loss of generality that $|z| > 1$. We need to estimate

$$\begin{aligned} &G(z - \mathbf{e}_d) - G(z + \mathbf{e}_d) \\ &\quad = \sum_{n=1}^{\infty} [p_n(z - \mathbf{e}_d) - p_n(z + \mathbf{e}_d)] = \sum_{n=1}^{\infty} \left[\overline{p}_n(z - \mathbf{e}_d) - \overline{p}_n(z + \mathbf{e}_d) \right] \\ &\qquad - \sum_{n=1}^{\infty} \left[p_n(z - \mathbf{e}_d) - \overline{p}_n(z - \mathbf{e}_d) - p_n(z + \mathbf{e}_d) + \overline{p}_n(z + \mathbf{e}_d) \right]. \end{aligned}$$

Note that $z - \mathbf{e}_d$ and $z + \mathbf{e}_d$ have the same "parity", so the above series converge absolutely even if $d = 2$. We will now show that

$$\frac{1}{2d} \sum_{n=1}^{\infty} \left[\overline{p}_n(z - \mathbf{e}_d) - \overline{p}_n(z + \mathbf{e}_d) \right] = \frac{2y}{\omega_d |z|^d} + O\left(\frac{y^2}{|z|^{d+2}} \right). \tag{8.5}$$

Indeed,

$$\overline{p}_n(z - \mathbf{e}_d) - \overline{p}_n(z + \mathbf{e}_d) = \frac{d^{d/2}}{(2\pi)^{d/2}} \frac{1}{n^{d/2}} e^{-\frac{|x|^2+(y-1)^2}{2n/d}} \left(1 - e^{-\frac{4y}{2n/d}}\right).$$

For $n \geq y$, we can use a Taylor approximation for $1 - e^{-\frac{4y}{2n/d}}$. The terms with $n < y$, do not contribute much. More specifically, the left-hand side of (8.5) equals

$$\frac{2d^{d/2+1}}{(2\pi)^{d/2}} \sum_{n=1}^{\infty} \frac{y}{n^{1+d/2}} e^{-\frac{|x|^2+(y-1)^2}{2n/d}} + O\left(\sum_{n=1}^{\infty} \frac{y^2}{n^{2+d/2}} e^{-\frac{|x|^2+(y-1)^2}{2n/d}}\right) + O(y^2 e^{-|z|}).$$

Lemma 4.3.2 then gives

$$\begin{aligned}
&\sum_{n=1}^{\infty} \left[\overline{p}_n(z - \mathbf{e}_d) - \overline{p}_n(z + \mathbf{e}_d)\right] \\
&\quad = \frac{2\,d\,\Gamma(d/2)}{\pi^{d/2}} \frac{y}{(|x|^2 + (y-1)^2)^{d/2}} + O\left(\frac{y^2}{|z|^{d+2}}\right) \\
&\quad = \frac{2\,d\,\Gamma(d/2)}{\pi^{d/2}} \frac{y}{|z|^d} + O\left(\frac{y^2}{|z|^{d+2}}\right) \\
&\quad = \frac{4d}{\omega_d} \frac{y}{|z|^d} + O\left(\frac{y^2}{|z|^{d+2}}\right).
\end{aligned}$$

The remaining work is to show that

$$\sum_{n=1}^{\infty} \left[p_n(z - \mathbf{e}_d) - \overline{p}_n(z - \mathbf{e}_d) - p_n(z + \mathbf{e}_d) + \overline{p}_n(z + \mathbf{e}_d)\right] = O(|z|^{-(d+1)}).$$

We mimic the argument used for (4.11); some details are left to the reader.

Again, the sum over $n < |z|$ is negligible. Due to the second (stronger) estimate in Theorem 2.3.6, the sum over $n > |z|^2$ is bounded by

$$\sum_{n > |z|^2} \frac{c}{n^{(d+3)/2}} = O\left(\frac{1}{|z|^{d+1}}\right).$$

For $n \in [|z|, |z|^2]$, apply Theorem 2.3.8 with $k = d + 5$ (for the case of symmetric increment distribution) to give

$$p_n(w) = \overline{p}_n(w) + \sum_{j=3}^{d+5} \frac{u_j(w/\sqrt{n})}{n^{(d+j-2)/2}} + O\left(\frac{1}{n^{(d+k-1)/2}}\right),$$

where $w = z \pm \mathbf{e}_d$. As remarked after Theorem 2.3.8, we then can estimate

$$|p_n(z - \mathbf{e}_d) - \overline{p}_n(z - \mathbf{e}_d) - p_n(z - \mathbf{e}_d) + \overline{p}_n(z - \mathbf{e}_d)|$$

up to an error of $O(n^{(-d+k-1)/2})$ by

$$I_{3,d+5}(n,z) := \sum_{j=3}^{d+5} \frac{1}{n^{(d+j-2)/2}} \left| u_j\left(\frac{z+\mathbf{e}_d}{\sqrt{n}}\right) - u_j\left(\frac{z-\mathbf{e}_d}{\sqrt{n}}\right)\right|.$$

Finally, due to Taylor expansion and the uniform estimate (2.29), one can obtain a bound on the sum $\sum_{n\in[|z|,|z|^2]} I_{3,d+5}(n,z)$ by imitating the final estimate in the proof of Theorem 4.3.1. We leave this to the reader. □

In Section 8.1.3 we give an exact expression for the Poisson kernel in $\mathcal{H}_2$ in terms of an integral. To motivate it, consider a random walk in $\mathbb{Z}^2$ starting at $\mathbf{e}_2$ stopped when it first reaches $\{x\mathbf{e}_1 : x \in \mathbb{Z}\}$. Then, the distribution of the first coordinate of the stopping position gives a probability distribution on $\mathbb{Z}$. In Corollary 8.1.7, we show that the characteristic function of this distribution is

$$\phi(\theta) = 2 - \cos\theta - \sqrt{(2-\cos\theta)^2 - 1}.$$

Using this and Proposition 2.2.2, we see that the probability that the first visit is to $x\mathbf{e}_1$ is

$$\frac{1}{2\pi}\int_{-\pi}^{\pi} e^{-ix\theta}\,\phi(\theta)\,d\theta = \frac{1}{2\pi}\int_{-\pi}^{\pi} \cos(x\theta)\,\phi(\theta)\,d\theta.$$

If, instead, the walk starts from $y\mathbf{e}_2$, then the position of its first visit to the origin can be considered as the sum of y independent random variables each with characteristic function ϕ. The sum has the characteristic function ϕ^y, and hence

$$H_{\mathcal{H}}(y\mathbf{e}_2, x\mathbf{e}_1) = \frac{1}{2\pi}\int_{-\pi}^{\pi} \cos(x\theta)\,\phi(\theta)^y\,d\theta.$$

8.1.2 Cube

In this subsection we give an explicit form for the Poisson kernel on a finite cube in $\mathbb{Z}^d$. Let $\mathcal{K}_n = \mathcal{K}_{n,d}$ be the cube

$$\mathcal{K}_n = \{(x^1,\ldots,x^d) \in \mathbb{Z}^d : 1 \le x^j \le n-1\}.$$

Note that $\#(\mathcal{K}_n) = (n-1)^d$ and $\partial\mathcal{K}_n$ consists of $2d$ copies of $\mathcal{K}_{n,d-1}$. Let S_j denote simple random walk and $\tau = \tau_n = \min\{j \geq 0 : S_j \notin \mathcal{K}_n\}$. Let $H_n = H_{\mathcal{K}_n}$ denote the Poisson kernel

$$H_n(x, y) = \mathbb{P}^x\{S_{\tau_n} = y\}.$$

If $d = 1$, the gambler's ruin estimate gives

$$H_n(x, n) = \frac{x}{n}, \qquad x = 0, 1, \ldots, n,$$

so we will restrict our consideration to $d \geq 2$. By symmetry, it suffices to determine $H_n(x, y)$ for y in one of the $(d-1)$-dimensional subcubes of $\partial\mathcal{K}_n$. We will consider

$$y \in \partial_n^1 := \{(n, \tilde{y}) \in \mathbb{Z}^d : \tilde{y} \in \mathcal{K}_{n,d-1}\}.$$

The set of functions on $\overline{\mathcal{K}}_n$ that are harmonic in $\mathcal{K}_n$ and equal zero on $\partial\mathcal{K}_n \setminus \partial_n^1$ is a vector space of dimension $\#(\partial_n^1)$, and one of its bases is $\{H_n(\cdot, y) : y \in \partial_n^1\}$. In the next proposition we will use another basis which is more explicit.

♣ The proposition below uses a discrete analogue of a technique from partial differential equations called *separation of variables*. We will then compare this to the Poisson kernel for Brownian motion that can be computed using the usual separation of variables.

Proposition 8.1.3 *If $x = (x^1, \ldots, x^d) \in \mathcal{K}_{n,d}$ and $y = (y^2, \ldots, y^d) \in \mathcal{K}_{n-1,d}$, then $H_n(x, (n, y))$ equals*

$$\left(\frac{2}{n}\right)^{d-1} \sum_{z \in \mathcal{K}_{n,d-1}} \frac{\sinh(\alpha_z x^1 \pi/n)}{\sinh(\alpha_z \pi)} \sin\left(\frac{z^2 x^2 \pi}{n}\right) \cdots$$
$$\sin\left(\frac{z^d x^d \pi}{n}\right) \sin\left(\frac{z^2 y^2 \pi}{n}\right) \cdots \sin\left(\frac{z^d y^d \pi}{n}\right),$$

where $z = (z^2, \ldots, z^d)$ and $\alpha_z = \alpha_{z,n}$ is the unique nonnegative number satisfying

$$\cosh\left(\frac{\alpha_z \pi}{n}\right) + \sum_{j=2}^{d} \cos\left(\frac{z^j \pi}{n}\right) = d. \tag{8.6}$$

Proof If $z = (z^2, \ldots, z^d) \in \mathbb{R}^{d-1}$, let f_z denote the function on $\mathbb{Z}^d$,

$$f_z(x^1, \ldots, x^d) = \sinh\left(\frac{\alpha_z x^1 \pi}{n}\right) \sin\left(\frac{z^2 x^2 \pi}{n}\right) \cdots \sin\left(\frac{z^d x^d \pi}{n}\right),$$

where α_z satisfies (8.6). It is straightforward to check that for any z, f_z is a discrete harmonic function on $\mathbb{Z}^d$ with

$$f_z(x) = 0, \quad x \in \partial \mathcal{K}_n \setminus \partial_n^1.$$

We now restrict our consideration to $z \in \mathcal{K}_{n,d-1}$. Let

$$\hat{f}_z = \frac{2^{(d-1)/2}}{n^{(d-1)/2} \sinh(\alpha_z \pi)} f_z, \quad z \in \mathcal{K}_{n,d-1},$$

and let $\hat{f}_z^*$ denote the restriction of $\hat{f}_z$ to ∂_n^1, considered as a function on $\mathcal{K}_{n,d-1}$,

$$\hat{f}_z^*(x) = \hat{f}_z((n,x)) = \left(\frac{2}{n}\right)^{(d-1)/2} \sin\left(\frac{z^2 x^2 \pi}{n}\right) \cdots \sin\left(\frac{z^d x^d \pi}{n}\right),$$
$$x = (x^2, \ldots, x^d) \in \mathcal{K}_{n,d-1}.$$

For integers $1 \leq j, k \leq n-1$, one can see (via the representation of sin in terms of exponentials) that

$$\sum_{l=1}^{n-1} \sin\left(\frac{jl\pi}{n}\right) \sin\left(\frac{kl\pi}{n}\right) = \begin{cases} 0 & j \neq k \\ n/2 & j = k. \end{cases} \tag{8.7}$$

Therefore, $\{\hat{f}_z^* : z \in \mathcal{K}_{n,d-1}\}$ forms an orthonormal basis for the set of functions on $\mathcal{K}_{n,d-1}$, in symbols,

$$\sum_{x \in \mathcal{K}_{n,d-1}} \hat{f}_z^*(x) \hat{f}_{\hat{z}}^*(x) = \begin{cases} 0, & z \neq \hat{z} \\ 1, & z = \hat{z} \end{cases}.$$

Hence, any function g on $\mathcal{K}_{n,d-1}$ can be written as

$$g(x) = \sum_{z \in \mathcal{K}_{n,d-1}} C(g,z) \hat{f}_z^*(x),$$

where

$$C(g,z) = \sum_{y \in \mathcal{K}_{n,d-1}} \hat{f}_z^*(y)\, g(y).$$

In particular, if $y \in \mathcal{K}_{n,d-1}$,

$$\delta_y(x) = \sum_{z \in \mathcal{K}_{n,d-1}} \hat{f}_z^*(y) \hat{f}_z^*(x).$$

Therefore, for each $y = (y_2, \ldots, y_n)$ the function

$$x \mapsto \sum_{z \in \mathcal{K}_{n,d-1}} \frac{\sinh(\alpha_z x^1 \pi/n)}{\sinh(\alpha_z \pi)} \hat{f}_z(y) \hat{f}_z((n, x_2, \ldots, x_n)),$$

is a harmonic function in $\mathcal{K}_{n,d}$ whose value on $\partial \mathcal{K}_{n,d}$ is $\delta_{(n,y)}$ and hence it must equal $x \mapsto H_n(x, (n, y))$. □

To simplify the notation, we will consider only the case $d = 2$ (but most of what we write extends to $d \geq 3$). If $d = 2$,

$$H_{\mathcal{K}_n}((x^1, x^2), (n, y)) = \frac{2}{n} \sum_{k=1}^{n} \frac{\sinh(a_k x^1 \pi/n)}{\sinh(a_k \pi)} \sin\left(\frac{k x^2 \pi}{n}\right) \sin\left(\frac{k y \pi}{n}\right), \tag{8.8}$$

where $a_k = a_{k,n}$ is the unique positive solution to

$$\cosh\left(\frac{a_k \pi}{n}\right) + \cos\left(\frac{k\pi}{n}\right) = 2.$$

Alternatively, we can write

$$a_k = \frac{n}{\pi} r\left(\frac{k\pi}{n}\right), \tag{8.9}$$

where r is the even function

$$r(t) = \cosh^{-1}(2 - \cos t). \tag{8.10}$$

Using $\cosh^{-1}(1 + x) = \sqrt{2x} + O(x^{3/2})$ as $x \to 0+$, we get

$$r(t) = |t| + O(|t|^3), \; t \in [-1, 1].$$

Now (8.9)–(8.10) imply that

$$a_k = k + O\left(\frac{k^3}{n^2}\right). \tag{8.11}$$

Since a_k increases with k, (8.11) implies that there is an $\epsilon > 0$ such that

$$a_k \geq \epsilon\, k, \quad 1 \leq k \leq n-1. \tag{8.12}$$

We will consider the scaling limit. Let B_t denote a two-dimensional Brownian motion. Let $\mathcal{K} = (0,1)^2$ and let

$$T = \inf\{t : B_t \notin \mathcal{K}\}.$$

The corresponding Poisson kernel $h_{\mathcal{K}}$ can be computed exactly in terms of an infinite series using the continuous analogue of the procedure above, giving

$$h((x^1,x^2),(1,y)) = 2\sum_{k=0}^{\infty} \frac{\sinh(kx^1\pi)}{\sinh(k\pi)} \sin(kx^2\pi)\, \sin(ky\pi) \tag{8.13}$$

(see Exercise 8.2).

Roughly speaking, we expect that

$$H_{\mathcal{K}_n}((nx^1,nx^2),(n,ny)) \approx \frac{1}{n} h((x_1,x_2),(1,y_2)),$$

and the next proposition gives a precise formulation of this.

Proposition 8.1.4 *There exists $c < \infty$ such that if $1 \leq j^1, j^2, l \leq n-1$ are integers, $x^i = j^i/n, y = l/n$,*

$$\left| n\, H_{\mathcal{K}_n}((j^1,j^2),(n,l)) - h((x^1,x^2),(1,y))\right| \leq \frac{c}{(1-x^1)^6\, n^2} \sin(x^2\pi)\, \sin(y\pi).$$

♣ A surprising fact about this proposition is how small the error term is. For fixed $x^1 < 1$, the error is $O(n^{-2})$ where one might only expect $O(n^{-1})$.

Proof Let $\rho = x^1$. Given $k \in \mathbb{N}$, note that $|\sin(kt)| \leq k \sin t$ for $0 < t < \pi$. (To see this, $t \mapsto k \sin t \pm \sin(kt)$ is increasing on $[0, t_k]$ where $t_k \in (0, \pi/2)$ solves $\sin(t_k) = 1/k$, and $\sin(\cdot)$ continues to increase up to $\pi/2$, while $\sin(k\,\cdot)$

stays bounded by 1. For $\pi/2 < t < \pi$ consider $\pi - t$ instead; details are left to the reader.) Therefore,

$$\frac{1}{\sin(x^2\pi)\,\sin(y\pi)}\left|\sum_{k\geq n^{2/3}}^{\infty}\frac{\sinh(kx^1\pi)}{\sinh(k\pi)}\sin(kx^2\pi)\,\sin(ky\pi)\right|$$

$$\leq \sum_{k\geq n^{2/3}}^{\infty} k^2\,\frac{\sinh(k\rho\pi)}{\sinh(k\pi)}$$

$$\leq \frac{1}{n^2}\sum_{k\geq n^{2/3}}^{\infty} k^5\,\frac{\sinh(k\rho\pi)}{\sinh(k\pi)}$$

$$\leq \frac{c}{n^2}\sum_{k\geq n^{2/3}}^{\infty} k^5\,e^{-k(1-\rho)\pi}. \tag{8.14}$$

Similarly, using (8.12),

$$\frac{1}{\sin(x^2\pi)\,\sin(y\pi)}\left|\sum_{k\geq n^{2/3}}^{\infty}\frac{\sinh(a_kx^1\pi)}{\sinh(a_k\pi)}\sin(kx^2\pi)\,\sin(ky\pi)\right|$$

$$\leq \frac{1}{n^2}\sum_{k\geq n^{2/3}} k^5\,e^{-k\epsilon(1-\rho)\pi}.$$

For $0 \leq x \leq 1$ and $k < n^{2/3}$,

$$\sinh(xa_k\pi) = \sinh(xk\pi)\left[1 + O\left(\frac{k^3}{n^2}\right)\right].$$

Therefore,

$$\frac{1}{\sin(x^2\pi)\,\sin(y\pi)}\left|\sum_{k<n^{2/3}}\left[\frac{\sinh(kx^1\pi)}{\sinh(k\pi)} - \frac{\sinh(a_kx^1\pi)}{\sinh(a_k\pi)}\right]\sin(kx^2\pi)\,\sin(ky\pi)\right|$$

$$\leq \frac{c}{n^2}\sum_{k\leq n^{2/3}} k^3\,e^{-k(1-\rho)\pi} \leq \frac{c}{n^2}\sum_{k\leq n^{2/3}} k^5\,e^{-k(1-\rho)\pi},$$

where the second to last inequality is obtained as in (8.14). Combining this with (8.8) and (8.13), we see that

$$\frac{|n\,H_{\mathcal{K}_n}((j^1,j^2),(n,nl)) - h((x^1,x^2),(1,y))|}{\sin(x^2\pi)\,\sin(y\pi)} \leq \frac{c}{n^2}\sum_{k=1}^{\infty} k^5\, e^{-\epsilon(1-\rho)k}$$
$$\leq \frac{c}{(1-\rho)^6\,n^2}.$$ □

♣ The error term in the last proposition is very good except for x^1 near 1. For x^1 close to 1, one can give good estimates for the Poisson kernel by using the Poisson kernel for a half plane (if x^2 is not near 0 or 1) or by a quadrant (if x^2 is near 0 or 1). These Poisson kernels are discussed in the next subsection.

8.1.3 Strips and quadrants in $\mathbb{Z}^2$

In the continuing discussion we think of $\mathbb{Z}^2$ as $\mathbb{Z} + i\mathbb{Z}$, and we will use complex numbers notation in this section. Recall r defined in (8.10) and note that

$$e^{r(t)} = 2 - \cos t + \sqrt{(2-\cos t)^2 - 1},$$
$$e^{-r(t)} = 2 - \cos t - \sqrt{(2-\cos t)^2 - 1}.$$

For each $t \geq 0$, the function

$$f_t(x+iy) = e^{xr(t)}\,\sin(yt), \qquad \hat{f}_t(x+iy) = e^{-xr(t)}\,\sin(yt),$$

is harmonic for simple random walk, and so is the function $\sinh(xr(t)) \cdot \sin(yt)$. The next proposition is an immediate generalization of Proposition 8.1.3 to rectangles that are not squares. The proof is the same and we omit it. We then take limits as the side lengths go to infinity to get expressions for other "rectangular" subsets of $\mathbb{Z}^2$.

Proposition 8.1.5 *If m, n are positive integers, let*

$$A_{m,n} = \{x + iy \in \mathbb{Z} \times i\mathbb{Z} : 1 \leq x \leq m-1, 1 \leq y \leq n-1\}.$$

Then

$$H_{A_{m,n}}(x+iy, iy_1) = H_{A_{m,n}}((m-x)+iy, m+iy_1)$$
$$= \frac{2}{n}\sum_{j=1}^{n-1}\frac{\sinh(r(\frac{j\pi}{n})(m-x))}{\sinh(r(\frac{j\pi}{n})m)}\,\sin\left(\frac{j\pi y}{n}\right)\,\sin\left(\frac{j\pi y_1}{n}\right). \tag{8.15}$$

Corollary 8.1.6 *If n is a positive integer, let*

$$A_{\infty,n} = \{x + iy \in \mathbb{Z} \times i\mathbb{Z} : 1 \le x < \infty, 1 \le y \le n-1\}.$$

Then

$$H_{A_{\infty,n}}(x+iy, iy_1) = \frac{2}{n}\sum_{j=1}^{n-1} \exp\left\{-r\left(\frac{j\pi}{n}\right)x\right\} \sin\left(\frac{j\pi y}{n}\right) \sin\left(\frac{j\pi y_1}{n}\right), \tag{8.16}$$

and

$$H_{A_{\infty,n}}(x+iy, x_1) = \frac{2}{\pi}\int_0^{\pi} \frac{\sinh(r(t)(n-y))}{\sinh(r(t)n)} \sin(tx)\, \sin(tx_1)\, dt. \tag{8.17}$$

Proof Note that

$$H_{A_{\infty,n}}(x+iy, iy_1) = \lim_{m\to\infty} H_{A_{m,n}}(x+iy, iy_1),$$

and

$$\lim_{m\to\infty} \frac{\sinh(r(\frac{j\pi}{n})(m-x))}{\sinh(r(\frac{j\pi}{n})m)} = \exp\left\{-r\left(\frac{j\pi}{n}\right)x\right\}.$$

This combined with (8.15) gives the first identity. For the second we write

$$\begin{aligned} H_{A_{\infty,n}}(x+iy, x_1) &= \lim_{m\to\infty} H_{A_{m,n}}(x+iy, x_1) \\ &= \lim_{m\to\infty} H_{A_{n,m}}(y+ix, ix_1) \\ &= \lim_{m\to\infty} \frac{2}{m}\sum_{j=1}^{m-1} \frac{\sinh(r(\frac{j\pi}{m})(n-y))}{\sinh(r(\frac{j\pi}{m})n)} \sin\left(\frac{j\pi x}{m}\right) \sin\left(\frac{j\pi x_1}{m}\right) \\ &= \frac{2}{\pi}\int_0^{\pi} \frac{\sinh(r(t)(n-y))}{\sinh(r(t)n)} \sin(tx)\, \sin(tx_1)\, dt. \end{aligned}$$

□

♣ We derived (8.16) as a limit of (8.15). We could also have derived it directly by considering the collection of harmonic functions

$$\exp\left\{-r\left(\frac{j\pi}{n}\right)x\right\} \sin\left(\frac{j\pi y}{n}\right), \qquad j = 1, \ldots, n-1.$$

Corollary 8.1.7 *Let*

$$A_+ = \{x + iy \in \mathbb{Z} \times i\mathbb{Z} : x > 0\}.$$

Then

$$H_{A_+}(x + iy, 0) = \frac{1}{2\pi} \int_{-\pi}^{\pi} e^{-xr(t)} \cos(yt)\, dt$$

Remark If $\mathcal{H}$ denotes the discrete upper half plane, then this corollary implies that

$$H_{\mathcal{H}}(iy, x) = H_{\mathcal{H}}(-x + iy, 0) = \frac{1}{2\pi} \int_{-\pi}^{\pi} e^{-yr(t)} \cos(xt)\, dt.$$

Proof Note that

$$\begin{aligned} H_{A_+}(x + iy, 0) &= \lim_{n\to\infty} H_{A_{\infty,2n}}(x + i(n+y), in) \\ &= \lim_{n\to\infty} \frac{1}{n} \sum_{j=1}^{2n-1} \exp\left\{-r\left(\frac{j\pi}{2n}\right)x\right\} \sin\left(\frac{j\pi}{2}\right) \sin\left(\frac{j\pi(n+y)}{2n}\right). \end{aligned}$$

Note that $\sin(j\pi/2) = 0$ if j is even. For odd j, we have $\sin^2(j\pi/2) = 1$ and $\cos(j\pi/2) = 0$, hence

$$\sin\left(\frac{j\pi}{2}\right) \sin\left(\frac{j\pi(n+y)}{2n}\right) = \cos\left(\frac{j\pi y}{2n}\right).$$

Therefore,

$$\begin{aligned} H_{A_+}(x + iy, 0) &= \lim_{n\to\infty} \frac{1}{n} \sum_{j=1}^{n} \exp\left\{-r\left(\frac{(2j-1)\pi}{2n}\right)x\right\} \cos\left(\frac{(2j-1)\pi y}{2n}\right) \\ &= \frac{1}{\pi} \int_{0}^{\pi} e^{-xr(t)} \cos(yt)\, dt. \end{aligned}$$

□

Remark As already mentioned, using the above expression for $(H_{A_+}(i, x), x \in \mathbb{Z})$, one can read off the characteristic function of the stopping position of simple random walk started from $\mathbf{e}_2$ and stopped at its first visit to the $\mathbb{Z} \times \{0\}$ (see also Exercise 8.1).

Corollary 8.1.8 *Let*

$$A_{\infty,\infty} = \{x + iy \in \mathbb{Z} + i\mathbb{Z} : x, y > 0\}.$$

Then,

$$H_{A_{\infty,\infty}}(x + iy, x_1) = \frac{2}{\pi} \int_0^\pi e^{-r(t)y} \sin(tx) \sin(tx_1)\, dt.$$

Proof Using (8.17),

$$\begin{aligned} H_{A_{\infty,\infty}}(x + iy, x_1) &= \lim_{n\to\infty} H_{A_{\infty,n}}(x + iy, x_1) \\ &= \lim_{n\to\infty} \frac{2}{\pi} \int_0^\pi \frac{\sinh(r(t)(n-y))}{\sinh(r(t)n)} \sin(tx) \sin(tx_1)\, dt \\ &= \frac{2}{\pi} \int_0^\pi e^{-r(t)y} \sin(tx) \sin(tx_1)\, dt. \end{aligned}$$

□

8.2 Eigenvalues for rectangles

In general it is hard to compute the eigenfunctions and eigenvectors for a finite subset A of $\mathbb{Z}^d$ with respect to simple random walk. One feasible case is that of a rectangle

$$A = \mathcal{R}(N_1, \ldots, N_d) := \{(x^1, \ldots, x^d) \in \mathbb{Z}^d : 0 < x^j < N_j\}.$$

If $\mathbf{k} = (k_1, \ldots, k_d) \in \mathbb{Z}^d$ with $1 \le k_j < N_j$, let

$$f_{\mathbf{k}}(x^1, \ldots, x^d) = f_{\mathbf{k},N_1,\ldots,N_d}(x^1, \ldots, x^d) = \prod_{j=1}^d \sin\left(\frac{x^j k_j \pi}{N_j}\right).$$

Note that $f_{\mathbf{k}} \equiv 0$ on $\partial\mathcal{R}(N_1, \ldots, N_d)$. A straightforward computation shows that

$$\mathcal{L} f_{\mathbf{k}}(x^1, \ldots, x^d) = \alpha(\mathbf{k}) f_{\mathbf{k}}.$$

where

$$\alpha(\mathbf{k}) = \alpha(\mathbf{k}; N_1, \ldots, N_d) = \frac{1}{d} \sum_{j=1}^d \left[\cos\left(\frac{k_j \pi}{N_j}\right) - 1\right].$$

Using (8.7) we can see that the functions

$$\{f_{\mathbf{k}} : 1 \leq k_j \leq N_j - 1\},$$

form an orthogonal basis for the set of functions on $\overline{\mathcal{R}(N_1, \ldots, N_d)}$ that vanish on $\partial\mathcal{R}(N_1, \ldots, N_d)$. Hence, this gives a complete set of eigenvalues and eigenvectors. We conclude that the (first) eigenvalue α of $\mathcal{R}(N_1, \ldots, N_d)$, defined in Section 6.9, is given by

$$\alpha = \frac{1}{d} \sum_{j=1}^{d} \cos\left(\frac{\pi}{N_j}\right).$$

In particular, as $n \to \infty$, the eigenvalue for $\mathcal{K}_n = \mathcal{K}_{n,d} = \mathcal{R}(n, \ldots, n)$, is given by

$$\alpha_{\mathcal{K}_n} = \cos\left(\frac{\pi}{n}\right) = 1 - \frac{\pi^2}{2n^2} + O\left(\frac{1}{n^4}\right).$$

8.3 Approximating continuous harmonic functions

It is natural to expect that discrete harmonic functions in $\mathbb{Z}^d$, when appropriately scaled, converge to (continuous) harmonic functions in $\mathbb{R}^d$. In this section we discuss some versions of this principle. We let $\mathcal{U} = \mathcal{U}_d = \{x \in \mathbb{R}^d : |x| < 1\}$ denote the unit ball in $\mathbb{R}^d$.

Proposition 8.3.1 *There exists $c < \infty$ such that the following is true for all positive integers n, m. Suppose that $f : (n+m)\mathcal{U} \to \mathbb{R}$ is a harmonic function. Then there is a function $\hat{f}$ on $\overline{\mathcal{B}}_n$ with $\mathcal{L}\hat{f}(x) = 0, x \in \mathcal{B}_n$ and such that*

$$|f(x) - \hat{f}(x)| \leq \frac{c\,\|f\|_\infty}{m^2}, \quad x \in \mathcal{B}_n. \tag{8.18}$$

In fact, one can choose (recall ξ_n from Section 6.3)

$$\hat{f}(x) = \mathbb{E}^x[f(S_{\xi_n})].$$

Proof Without loss of generality, assume that $\|f\|_\infty = 1$. Since f is defined on $(n+1)\mathcal{U}_d$, $\hat{f}$ is well defined. By definition we know that $\mathcal{L}\hat{f}(x) = 0, x \in \mathcal{B}_n$. We need to prove (8.18). By (6.9), if $x \in \mathcal{B}_n$,

$$f(x) = \mathbb{E}^x\left[f(S_{\xi_n}) - \sum_{j=0}^{\xi_n - 1} \mathcal{L}f(S_j)\right] = \hat{f}(x) - \phi(x),$$

where

$$\phi(x) = \sum_{z \in \mathcal{B}_n} G_{\mathcal{B}_n}(x,z)\, \mathcal{L}f(z).$$

In Section 6.2, we observed that there is a c such that all fourth-order derivatives of f at x are bounded above by $c\,(n+m-|x|)^{-4}$. By expanding in a Taylor series, using the fact that f is harmonic, and also using the symmetry of the random walk, this implies that

$$|\mathcal{L}f(x)| \le \frac{c}{(n+m-|x|)^4}.$$

Therefore, we have

$$|\phi(x)| \le c \sum_{k=0}^{n-1} \sum_{k \le |z| < k+1} \frac{G_{\mathcal{B}_n}(x,z)}{(n+m-k)^4}. \tag{8.19}$$

We claim that there is a c_1 such that for all x,

$$\sum_{n-l \le |z| \le n-1} G_{\mathcal{B}_n}(x,z) \le c\, l^2.$$

Indeed, the proof of this is essentially the same as the proof of (5.5). Once we have the last estimate, summing by parts the right-hand side of (8.19) gives that $|\phi(x)| \le c/m^2$. □

The next proposition can be considered as a converse of the last proposition. If $f : \mathbb{Z}^d \to \mathbb{R}$ is a function, we will also write f for the piecewise constant function on $\mathbb{R}^d$ defined as follows. For each $x = (x^1, \ldots, x^j) \in \mathbb{Z}^d$, let $\square_x$ denote the cube of side length 1 centered at x,

$$\square_x = \left\{ (y^1, \ldots, y^d) \in \mathbb{R}^d : -\frac{1}{2} \le y^j - x^j < \frac{1}{2} \right\}.$$

The sets $\{\square_x : x \in \mathbb{Z}^d\}$ partition $\mathbb{R}^d$.

Proposition 8.3.2 *Suppose that f_n is a sequence of functions on $\mathbb{Z}^d$ satisfying $\mathcal{L}f_n(x) = 0, x \in \mathcal{B}_n$ and $\sup_x |f_n(x)| \le 1$. Let $g_n : \mathbb{R}^d \to \mathbb{R}$ be defined by $g_n(y) = f_n(ny)$. Then there exists a subsequence n_j and a function g that is harmonic on $\mathcal{U}$ such that $g_{n_j} \longrightarrow g$ uniformly on every compact $K \subset \mathcal{U}$.*

Proof Let J be a countable dense subset of $\mathcal{U}$. For each $y \in J$, the sequence $g_n(y)$ is bounded and hence has a subsequential limit. By a standard diagonalization procedure, we can find a function g on J such that

$$g_{n_j}(y) \longrightarrow g(y), \qquad y \in J.$$

For notational convenience, for the rest of this proof we will assume that in fact $g_n(y) \longrightarrow g(y)$, but the proof works equally well if there is only a subsequence.

Given that $r < 1$, let $r\mathcal{U} = \{y \in \mathcal{U} : |y| < r\}$. Using Theorem 6.3.8, we can see that there is a $c_r < \infty$ such that for all n, $|g_n(y_1) - g_n(y_2)| \leq c_r\,[|y_1 - y_2| + n^{-1}]$ for $y_1, y_2 \in r\mathcal{U}$. In particular, $|g(y_1) - g(y_2)| \leq c_r\,|y_1 - y_2|$ for $y_1, y_2 \in J \cap r\mathcal{U}$. Hence, we can extend g continuously to $r\mathcal{U}$ such that

$$|g(y_1) - g(y_2)| \leq c_r\,|y_1 - y_2|, \qquad y_1, y_2 \in r\mathcal{U}, \tag{8.20}$$

and a standard 3ϵ-argument shows that g_n converges to g uniformly on $r\mathcal{U}$.

Since g is continuous, in order to show that g is harmonic, it suffices to show that it has the spherical mean value property, i.e. if $y \in \mathcal{U}$ and $|y| + \epsilon < 1$,

$$\int_{|z-y|=\epsilon} g(z)\, ds_\epsilon(z) = g(y).$$

Here, s_ϵ denotes surface measure normalized to have measure one. This can be established from the discrete mean value property for the functions f_n, using Proposition 7.7.1 and (8.20). We omit the details. □

8.4 Estimates for the ball

One is often interested in comparing quantities for the simple random walk on the discrete ball $\mathcal{B}_n$ with corresponding quantities for Brownian motion. Since the Brownian motion is rotationally invariant, balls are very natural domains to consider. However, the lattice effects at the boundary mean that it is harder to control the rate of convergence of the simple random walk. This section presents some basic comparison estimates.

We first consider the Green's function $G_{\mathcal{B}_n}(x, y)$. If $x = 0$, Proposition 6.3.5 gives sharp estimates. It is trickier to estimate this for other x, y. We will let g denote the Green's function for Brownian motion in $\mathbb{R}^d$ with covariance matrix $d^{-1}I$,

$$g(x, y) = C_d\,|x - y|^{2-d}, \quad d \geq 3,$$

$$g(x, y) = -C_d\,\log|x - y|, \quad d = 2,$$

and define $g_n(x,y)$ by

$$g_n(x,y) = g(x,y) - \mathbb{E}^x[g(B_{T_n},y)].$$

Here, B is a d-dimensional Brownian motion with covariance $d^{-1}I$ and $T_n = \inf\{t : |B_t| = n\}$. The reader should compare the formula for $g_n(x,y)$ to corresponding formulas for $G_{\mathcal{B}_n}(x,y)$ in Proposition 4.6.2. The Green's function for standard Brownian motion is $g_n(x,y)/d$.

Proposition 8.4.1 *If $d \geq 2$ and $x, y \in \mathcal{B}_n$,*

$$G_{\mathcal{B}_n}(x,y) = g_n(x,y) + O\left(\frac{1}{|x-y|^d}\right) + O\left(\frac{\log^2 n}{(n-|y|)^{d-1}}\right).$$

♣ This estimate is not optimal, but improvements will not be studied in this book. Note that it follows that for every $\epsilon > 0$, if $|x|, |y| \leq (1-\epsilon)n$,

$$G_{\mathcal{B}_n}(x,y) = g_n(x,y)\left[1 + O\left(\frac{1}{|x-y|^2}\right) + O_\epsilon\left(\frac{\log^2 n}{n}\right)\right],$$

where we write O_ϵ to indicate that the implicit constants depend on ϵ but are uniform in x, y, n. In particular, we have uniform convergence on compact subsets of the open unit ball.

Proof We will do the $d \geq 3$ case; the $d = 2$ case is done similarly. By Proposition 4.6.2,

$$G_{\mathcal{B}_n}(x,y) = G(x,y) - \mathbb{E}^x\left[G(S_{\xi_n},y)\right].$$

Therefore,

$$|g_n(x,y) - G_{\mathcal{B}_n}(x,y)| \leq |g(x,y) - G(x,y)| + \left|\mathbb{E}^x[g(B_{T_n},y)] - \mathbb{E}^x[G(S_{\xi_n},y)]\right|.$$

By Theorem 4.3.1,

$$|g(x,y) - G(x,y)| \leq \frac{c}{|x-y|^d}$$

$$G(S_{\xi_n},y) = \frac{C_d}{|S_{\xi_n} - y|^{d-2}} + O\left(\frac{1}{(1+|S_{\xi_n}-y|)^d}\right),$$

Note that

$$\begin{aligned}\mathbb{E}^x[(1+|S_{\xi_n}-y|)^{-d}] &\leq |n+1-|y||^{-2}\,\mathbb{E}^x[G(S_{\xi_n},y)]\\ &\leq c\,|n+1-|y||^{-2}\,G(x,y)\\ &\leq c\,|n+1-|y||^{-2}\,|x-y|^{2-d}\\ &\leq c\left[|x-y|^{-d}+(n+1-|y|)^{1-d}\right].\end{aligned}$$

We can define a Brownian motion B and a simple random walk S on the same probability space such that for each r,

$$\mathbb{P}\left\{|B_{T_n}-S_{\xi_n}|\geq r\,\log n\right\}\leq \frac{c}{r};$$

see Proposition 7.7.1. Since $|B_{T_n}-S_{\xi_n}|\leq c\,n$, we see that

$$\mathbb{E}\left[|B_{T_n}-S_{\xi_n}|\right]\leq\sum_{k=1}^{cn}\mathbb{P}(|B_{T_n}-S_{\xi_n}|\geq k)\leq c\,\log^2 n.$$

Also,

$$\left|\frac{C_d}{|x-y|^{d-2}}-\frac{C_d}{|z-y|^{d-2}}\right|\leq\frac{c\,|x-z|}{[n-|y|]^{d-1}}.$$

□

Let $\alpha_{\mathcal{B}_n}$ denote the eigenvalue for the ball as in Section 6.9 and define λ_n by $\alpha_{\mathcal{B}_n}=e^{-\lambda_n}$. Let $\lambda=\lambda(d)$ be the eigenvalue of the unit ball for a standard d-dimensional Brownian motion B_t, i.e.

$$\mathbb{P}\{|B_s|<1,\ 0\leq s\leq t\}\sim c\,e^{-\lambda t},\qquad t\to\infty.$$

Since the random walk suitably normalized converges to Brownian motion, one would conjecture that $dn^2\lambda_n$ is approximately λ for large n. The next proposition establishes this but again not with the optimal error bound.

Proposition 8.4.2

$$\lambda_n=\frac{\lambda}{dn^2}\left[1+O\left(\frac{1}{\log n}\right)\right].$$

Proof By Theorem 7.1.1, we can find a $b>0$ such that a simple random walk S_n and a standard Brownian motion B_t can be defined on the same probability space so that

$$\mathbb{P}\left\{\max_{0\le j\le n^3} |S_j - B_{j/d}| \ge b\,\log n\right\} \le b\,n^{-1}.$$

By Corollary 6.9.6, there is a c_1 such that for all n and all j,

$$\mathbb{P}\{|S_j| < n,\ \ j \le kn^2\} \ge c_1\,e^{-\lambda_n k\,n^2}. \tag{8.21}$$

For Brownian motion, we know that there is a c_2 such that

$$\mathbb{P}\{|B_t| < 1, 0 \le t \le k\} \le c_2\,e^{-\lambda k}.$$

By the coupling, we know that for all n sufficiently large

$$\begin{aligned}\mathbb{P}\{|S_j| < n,\ \ j \le d\,\lambda^{-1} n^2\,\log n\} \le \mathbb{P}\{&|B_t| < n + b\log n,\\ &t \le \lambda^{-1}\,n^2\,\log n\} + b\,n^{-1},\end{aligned}$$

and due to Brownian scaling, we obtain

$$\begin{aligned}c_1\,\exp\left\{-\frac{dn^2\lambda_n}{\lambda}\,\log n\right\} &\le c_2\,\exp\left\{-\frac{n^2\,\log n}{(n+b\log n)^2}\right\} + b\,n^{-1}\\ &\le c_3\,\exp\left\{-\log n + O\left(\frac{\log^2 n}{n}\right)\right\}.\end{aligned}$$

Taking logarithms, we get

$$\frac{dn^2\lambda_n}{\lambda} \ge 1 - O\left(\frac{1}{\log n}\right).$$

A similar argument, reversing the roles of the Brownian motion and the random walk, gives

$$\frac{dn^2\lambda_n}{\lambda} \le 1 + O\left(\frac{1}{\log n}\right).$$

□

Exercises

Exercise 8.1 Suppose that S_n is simple random walk in $\mathbb{Z}^2$ started at the origin and

$$T = \min\left\{j \ge 1 : S_j \in \{x\mathbf{e}_1 : x \in \mathbb{Z}\}\right\}.$$

Let X denote the first component of S_T. Show that the characteristic function of X is

$$\phi(t) = 1 - \sqrt{(2 - \cos t)^2 - 1}.$$

Exercise 8.2 Let $V = \{(x, y) \in \mathbb{R}^2 : 0 < x, y < 1\}$ and let $\partial_1 V = \{(1, y) : 0 \leq y \leq 1\}$. Suppose that $g : \partial V \to \mathbb{R}$ is a continuous function that vanishes on $\partial V \setminus \partial_1 V$. Show that the unique continuous function on $\overline{V}$ that is harmonic in V and agrees with g on ∂V is

$$f(x, y) = 2 \sum_{k=1}^{\infty} c_k \frac{\sinh(kx\pi)}{\sinh(k\pi)} \sin(ky\pi),$$

where

$$c_k = \int_0^1 \sin(tk\pi)\, g(1, t)\, dt.$$

Use this to derive (8.13).

Exercise 8.3 Let $A_{\infty,\infty}$ be as in Corollary 8.1.8. Suppose that x_n, y_n, k_n are sequences of positive integers with

$$\lim_{n \to \infty} \frac{x_n}{n} = x, \qquad \lim_{n \to \infty} \frac{y_n}{n} = y, \qquad \lim_{n \to \infty} \frac{k_n}{n} = k,$$

where x, y, k are positive real numbers. Find

$$\lim_{n \to \infty} n\, H_{A_{\infty,\infty}}(x_n + iy_n, k_n).$$

Exercise 8.4 Let f_n be the eigenfunction associated to the d-dimensional simple random walk in $\mathbb{Z}^d$ on $\mathcal{B}_n$, i.e.

$$\mathcal{L} f_n(x) = (1 - e^{-\lambda_n}) f_n(x), \quad x \in \mathcal{B}_n,$$

with $f_n \equiv 0$ on $\mathbb{Z}^d \setminus \mathcal{B}_n$ or equivalently,

$$f_n(x) = (1 - e^{-\lambda_n}) \sum_{y \in \mathcal{B}_n} G_{\mathcal{B}_n}(x, y) f(y).$$

This defines the function up to a multiplicative constant; fix the constant by asserting that $f_n(0) = 1$. Extend f_n to be a function of $\mathbb{R}^d$ as in Section 8.3 and let

$$F_n(x) = f_n(nx).$$

The goal of this problem is to show that the limit

$$F(x) = \lim_{n\to\infty} F_n(x),$$

exists and satisfies

$$F(x) = \lambda \int_{|y|\le 1} g(x,y)\, F(y)\, d^d y, \tag{8.22}$$

where g is the Green's function for Brownian motion with a constant chosen as in Section 8.4. In other words, F is the eigenfunction for Brownian motion. (The eigen*function* is the same, whether we choose covariance matrix I or $d^{-1}I$.) Useful tools for this exercise are Proposition 6.9.4, Exercise 6.6, Proposition 8.4.1, and Proposition 8.4.2. In particular,

(i) Show that there exist c_1, c_2 such that

$$c_1[1 - |x|] \le F_n(x) \le c_2[1 - |x| + n^{-1}].$$

(ii) Use a diagonalization argument to find a subsequence n_j such that for all x with rational coordinates, the limit

$$F(x) = \lim_{j\to\infty} F_{n_j}(x)$$

exists.

(iii) Show that for every $r < 1$, there is a c_r such that for $|x|, |y| \le r$,

$$|F_n(x) - F_n(y)| \le c_r\, [|x - y] + n^{-1}].$$

(iv) Show that F is uniformly continuous on the set of points in the unit ball with rational coordinates and hence can be defined uniquely on $\{|z| \le 1\}$ by continuity.

(v) Show that if $|x|, |y| < c_r$, then

$$|F(x) - F(y)| \le c_r\, |x - y|.$$

(vi) Show that F satisfies (8.22).

(vii) You may take it as a given that there is a unique solution to (8.22) with $F(0) = 1$ and $F(x) = 0$ for $|x| = 1$. Use this to show that F_n converges to F uniformly.

9
Loop measures

9.1 Introduction

Problems in random walks are closely related to problems on loop measures, spanning trees, and determinants of Laplacians. In this chapter we will give some of the relations. Our basic viewpoint will be different from that normally taken in probability. Instead of concentrating on probability measures, we consider arbitrary (positive) measures on paths and loops.

Considering measures on paths or loops that are not probability measures is standard in statistical physics. Typically, one considers weights on paths of the form $e^{-\beta E}$ where β is a parameter and E is the "energy" of a configuration. If the total mass is finite (say, if there are only a finite number of configurations) such weights can be made into probability measures by normalizing. There are times when it is more useful to think of the probability measures and other times when the unnormalized measure is important. In this chapter we take the configurational view.

9.2 Definitions and notations

Throughout this chapter we will assume that

$$\mathcal{X} = \{x_0, x_1, \ldots, x_{n-1}\}, \qquad \text{or} \qquad \mathcal{X} = \{x_0, x_1, \ldots\}$$

is a finite or countably infinite set of *points* or *vertices* with a distinguished vertex x_0 called the *root*.

- A finite sequence of points $\omega = [\omega_0, \omega_1, \ldots, \omega_k]$ in $\mathcal{X}$ is called a *path* of *length* k. We write $|\omega|$ for the length of ω.

- A path is called a *cycle* if $\omega_0 = \omega_k$. If $\omega_0 = x$, we call the cycle an *x-cycle* and call x the *root* of the cycle.

We allow the trivial cycles of length zero consisting of a single point.

- If $x \in \mathcal{X}$, we write $x \in \omega$ if $x = \omega_j$ for some $j = 0, \ldots, |\omega|$.
- If $A \subset \mathcal{X}$, we write $\omega \subset A$ if all the vertices of ω are in A.

- A *weight* q is a nonnegative function $q : \mathcal{X} \times \mathcal{X} \to [0, \infty)$ that induces a weight on paths

$$q(\omega) = \prod_{j=1}^{|\omega|} q(\omega_{j-1}, \omega_j).$$

  By convention $q(\omega) = 1$ if $|\omega| = 0$.
- q is *symmetric* if $q(x, y) = q(y, x)$ for all x, y

♣ Although we are doing this in generality, one good example to have in mind is $\mathcal{X} = \mathbb{Z}^d$ or $\mathcal{X}$ equal to a finite subset of $\mathbb{Z}^d$ containing the origin with $x_0 = 0$. The weight q is that obtained from simple random walk, i.e. $q(x, y) = 1/2d$ if $|x - y| = 1$ and $q \equiv 0$ otherwise.

- We say that $\mathcal{X}$ is *q-connected* if for every $x, y \in \mathcal{X}$ there exists a path

$$\omega = [\omega_0, \omega_1, \ldots, \omega_k]$$

  with $\omega_0 = x, \omega_k = y$ and $q(\omega) > 0$.
- q is called a *(Markov) transition probability* if for each x

$$\sum_y q(x, y) = 1.$$

  In this case $q(\omega)$ denotes the probability that the chain starting at ω_0 enters states $\omega_1, \ldots, \omega_k$ in that order. If $\mathcal{X}$ is q-connected, q is called *irreducible*.
- q is called a *subMarkov transition probability* if for each x

$$\sum_y q(x, y) \leq 1,$$

  and it is called *strictly subMarkov* if the sum is strictly less than one for at least one x. Again, q is called *irreducible* if $\mathcal{X}$ is q-connected. A subMarkov transition probability q on $\mathcal{X}$ can be made into a transition probability on $\mathcal{X} \cup \{\Delta\}$ by setting $q(\Delta, \Delta) = 1$ and

$$q(x, \Delta) = 1 - \sum_y q(x, y).$$

The first time that this Markov chain reaches Δ is called the *killing time* for the subMarkov chain.

♣ If q is the weight corresponding to simple random walk in $\mathbb{Z}^d$, then q is a transition probability if $\mathcal{X} = \mathbb{Z}^d$ and q is a strictly subMarkov transition probability if $\mathcal{X}$ is a proper subset of $\mathbb{Z}^d$.

- If q is a transition probability on $\mathcal{X}$, two important ways to get subMarkov transition probabilities are:
  - Take $A \subset \mathcal{X}$ and consider $q(x, y)$ restricted to A. This corresponds to the Markov chain *killed when it leaves A*.
  - Let $0 < \lambda < 1$ and consider λq. This corresponds to the Markov chain *killed at geometric rate* $(1 - \lambda)$.
- The *rooted loop measure* $m = m_q$ is the measure on cycles defined by $m(\omega) = 0$ if $|\omega| = 0$ and

$$m(\omega) = m_q(\omega) = \frac{q(\omega)}{|\omega|}, \qquad |\omega| \geq 1.$$

- An *unrooted loop or cycle* is an equivalence class of cycles under the equivalence

$$[\omega_0, \omega_1, \ldots, \omega_k] \sim [\omega_j, \omega_{j+1}, \ldots, \omega_k, \omega_1, \ldots, \omega_j]. \tag{9.1}$$

We denote unrooted loops by $\overline{\omega}$ and write $\omega \sim \overline{\omega}$ if ω is a cycle that produces the unrooted loop $\overline{\omega}$.

The lengths and weights of all representatives of $\overline{\omega}$ are the same, so it makes sense to write $|\overline{\omega}|$ and $q(\overline{\omega})$.

- If $\overline{\omega}$ is an unrooted loop, let

$$K(\overline{\omega}) = \#\{\omega : \omega \sim \overline{\omega}\}$$

be the number of representatives of the equivalence class. The reader can easily check that $K(\overline{\omega})$ divides $|\overline{\omega}|$ but can be smaller. For example, if $\overline{\omega}$

is the unrooted loop corresponding to a rooted loop $\omega = [x, y, x, y, x]$ with distinct vertices x, y, then $|\overline{\omega}| = 4$ but $K(\overline{\omega}) = 2$.

- The *unrooted loop measure* is the measure $\overline{m} = \overline{m}_q$ obtained from m by "forgetting the root," i.e.

$$\overline{m}(\overline{\omega}) = \sum_{\omega \sim \overline{\omega}} \frac{q(\omega)}{|\omega|} = \frac{K(\overline{\omega})\, q(\overline{\omega})}{|\overline{\omega}|}.$$

- A weight q generates a directed graph with vertices $\mathcal{X}$ and directed edges $= \{(x, y) \in \mathcal{X} \times \mathcal{X} : q(x, y) > 0\}$. Note that this allows "self-loops" of the form (x, x). If q is symmetric, then this is an undirected graph. In this chapter, *graph* will mean undirected graph.
- If $\#(\mathcal{X}) = n < \infty$, a *spanning tree* $\mathcal{T}$ (of the complete graph) on vertices $\mathcal{X}$ is a collection of $n - 1$ edges in $\mathcal{X}$ such that $\mathcal{X}$ with these edges is a connected graph.
- Given q, the weight of a tree $\mathcal{T}$ (with respect to root x_0) is

$$q(\mathcal{T}; x_0) = \prod_{(x,x') \in \mathcal{T}} q(x, x'),$$

where the product is over all directed edges $(x, x') \in \mathcal{T}$ and the direction is chosen so that the unique self-avoiding path from x to x_0 in $\mathcal{T}$ goes through x'.
- If q is symmetric, then $q(\mathcal{T}; x_0)$ is independent of the choice of the root x_0 and we will write $q(\mathcal{T})$ for $(\mathcal{T}; x_0)$. Any tree with positive weight is a subgraph of the graph generated by q.
- If q is a weight and $\lambda > 0$, we write q_λ for the weight λq. Note that $q_\lambda(\omega) = \lambda^{|\omega|}\, q(\omega), q_\lambda(\mathcal{T}) = \lambda^{n-1}\, q(\mathcal{T})$. If q is a subMarkov transition probability and $\lambda \leq 1$, then q_λ is also a subMarkov transition probability for a chain moving as q with an additional geometric killing.
- Let $\mathcal{L}_j$ denote the set of (rooted) cycles of length j and

$$\mathcal{L}_j(A) = \{\omega \in \mathcal{L}_j : \omega \subset A\},$$

$$\mathcal{L}_j^x(A) = \{\omega \in \mathcal{L}_j(A) : x \in \omega\}, \quad \mathcal{L}_j^x = \mathcal{L}_j^x(\mathcal{X}).$$

$$\mathcal{L} = \bigcup_{j=0}^{\infty} \mathcal{L}_j, \quad \mathcal{L}(A) = \bigcup_{j=0}^{\infty} \mathcal{L}_j(A), \quad \mathcal{L}^x(A) = \bigcup_{j=0}^{\infty} \mathcal{L}_j^x(A), \quad \mathcal{L}^x = \bigcup_{j=0}^{\infty} \mathcal{L}_j^x.$$

We also write $\overline{\mathcal{L}}_j, \overline{\mathcal{L}}_j(A)$, etc., for the analogous sets of unrooted cycles.

9.2.1 Simple random walk on a graph

An important example is simple random walk on a graph. There are two different definitions that we will use. Suppose that $\mathcal{X}$ is the set of vertices of an (undirected) graph. We write $x \sim y$ if x is *adjacent* to y, i.e. if $\{x, y\}$ is an edge. Let

$$\deg(x) = \#\{y : x \sim y\}$$

be the *degree* of x. We assume that the graph is connected.

- *Simple random walk* on the graph is the Markov chain with transition probability

$$q(x, y) = \frac{1}{\deg(x)}, \qquad x \sim y.$$

  If $\mathcal{X}$ is finite, the invariant probability measure for this Markov chain is proportional to $\deg(x)$.
- Suppose that

$$d = \sup_{x \in \mathcal{X}} \deg(x) < \infty.$$

  The *lazy (simple random) walk* on the graph, is the Markov chain with symmetric transition probability

$$q(x, y) = \frac{1}{d}, \qquad x \sim y,$$

$$q(x, x) = \frac{d - \deg(x)}{d}.$$

  We can also consider this as simple random walk on the augmented graph that has added $d - \deg(x)$ directed self-loops at each vertex x. If $\mathcal{X}$ is finite, the invariant probability measure for this Markov chain is uniform.
- A graph is *regular* (or *d-regular*) if $\deg(x) = d$ for all x. For regular graphs, the lazy walk is the same as the simple random walk.
- A graph is *transitive* if "all the vertices look the same," i.e. if for each $x, y \in \mathcal{X}$ there is a graph isomorphism that takes x to y. Any transitive graph is regular.

9.3 Generating functions and loop measures

In this section, we fix a set of vertices $\mathcal{X}$ and a weight q on $\mathcal{X}$.

- If $x \in \mathcal{X}$, the *x-cycle generating function* is given by

$$g(\lambda; x) = \sum_{\omega \in \mathcal{L}, \omega_0 = x} \lambda^{|\omega|} q(\omega) = \sum_{\omega \in \mathcal{L}, \omega_0 = x} q_\lambda(\omega).$$

  If q is a subMarkov transition probability and $\lambda \leq 1$, then $g(\lambda; x)$ denotes the expected number of visits of the chain to x before being killed for a subMarkov chain with weight q_λ started at x.

- For any cycle ω we define

$$d(\omega) = \#\{j : 1 \leq j \leq |\omega|, \omega_j = \omega_0\},$$

  and we call ω an *irreducible cycle* if $d(\omega) = 1$.
- The *first return to x generating function* is defined by

$$f(\lambda; x) = \sum_{\omega \in \mathcal{L}, |\omega| \geq 1, \omega_0 = x, d(\omega) = 1} q(\omega) \, \lambda^{|\omega|}.$$

  If λq is a subMarkov transition probability, then $f(\lambda; x)$ is the probability that the chain starting at x returns to x before being killed.

  One can check as in (4.6), that

$$g(\lambda; x) = 1 + f(\lambda; x) \, g(\lambda, x),$$

which yields

$$g(\lambda; x) = \frac{1}{1 - f(\lambda; x)}. \tag{9.2}$$

- If $\mathcal{X}$ is finite, the *cycle generating function* is

$$g(\lambda) = \sum_{x \in \mathcal{X}} g(\lambda; x) = \sum_{\omega \in \mathcal{L}} \lambda^{|\omega|} q(\omega) = \sum_{\omega \in \mathcal{L}} q_\lambda(\omega).$$

  Since each $x \in \mathcal{X}$ has a unique cycle of length 0 rooted at x,

$$g(0; x) = 1, \quad g(0) = \#(\mathcal{X}).$$

- If $\mathcal{X}$ is finite, the *loop measure generating function* is

$$\Phi(\lambda) = \sum_{\omega \in \mathcal{L}} \lambda^{|\omega|}\, m_q(\omega) = \sum_{\overline{\omega} \in \overline{\mathcal{L}}} \lambda^{|\overline{\omega}|}\, \overline{m}_q(\overline{\omega}) = \sum_{\omega \in \mathcal{L},\, |\omega| \geq 1} \frac{\lambda^{|\omega|}}{|\omega|}\, q(\omega).$$

Note that if $\#(\mathcal{X}) = n < \infty$,

$$\Phi(0) = 0, \qquad g(\lambda) = \lambda\, \Phi'(\lambda) + n, \qquad \Phi(\lambda) = \int_0^\lambda \frac{g(s) - n}{s}\, ds.$$

- If $A \subset \mathcal{X}$ is finite, we write

$$F(A;\lambda) = \exp\left\{ \sum_{\omega \in \mathcal{L}(A), |\omega| \geq 1} \frac{q(\omega)\, \lambda^{|\omega|}}{|\omega|} \right\}$$
$$= \exp\left\{ \sum_{\overline{\omega} \in \overline{\mathcal{L}}(A), |\overline{\omega}| \geq 1} \frac{q(\overline{\omega})\, K(\overline{\omega})\, \lambda^{|\overline{\omega}|}}{|\overline{\omega}|} \right\}.$$

In other words, $\log F(A;\lambda)$ is the loop measure (with weight q_λ) of the set of loops in A. In particular, $F(\mathcal{X};\lambda) = e^{\Phi(\lambda)}$.

- If $x \in A$ (A not necessarily finite), let $\log F_x(A;\lambda)$ denote the loop measure (with weight q_λ) of the set of loops in A that include x, i.e.

$$F_x(A;\lambda) = \exp\left\{ \sum_{\omega \in \mathcal{L}^x(A), |\omega| \geq 1} \frac{q(\omega)\, \lambda^{|\omega|}}{|\omega|} \right\}$$
$$= \exp\left\{ \sum_{\overline{\omega} \in \overline{\mathcal{L}}^x(A), |\overline{\omega}| \geq 1} \frac{q(\overline{\omega})\, K(\overline{\omega})\, \lambda^{|\overline{\omega}|}}{|\overline{\omega}|} \right\}.$$

More generally, if $V \subset A$, $\log F_V(A;\lambda)$ denotes the loop measure of loops in A that intersect V,

$$F_V(A;\lambda) = \exp\left\{ \sum_{\omega \in \mathcal{L}(A), |\omega| \geq 1, V \cap \omega \neq \emptyset} \frac{q(\omega)\, \lambda^{|\omega|}}{|\omega|} \right\}$$
$$= \exp\left\{ \sum_{\overline{\omega} \in \overline{\mathcal{L}}(A), |\overline{\omega}| \geq 1, V \cap \overline{\omega} \neq \emptyset} \frac{q(\overline{\omega})\, K(\overline{\omega})\, \lambda^{|\overline{\omega}|}}{|\overline{\omega}|} \right\}.$$

If η is a path, we write F_η for F_V where V denotes the vertices in η. Note that $F(A;\lambda) = F_A(A;\lambda)$.

- We write $F(A) = F(A;1), F_x(A) = F_x(A;1)$.

Proposition 9.3.1 *If $A = \{y_1, \ldots, y_k\}$, then*

$$F(A;\lambda) = F_A(A;\lambda) = F_{y_1}(A;\lambda)\, F_{y_2}(A_1;\lambda) \cdots F_{y_k}(A_{k-1};\lambda), \tag{9.3}$$

where $A_i = A \setminus \{y_1, \ldots, y_i\}$. More generally, if $V = \{y_1, \ldots, y_j\} \subset A$ then

$$F_V(A;\lambda) = F_{y_1}(A;\lambda)\, F_{y_2}(A_1;\lambda) \cdots F_{y_j}(A_{j-1};\lambda). \tag{9.4}$$

In particular, the products on the right-hand side of (9.3) and (9.4) are independent of the ordering of the vertices.

Proof This follows from the definition and the observation that the collection of loops that intersect V can be partitioned into those that intersect y_1, those that do not intersect y_1 but intersect y_2, etc. □

The next lemma is an important relationship between one generating function and the exponential of another generating function.

Lemma 9.3.2 *Suppose that $x \in A \subset \mathcal{X}$. Let*

$$g_A(\lambda;x) = \sum_{\omega \in \mathcal{L}(A), \omega_0 = x} q(\omega)\, \lambda^{|\omega|}.$$

Then,

$$F_x(A;\lambda) = g_A(\lambda;x).$$

Remark If $\lambda = 1$ and q is a transition probability, then $g_A(1;x)$ (and hence by the lemma $F_x(A) = F_x(A;1)$) is the expected number of visits to x by a random walk starting at x before its first visit to $\mathcal{X} \setminus A$. In other words, $F_x(A)^{-1}$ is the probability that a random walk starting at x reaches $\mathcal{X} \setminus A$ before its first return to x. Using this interpretation for $F_x(A)$, the fact that the product on the right-hand side of (9.3) is independent of the ordering is not so obvious. See Exercise 9.2 for a more direct proof in this case.

Proof Suppose that $\overline{\omega} \in \overline{\mathcal{L}}^x(A)$. Let $d_x(\overline{\omega})$ be the number of times that a representative ω of $\overline{\omega}$ visits x (this is the same for all representatives ω). For representatives ω with $\omega_0 = x$, $d(\omega) = d_x(\overline{\omega})$. It is easy to verify that the number of representatives ω of $\overline{\omega}$ with $\omega_0 = x$ is $K(\overline{\omega})d_x(\overline{\omega})/|\overline{\omega}|$. From this, we see that

$$\overline{m}(\overline{\omega}) = \sum_{\omega\sim\overline{\omega}} \frac{q(\omega)}{|\omega|} = \frac{K(\overline{\omega})\, q(\overline{\omega})}{|\overline{\omega}|} = \sum_{\omega\sim\overline{\omega},\omega_0=x} \frac{q(\omega)}{d(\omega)}.$$

Therefore,

$$\sum_{\overline{\omega}\in\mathcal{L}^x(A)} \overline{m}(\overline{\omega})\,\lambda^{|\overline{\omega}|} = \sum_{\omega\in\mathcal{L}(A),\omega_0=x} \frac{q(\omega)\,\lambda^{|\omega|}}{d(\omega)}$$
$$= \sum_{j=1}^{\infty} \frac{1}{j} \sum_{\omega\in\mathcal{L}(A),\omega_0=x,d(\omega)=j} q(\omega)\,\lambda^{|\omega|}.$$

An x-cycle ω with $d_x(\omega) = j$ can be considered as a concatenation of j x-cycles ω' with $d(\omega') = 1$. Using this, we can see that

$$\sum_{\omega\in\mathcal{L}(A),\omega_0=x,d(\omega)=j} q(\omega)\,\lambda^{|\omega|} = \left[\sum_{\omega\in\mathcal{L}(A),\omega_0=x,d(\omega)=1} q(\omega)\,\lambda^{|\omega|}\right]^j = f_A(\lambda;x)^j.$$

Therefore,

$$\log F_x(A;\lambda) = \sum_{j=1}^{\infty} \frac{f_A(x;\lambda)^j}{j} = -\log[1 - f_A(\lambda;x)] = \log g_A(\lambda;x).$$

The last equality uses (9.2). □

Proposition 9.3.3 *Suppose that* $\#(\mathcal{X}) = n < \infty$ *and* $\lambda > 0$ *satisfies* $F(\mathcal{X};\lambda) < \infty$. *Then*

$$F(\mathcal{X};\lambda) = \frac{1}{\det[I - \lambda Q]},$$

where Q *denotes the* $n \times n$ *matrix* $[q(x,y)]_{x,y\in\mathcal{X}}$.

Proof Without loss of generality we may assume that $\lambda = 1$. We prove by induction on n. If $n = 1$ and $Q = (r)$, then $F(\mathcal{X};1) = 1/(1-r)$. To do the inductive step, suppose that $n > 1$ and $x \in \mathcal{X}$, then

$$g(1;x) = \left[\sum_{j=0}^{\infty} Q^j\right]_{x,x} = \left[(I-Q)^{-1}\right]_{x,x} = \frac{\det[I - Q_x]}{\det[I - Q]},$$

where Q_x denotes the matrix Q with the row and column corresponding to x removed. The last equality follows from the adjoint form of the inverse. Using (9.3) and the inductive hypothesis on $\mathcal{X} \setminus \{x\}$, we get the result. □

Remark The matrix $I - \lambda Q$ is often called the (negative of the) Laplacian. The last proposition and others below relate the determinant of the Laplacian to loop measures and trees.

- Let $\lambda_{0,x}$ denote the radius of convergence of $g(\lambda; x)$.
  - If $\mathcal{X}$ is q-connected, then $\lambda_{0,x}$ is independent of x and we write just λ_0.
  - If $\mathcal{X}$ is q-connected and finite, then λ_0 is also the radius of convergence for $g(\lambda)$ and $F(\mathcal{X}; \lambda)$ and $1/\lambda_0$ is the largest eigenvalue for the matrix $Q = (q(x,y))$. If q is a transition probability, then $\lambda_0 = 1$. If q is an irreducible, strictly subMarkov transition probability, then $\lambda_0 > 1$.

If $\mathcal{X}$ is q-connected and finite, then $g(\lambda_0) = g(\lambda_0; x) = F(\mathcal{X}; \lambda_0) = \infty$. However, one can show easily that $F(\mathcal{X} \setminus \{x\}; \lambda_0) < \infty$. The next proposition shows how to compute the last quantity from the generating functions.

Proposition 9.3.4 *Let λ_0 be the radius of convergence of g. Then, if $x \in \mathcal{X}$,*

$$\log F(\mathcal{X} \setminus \{x\}; \lambda_0) = \lim_{\lambda \to \lambda_0 -} [\log F(\mathcal{X}; \lambda) - \log g(\lambda; x)].$$

Proof

$$\begin{aligned}\log F(\mathcal{X} \setminus \{x\}; \lambda_0) &= \lim_{\lambda \to \lambda_0 -} \log F(\mathcal{X} \setminus \{x\}; \lambda) \\ &= \lim_{\lambda \to \lambda_0 -} [\log F(\mathcal{X}; \lambda) - \log F_x(\mathcal{X}; \lambda)].\end{aligned}$$

□

If $\#(\mathcal{X}) < \infty$ and a subset $\mathcal{E}$ of edges is given, then simple random walk on the graph $(\mathcal{X}, \mathcal{E})$ is the Markov chain corresponding to

$$q(x,y) = [\#\{z : (x,z) \in \mathcal{E}\}]^{-1}, \qquad (x,y) \in \mathcal{E}.$$

If $\#(\mathcal{X}) = n < \infty$ and $(\mathcal{X}, \mathcal{E})$ is transitive, then

$$g(\lambda; x) = n^{-1} g(\lambda),$$

and hence we can write Proposition 9.3.4 as

$$\log F(\mathcal{X} \setminus \{x\}, \lambda_0) = \log n + \lim_{\lambda \to \lambda_0 -} [\Phi(\lambda) - \log g(\lambda)].$$

Proposition 9.3.5 *Suppose that $\mathcal{X}$ is finite and q is an irreducible transition probability, reversible with respect to the invariant probability π. Let $\alpha_1 = 1, \alpha_2, \ldots, \alpha_n$ denote the eigenvalues of $Q = [q(x, y)]$. Then, for every $x \in \mathcal{X}$,*

$$\frac{1}{F(\mathcal{X} \setminus \{x\})} = \pi(x) \prod_{j=2}^{n} (1 - \alpha_j)$$

Proof Since the eigenvalues of $I - \lambda Q$ are $1 - \lambda\,\alpha_1, \ldots, 1 - \lambda\,\alpha_n$, we see that

$$\lim_{\lambda \to 1-} \frac{\det[I - \lambda Q]}{1 - \lambda} = \prod_{j=2}^{n} (1 - \alpha_j).$$

If $\lambda < 1$, then Proposition 9.3.3 states that

$$F(\mathcal{X}; \lambda) = \frac{1}{\det[I - \lambda Q]}.$$

Also, as $\lambda \to 1-$,

$$g(\lambda; x) \sim \pi(x)\,(1 - \lambda)^{-1},$$

where π denotes the invariant probability. (This can be seen by recalling that $g(\lambda; x)$ is the number of visits to x by a chain starting at x before a geometric killing time with rate $(1 - \lambda)$. The expected number of steps before killing is $1/(1-\lambda)$, and since the killing is independent of the chain, the expected number of visits to x before being killed is asymptotic to $\pi(x)/(1-\lambda)$.) Therefore, using Proposition 9.3.4,

$$\begin{aligned} \log F(\mathcal{X} \setminus \{x\}) &= \lim_{\lambda \to 1-} [\log F(\mathcal{X}; \lambda) - \log g(\lambda; x)] \\ &= -\log \pi(x) - \sum_{j=2}^{n} \log(1 - \alpha_j). \end{aligned}$$

□

9.4 Loop soup

- If V is a countable set and $\nu : V \to [0, \infty)$ is a measure, then a *(Poisson) soup* from ν is a collection of independent Poisson processes

$$N_t^x, \quad x \in V,$$

where N_t^x has parameter $\nu(x)$. A *soup realization* is the corresponding collection of multisets[†] $\mathcal{A}_t$ where the number of times that x appears in $\mathcal{A}_t$ is N_t^x. This can be considered as a stochastic process taking values in multisets of elements of V.

- The *rooted loop soup* $\mathcal{C}_t$ is a soup realization from m.
- The *unrooted loop soup* $\overline{\mathcal{C}}_t$ is a soup realization from $\overline{m}$.

From the definitions of m and $\overline{m}$ we can see that we can obtain an unrooted loop soup $\overline{\mathcal{C}}_t$ from a rooted loop soup $\mathcal{C}_t$ by "forgetting the roots" of the loops in $\mathcal{C}_t$. To obtain $\mathcal{C}_t$ from $\overline{\mathcal{C}}_t$, we need to add some randomness. More specifically, if $\overline{\omega}$ is a loop in an unrooted loop soup $\overline{\mathcal{C}}_t$, we choose a rooted loop ω by choosing uniformly among the $K(\overline{\omega})$ representatives of $\overline{\omega}$. It is not hard to show that with probability one, for each t, there is at most one loop in

$$\mathcal{C}_t \setminus \left[\bigcup_{s<t} \mathcal{C}_s\right].$$

Hence, we can order the loops in $\mathcal{C}_t$ (or $\overline{\mathcal{C}}_t$) according to the "time" at which they were created; we call this the chronological order.

Proposition 9.4.1 *Suppose that $x \in A \subset \mathcal{X}$ with $F_x(A) < \infty$. Let $\overline{\mathcal{C}}_t(A;x)$ denote an unrooted loop soup $\overline{\mathcal{C}}_t$ restricted to $\bar{\mathcal{L}}^x(A)$. Then, with probability one, $\overline{\mathcal{C}}_1(A;x)$ contains a finite number of loops which we can write in chronological order*

$$\overline{\omega}_1, \ldots, \overline{\omega}_k.$$

Suppose that independently for each unrooted loop $\overline{\omega}_j$, a rooted loop ω_j with $\omega_0 = x$ is chosen uniformly among the $K(\overline{\omega})\, d_x(\overline{\omega})/|\overline{\omega}|$ representatives of $\overline{\omega}$ rooted at x, and these loops are concatenated to form a single loop

$$\eta = \omega_1 \oplus \omega_2 \oplus \cdots \oplus \omega_k.$$

Then, for any loop $\eta' \subset A$ rooted at x,

$$\mathbb{P}\{\eta = \eta'\} = \frac{q(\eta')}{F_x(A)}.$$

[†] A multiset is a generalization of a set where elements can appear multiple times in the collection.

Proof We first note that $\omega_1, \ldots, \omega_k$ as given above is the realization of the loops in a Poissonian realization corresponding to the measure

$$m_x^*(\omega) = \frac{q(\omega)}{d_x(\omega)},$$

up to time one listed in chronological order, restricted to loops $\omega \in \mathcal{L}(A)$ with $\omega_0 = x$. Using the argument of Lemma 9.3.2, the probability that no loop appears is

$$\exp\left\{-\sum_{\omega\in\mathcal{L}(A);\omega_0=x} m_x^*(\omega)\right\} = \exp\left\{-\sum_{\omega\in\mathcal{L}(A);\omega_0=x} \frac{q(\omega)}{d_x(\omega)}\right\} = \frac{1}{F_x(A)}.$$

More generally, suppose that $\eta' \in \mathcal{L}(A)$ is given with $\eta'_0 = x$ and $d(\eta') = k$. For any choice of positive $j_1, \ldots, j_r$ integers summing to k, we have a decomposition of η',

$$\eta' = \omega_1 \oplus \cdots \oplus \omega_r$$

where ω_i is a loop rooted at x with $d_x(\omega_i) = j_i$. The probability that $\omega_1, \ldots, \omega_r$ (and no other loops) appear in the realization up to time 1 in this order is

$$\begin{aligned}\frac{\exp\left\{-\sum_{\omega\in\mathcal{L}(A);\omega_0=x} m_x^*(\omega)\right\}}{r!}\, m_x^*(\omega_1)\cdots m_x^*(\omega_r) &= \frac{1}{r!\, F_x(A)}\,\frac{q(\omega_1)\cdots q(\omega_r)}{j_1,\ldots,j_r}\\ &= \frac{q(\eta')}{F_x(A)}\,\frac{1}{r!\,(j_1,\ldots,j_r)}.\end{aligned}$$

The proposition then follows from the following combinatorial fact that we leave as Exercise 9.1:

$$\sum_{j_1+\cdots+j_r=k} \frac{1}{r!\,(j_1,\ldots,j_r)} = 1.$$

□

9.5 Loop erasure

- A path $\omega = [\omega_0, \ldots, \omega_n]$ is *self-avoiding* if $\omega_j \neq \omega_k$ for $0 \leq j < k \leq n$.

Given that a path $\omega = [\omega_0, \ldots, \omega_n]$, there are a number of ways to obtain a self-avoiding subpath of ω that goes from ω_0 to ω_n. The next definition gives one way.

- If $\omega = [\omega_0, \ldots, \omega_n]$ is a path, $LE(\omega)$ denotes its *(chronological) loop-erasure* defined as follows.
  - Let $\sigma_0 = \max\{j \leq n : \omega_j = 0\}$. Set $\eta_0 = \omega_0 = \omega_{\sigma_0}$.
  - Suppose that $\sigma_i < n$. Let $\sigma_{i+1} = \max\{j \leq n : \omega_j = \omega_{\sigma_i+1}\}$. Set $\eta_{i+1} = \omega_{\sigma_{i+1}} = \omega_{\sigma_i+1}$.
  - If $i_\omega = \min\{i : \sigma_i = n\} = \min\{i : \eta_i = \omega_n\}$, then $LE(\omega) = [\eta_0, \ldots, \eta_i]$.

A weight q on paths induces a new weight $\hat{q}_A$ on self-avoiding paths by specifying that the weight of a self-avoiding path η is the sum of the weights of all the paths ω in A for which $\eta = L(\omega)$. The next proposition describes this weight.

Proposition 9.5.1 *Suppose that $A \subset \mathcal{X}, i \geq 1$ and $\eta = [\eta_0, \ldots, \eta_i]$ is a self-avoiding path whose vertices are in A. Then,*

$$\hat{q}_A(\eta) := \sum_{\omega \in \mathcal{L}(A); LE(\omega)=\eta} q(\omega) = q(\eta)\, F_\eta(A), \tag{9.5}$$

where, as before,

$$F_\eta(A) = \exp\left\{ \sum_{\omega \in \mathcal{L}(A), |\omega| \geq 1, \eta \cap \omega \neq \emptyset} \frac{q(\omega)}{|\omega|} \right\}.$$

Proof Let $A_{-1} = A, A_j = A \setminus \{\eta_0, \ldots, \eta_j\}$. Given any ω with $LE(\omega) = \eta$, we can decompose ω as

$$\omega = \omega^0 \oplus [\eta_0, \eta_1] \oplus \omega^1 \oplus [\eta_1, \eta_2] \oplus \cdots \oplus [\eta_{i-1}, \eta_i] \oplus \omega^i,$$

where ω^j denotes the loop

$$[\omega_{\sigma_{j-1}+1}, \ldots, \omega_{\sigma_j}]$$

(here, $\sigma_{-1} = -1$). The loop ω^j can be any loop rooted at η_j contained in A_{j-1}. The total measure of such loops is $F_{\eta_j}(A_{j-1})$; see Lemma 9.3.2. The result then follows from (9.4). □

In particular, $\hat{q}_A(\eta)$ depends on A. The next proposition discusses the "Radon–Nikodym derivative" of $\hat{q}_{A_1}$ with respect to $\hat{q}_A$ for $A_1 \subset A$.

- If $V_1, V_2 \subset A$, Let

$$F_{V_1,V_2}(A) = \exp\left\{ \sum_{\omega \in \mathcal{L}(A), \omega \cap V_1 \neq \emptyset, \omega \cap V_2 \neq \emptyset} \frac{q(\omega)}{|\omega|} \right\}.$$

Proposition 9.5.2 *Suppose that $A_1 \subset A$ and $\eta = [\eta_0, \ldots, \eta_i]$ is a self-avoiding path whose vertices are in A_1. Then*

$$\hat{q}_A(\eta) = \hat{q}_{A_1}(\eta)\, F_{\eta, A \setminus A_1}(A).$$

Proof This follows immediately from the relation $F_\eta(A) = F_\eta(A_1)\, F_{\eta, A\setminus A_1}(A)$. □

The "inverse" of loop erasing is loop addition. Suppose that $\eta = [\eta_0, \ldots, \eta_k]$ is a self-avoiding path. We define a random variable Z_η taking values in the set of paths ω with $LE(\omega) = \eta$ as follows. Let $\overline{\mathcal{C}}_t$ be a realization of the unrooted loop soup in A as in Proposition 9.4.1. For each $0 \le j \le k$, let

$$\overline{\omega}_{1,j},\ \overline{\omega}_{2,j},\ \ldots,\ \overline{\omega}_{s_j,j},$$

denote the loops in $\overline{\mathcal{C}}_1$ that intersect η_j but do not intersect $\{\eta_0, \ldots, \eta_{j-1}\}$. These loops are listed in the order that they appear in the soup. For each such loop $\overline{\omega}_{i,j}$, choose a representative $\omega_{i,j}$ roooted at η_j; if there is more than one choice for the representative, choose it uniformly. We then concatenate these loops to give

$$\tilde{\omega}_j = \omega_{1,j} \oplus \cdots \oplus \omega_{s_j,j}.$$

If $s_j = 0$, define $\tilde{\omega}_j$ to be the trivial loop $[\eta_j]$. We then concatenate again to define

$$\omega = Z(\eta) = \tilde{\omega}_0 \oplus [\eta_0, \eta_1] \oplus \tilde{\omega}_1 \oplus [\eta_1, \eta_2] \oplus \cdots \oplus [\eta_{k-1}, \eta_k] \oplus \tilde{\omega}_k.$$

Proposition 9.4.1 tells us that there is another way to construct a random variable with the distribution of Z_η. Suppose that $\tilde{\omega}_0, \ldots, \tilde{\omega}_k$ are chosen independently (given η) with $\tilde{\omega}_j$ having the distribution of a cycle in $A \setminus \{\eta_0, \ldots, \eta_{j-1}\}$ rooted at η_j. In other words if $\omega' \in \mathcal{L}(A \setminus \{\eta_0, \ldots, \eta_{j-1}\})$ with $\omega'_0 = \eta_j$, then the probability that $\tilde{\omega}_j = \omega'$ is $q(\omega')/F_{\eta_j}(A \setminus \{\eta_0, \ldots, \eta_{j-1}\})$.

9.6 Boundary excursions

Boundary excursions in a set A are paths that begin and end on the boundary and otherwise stay in A. Suppose that $\mathcal{X}, q$ are given. If $A \subset \mathcal{X}$ we define

$$\partial A = (\partial A)_q = \{y \in \mathcal{X} \setminus A : q(x,y) + q(y,x) > 0 \text{ for some } x \in A\}.$$

- A *(boundary) excursion* in A is a path $\omega = [\omega_0, \ldots, \omega_n]$ with $n \geq 2$ such that $\omega_0, \omega_n \in \partial A$ and $\omega_1, \ldots, \omega_{n-1} \in A$.
- The set of boundary excursions with $\omega_0 = x$ and $\omega_{|\omega|} = y$ is denoted as $\mathcal{E}_A(x, y)$, and

$$\mathcal{E}_A = \bigcup_{x.y \in \partial A} \mathcal{E}_A(x, y).$$

- Let $\hat{\mathcal{E}}_A(x, y), \hat{\mathcal{E}}_A$ denote the subsets of $\mathcal{E}_A(x, y), \mathcal{E}_A$, respectively, consisting of the self-avoiding paths. If $x = y$, the set $\hat{\mathcal{E}}_A(x, y)$ is empty.
- The measure q restricted to $\mathcal{E}_A$ is called the *excursion measure* on A.
- The measure q restricted to $\hat{\mathcal{E}}_A$ is called the *self-avoiding excursion measure* on A.
- The *loop-erased excursion measure* on A, is the measure on $\hat{\mathcal{E}}_A$ given by

$$\hat{q}(\eta) = q\{\omega \in \mathcal{E}_A : LE(\omega) = \eta\}.$$

  As in (9.5), we can see that

$$\hat{q}(\eta) = q(\eta)\, F_\eta(A). \tag{9.6}$$

- If $x, y \in \partial A$, $q, \hat{q}$ can also be considered as measures on $\mathcal{E}_A(x, y)$ or $\hat{\mathcal{E}}_A(x, y)$ by restricting to those paths that begin at x and end at y. If $x = y$, these measures are trivial for the self-avoiding and loop-erased excursion measures.

♣ If μ is a measure on a set K and $K_1 \subset K$, then the restriction of μ to K_1 is the measure ν defined by $\nu(V) = \mu(V \cap K_1)$. If μ is a probability measure, this is related to but not the same as the conditional measure, given that K_1; the conditional measure normalizes to make the measure a probability measure. A family of measures μ_A, indexed by subsets A, supported on $\mathcal{E}_A$ (or $\mathcal{E}_A(x, y)$) is said to have the *restriction property* if, whenever $A_1 \subset A$, then μ_{A_1} is μ_A restricted to $\mathcal{E}_{A_1}$ ($\mathcal{E}_{A_1}(x, y)$). The excursion measure and the self-avoiding excursion measure have the restriction property. However, the loop-erased excursion measure does not have the restriction property. This can be seen from (9.6), since it is possible that $F_\eta(A_1) \neq F_\eta(A)$.

The loop-erased excursion measure $\hat{q}$ is obtained from the excursion measure q by a deterministic function on paths (loop erasure). Since this function is not one-to-one, we cannot obtain q from $\hat{q}$ without adding some extra randomness.

However, one can obtain q from $\hat{q}$ by adding random loops as described at the end of Section 9.5.

The next definition is a generalization of the boundary Poisson kernel defined in Section 6.7.

- The *boundary Poisson kernel* is the function $H_{\partial A} : \partial A \times \partial A \to [0, \infty)$ given by

$$H_{\partial A}(x, y) = \sum_{\omega \in \mathcal{E}_A(x,y)} q(\omega).$$

Note that if $\omega \in \mathcal{E}_A(x, y)$, then $LE(\omega) \in \hat{\mathcal{E}}_A(x, y)$. In particular, if $x \neq y$,

$$H_{\partial A}(x, y) = \sum_{\eta \in \hat{\mathcal{E}}_A(x,y)} \hat{q}(\eta).$$

Suppose that k is a positive integer and $x_1, \ldots, x_k, y_1, \ldots, y_k$ are distinct points in ∂A. We write $\mathbf{x} = (x_1, \ldots, x_k), \mathbf{y} = (y_1, \ldots, y_k)$. We let

$$\mathcal{E}_A(\mathbf{x}, \mathbf{y}) = \mathcal{E}_A(x_1, y_1) \times \cdots \times \mathcal{E}_A(x_k, y_k),$$

and we write $[\omega] = (\omega^1, \ldots, \omega^k)$ for an element of $\mathcal{E}_A(\mathbf{x}, \mathbf{y})$ and

$$q([\omega]) = (q \times \cdots \times q)([\omega]) = q(\omega^1)\, q(\omega^2) \cdots q(\omega^k).$$

We can consider $q \times \cdots \times q$ as a measure on $\mathcal{E}_A(\mathbf{x}, \mathbf{y})$. We define $\hat{\mathcal{E}}_A(\mathbf{x}, \mathbf{y})$ similarly.

- The *nonintersecting excursion measure* $q_A(\mathbf{x}, \mathbf{y})$ at $(\mathbf{x}, \mathbf{y})$ is the restriction of the measure $q \times \cdots \times q$ to the set of $[\omega] \in \mathcal{E}_A(\mathbf{x}, \mathbf{y})$ that do not intersect, i.e. $\omega^i \cap \omega^j = \emptyset, 1 \leq i < j \leq k$.
- The *nonintersecting self-avoiding excursion measure* at $(\mathbf{x}, \mathbf{y})$ is the restriction of the measure $q \times \cdots \times q$ to the set of $[\omega] \in \hat{\mathcal{E}}_A(\mathbf{x}, \mathbf{y})$ that do not intersect. Equivalently, it is the restriction of the nonintersecting excursion measure to $\hat{\mathcal{E}}_A(\mathbf{x}, \mathbf{y})$.

There are several ways to define the nonintersecting loop-erased excursion measure. It turns out that the most obvious way (restricting the loop-erased excursion measure to k-tuples of walks that do not intersect) is neither the most important nor the most natural. To motivate our definition, let us consider the nonintersecting excursion measure with $k = 2$. This is the measure on pairs of excursions (ω^1, ω^2) that give measures $q(\omega^1)\, q(\omega^2)$ to each (ω^1, ω^2) satisfying $\omega^1 \cap \omega^2 = \emptyset$. Another way of saying this is as follows.

- Given that ω^1, the measure on ω^2 is q restricted to those excursions $\omega \in \mathcal{E}_A(x_2, y_2)$ such that $\omega \cap \omega^1 = \emptyset$. In other words, the measure is q restricted to $\mathcal{E}_{A\setminus\omega_1}(x_2, y_2)$.

More generally, if $k \geq 2$ and $1 \leq j \leq k-1$, the following holds.

- Given that $\omega^1, \ldots, \omega^j$, the measure on ω^{j+1} is q restricted to excursions in $\mathcal{E}_A(x_{j+1}, y_{j+1})$ that do not intersect $\omega^1 \cup \cdots \cup \omega^j$. In other words, the measure is q restricted to $\mathcal{E}_{A\setminus(\omega^1\cup\cdots\cup\omega^j)}(x_{j+1}, y_{j+1})$.

The nonintersecting self-avoiding excursion measure satisfies the analogous property. We will use this as the basis for our definition of the *nonintersecting loop-erased measure* $\hat{q}_A(\mathbf{x}, \mathbf{y})$ at $(\mathbf{x}, \mathbf{y})$. We want our definition to satisfy the following.

- Given that $\eta^1, \ldots, \eta^j$, the measure on η^{j+1} is the same as $\hat{q}_{A\setminus(\eta^1\cup\cdots\cup\eta^j)}(x_{j+1}, y_{j+1})$.

  This leads to the following definition.

- The measure $\hat{q}_A(\mathbf{x}, \mathbf{y})$ is the measure on $\hat{\mathcal{E}}_A(\mathbf{x}, \mathbf{y})$ obtained by restricting $q_A(\mathbf{x}, \mathbf{y})$ to the set V of k-tuples $[\omega] \in \mathcal{E}_A(\mathbf{x}, \mathbf{y})$ that satisfy

$$\omega^{j+1} \cap [\eta^1 \cup \cdots \cup \eta^j] = \emptyset, \quad j = 1, \ldots, k-1, \tag{9.7}$$

  where $\eta^j = LE(\omega^j)$, and then considering it as a measure on the loop erasures. In other words,

$$\hat{q}_A(\eta^1, \ldots, \eta^k) = q\{(\omega^1, \ldots, \omega^k) \in V : LE(\omega^j) = \eta^j,$$
$$j = 1, \ldots, k, \text{ satisfying (9.7)}\}.$$

This definition may look unnatural because it seems that it might depend on the order of the pairs of vertices. However, the next proposition shows that this is not the case.

Proposition 9.6.1 *The $\hat{q}_A(\mathbf{x}, \mathbf{y})$-measure of a k-tuple $(\eta^1, \ldots, \eta^k)$ is*

$$\left[\prod_{j=1}^{k} \hat{q}_A(\eta^j)\right] 1\{\eta^i \cap \eta^j \neq \emptyset, 1 \leq i < j \leq n\} F_{\eta^1,\ldots,\eta^k}(A)^{-1},$$

where

$$F_{\eta^1,\ldots,\eta^k}(A) = \exp\left\{\sum_{\omega\in\mathcal{L}(A)} \frac{q(\omega)}{|\omega|} J(\omega;\eta^1,\ldots,\eta^k)\right\},$$

and $J(\omega;\eta^1,\ldots,\eta^k) = \max\{0, s-1\}$, *where* s *is the number of paths* $\eta^1,\ldots,\eta^k$ *intersected by* ω.

Proof Proposition 9.5.1 implies that

$$\prod_{j=1}^{k} \hat{q}_A(\eta^j) = \prod_{j=1}^{k} q(\eta^j) \prod_{j=1}^{k} \exp\left\{\sum_{\omega\in\mathcal{L}(A),|\omega|\ge 1,\omega\cap\eta_j\neq\emptyset} \frac{q(\omega)}{|\omega|}\right\}. \tag{9.8}$$

However, assuming that $\eta^i \cap \eta^j = \emptyset$ for $i \neq j$,

$$\hat{q}_A(\eta^1,\ldots,\eta^j) = \prod_{j=1}^{k} q(\eta^j) \prod_{j=1}^{k} \exp\left\{\sum_{\omega\in\mathcal{L}(A\setminus(\eta_1\cup\cdots\eta_{j-1})),|\omega|\ge 1,\omega\cap\eta_j\neq\emptyset} \frac{q(\omega)}{|\omega|}\right\}. \tag{9.9}$$

If a loop ω intersects s of the η^j, where $s \ge 2$, then it appears s times in (9.8) but only one time in (9.9). □

- Let $\hat{H}_{\partial A}(\mathbf{x},\mathbf{y})$ denote the total mass of the measure $\hat{q}_A(\mathbf{x},\mathbf{y})$.

If $k = 1$, we know that $\hat{H}_{\partial A}(x,y) = H_{\partial A}(x,y)$. The next proposition shows that for $k > 1$, we can describe $\hat{H}_{\partial A}(\mathbf{x},\mathbf{y})$ in terms of the quantities $H_{\partial A}(x_i,y_j)$. The identity is a generalization of a result of Karlin and McGregor on Markov chains (see Exercise 9.3). If π is a permutation of $\{1,\ldots,k\}$, we also write $\pi(\mathbf{y})$ for $(y_{\pi(1)},\ldots,y_{\pi(k)})$.

Proposition 9.6.2 (Fomin's identity)

$$\sum_{\pi} (-1)^{\operatorname{sgn}\pi}\, \hat{H}_{\partial A}(\mathbf{x},\pi(\mathbf{y})) = \det\left[H_{\partial A}(x_i,y_j)\right]_{1\le i,j\le k}. \tag{9.10}$$

Remark If A is a simply connected subset of $\mathbb{Z}^2$ and q comes from simple random walk, then topological considerations tell us that $\hat{H}_{\partial A}(\mathbf{x},\pi(\mathbf{y}))$ is nonzero for at most one permutation π. If we order the vertices so that this permutation is the identity, Fomin's identity becomes

$$\hat{H}_{\partial A}(\mathbf{x},\mathbf{y}) = \det\left[H_{\partial A}(x_i,y_j)\right]_{1\le i,j\le k}.$$

Proof We will say that $[\omega]$ is *nonintersecting* if (9.7) holds and otherwise we call it *intersecting*. Let

$$\mathcal{E}^* = \bigcup_{\pi} \mathcal{E}_A(\mathbf{x}, \pi(\mathbf{y})). \tag{9.11}$$

Let $\mathcal{E}^*_{NI}$ be the set of nonintersecting $[\omega] \in \mathcal{E}^*$, and let $\mathcal{E}^*_I = \mathcal{E}^* \setminus \mathcal{E}^*_{NI}$ be the set of intersecting $[\omega]$. We will define a function $\phi : \mathcal{E}^* \to \mathcal{E}^*$ with the following properties.

- ϕ is the identity on $\mathcal{E}^*_{NI}$.
- $q([\omega]) = q(\phi([\omega]))$.
- If $[\omega] \in \mathcal{E}^*_I \cap \mathcal{E}_A(\mathbf{x}, \pi(\mathbf{y}))$, then $\phi([\omega]) \in \mathcal{E}_A(\mathbf{x}, \pi_1(\mathbf{y}))$ where $\mathrm{sgn}\pi_1 = -\mathrm{sgn}\pi$. In fact, π_1 is the composition of π and a transposition.
- $\phi \circ \phi$ is the identity. In particular, ϕ is a bijection.

To show that existence of such a ϕ proves the proposition, first note that

$$\det\left[H_{\partial A}(x_i, y_j)\right]_{1 \le i,j \le k} = \sum_{\pi} (-1)^{\mathrm{sgn}\pi} \prod_{i=1}^{k} H_{\partial A}(x_i, y_{\pi(i)}).$$

Also,

$$H_{\partial A}(x_i, y_{\pi(i)}) = \sum_{\omega \in \mathcal{E}_A(x_i, y_{\pi(i)})} q(\omega).$$

Therefore, by expanding the product, we have

$$\det\left[H_{\partial A}(x_i, y_j)\right]_{1 \le i,j \le k} = \sum_{[\omega] \in \mathcal{E}^*} (-1)^{\mathrm{sgn}\pi}\, q([\omega]) = \sum_{[\omega] \in \mathcal{E}^*} (-1)^{\mathrm{sgn}\pi_1}\, q([\phi(\omega)]).$$

In the first summation the permutation π is as in (9.11). Hence the sum of all the terms that come from $\omega \in \mathcal{E}^*_I$ is zero, and

$$\det\left[H_{\partial A}(x_i, y_j)\right]_{1 \le i,j \le k} = \sum_{[\omega] \in \mathcal{E}^*_{NI}} (-1)^{\mathrm{sgn}\pi}\, q([\omega]).$$

But the right-hand side is the same as the left-hand side of (9.10). We define ϕ to be the identity on $\mathcal{E}^*_{NI}$, and we now proceed to define the bijection ϕ on $\mathcal{E}^*_I$.

Let us first consider the $k = 2$ case. Let $[\omega] \in \mathcal{E}^*_I$,

$$\xi = \omega^1 = [\xi_0, \ldots, \xi_m] \in \mathcal{E}_A(x_1, y), \qquad \omega = \omega^2 = [\omega_0, \ldots, \omega_n] \in \mathcal{E}_A(x_2, y'),$$
$$\eta = [\eta_0, \ldots, \eta_l] = LE(\xi),$$

where $y = y_1, y' = y_2$ or $y = y_2, y' = y_1$. Since $[\omega] \in \mathcal{E}_I^*$, we know that

$$\eta \cap \omega = L(\xi) \cap \omega \neq \emptyset.$$

Define

$$s = \min\{l : \eta_l \in \omega\}, \quad t = \max\{l : \xi_l = \eta_s\}, \quad u = \max\{l : \omega_l = \eta_s\}.$$

Then, we can write $\xi = \xi^- \oplus \xi^+, \omega = \omega^- \oplus \omega^+$
where

$$\xi^- = [\xi_0, \ldots, \xi_t], \quad \xi^+ = [\xi_t, \ldots, \xi_m],$$
$$\omega^- = [\omega_0, \ldots, \omega_u], \quad \omega^+ = [\omega_u, \ldots, \omega_n].$$

We define

$$\phi([\omega]) = \phi((\xi^- \oplus \xi^+, \omega^- \oplus \omega^+)) = (\xi^- \oplus \omega^+, \omega^- \oplus \xi^+).$$

Note that $\xi^- \oplus \omega^+ \in \mathcal{E}_A(x_1, y')$, $\omega^- \oplus \xi^+ \in \mathcal{E}_A(x_2, y)$, and $q(\phi([\omega])) = q([\omega])$. A straightforward check shows that $\phi \circ \phi$ is the identity.

Suppose that $k > 2$ and $[\omega] \in \mathcal{E}_I^*$. We will change two paths as in the $k = 2$ case and leave the others fixed, being careful in our choice of paths to make

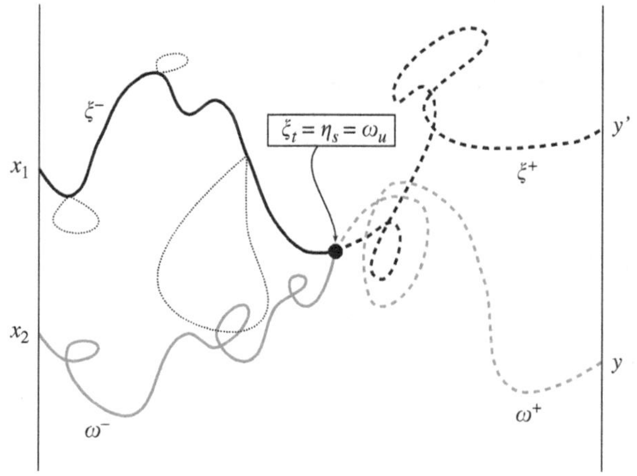

Figure 9.1 The paths $\xi^-, \xi^+, \omega^-, \omega^+$ in the proof of Fomin's identity.

sure that $\phi(\phi([\omega])) = [\omega]$. Let $\eta^i = LE(\omega^i)$. We define

$$r = \min\{i : \eta^i \cap \omega^j \neq \emptyset \text{ for some } j > i\},$$
$$s = \min\{l : \eta^r_l \in \omega^{i+1} \cup \cdots \cup \omega^k\},$$
$$b = \min\{j > r : \eta^r_s \in \omega^j\},$$
$$t = \max\{l : \omega^r_l = \eta^r_s\}, \qquad u = \max\{l : \omega^b_l = \eta^r_s\}.$$

We make the interchange

$$(\omega^{r,-} \oplus \omega^{r,+}, \omega^{b,-} \oplus \omega^{b,+}) \longleftrightarrow (\omega^{r,-} \oplus \omega^{b,+}, \omega^{b,-} \oplus \omega^{r,+})$$

as in the previous paragraph (with $(\omega^r, \omega^b) = (\xi, \omega)$) leaving the other paths fixed. This defines ϕ, and it is then straightforward to check that $\phi \circ \phi$ is the identity. □

9.7 Wilson's algorithm and spanning trees

Kirchhoff was the first to relate the number of spanning trees of a graph to a determinant. Here, we derive a number of these results. We use a more recent technique, *Wilson's algorithm*, to establish the results. This algorithm is an efficient method to produce spanning trees from the uniform distribution using loop-erased random walk. We describe it in the proof of the next proposition. The basic reason why this algorithm works is that the product on the right-hand side of (9.3) is independent of the ordering of the vertices.

Proposition 9.7.1 *Suppose that* $\#(\mathcal{X}) = n < \infty$ *and* q *are transition probabilities for an irreducible Markov chain on* $\mathcal{X}$. *Then*

$$\sum_{\mathcal{T}} q(\mathcal{T}; x_0) = \frac{1}{F(\mathcal{X} \setminus \{x_0\})}. \tag{9.12}$$

Proof We will describe an algorithm due to David Wilson that chooses a spanning tree at random. Let $\mathcal{X} = \{x_0, \ldots, x_{n-1}\}$.

- Start the Markov chain at x_1 and let it run until it reaches x_0. Take the loop-erasure of the set of points visited $[\eta_0 = x_1, \eta_1, \ldots, \eta_i = x_0]$. Add the edges $[\eta_0, \eta_1], [\eta_1, \eta_2], \ldots, [\eta_{i-1}, \eta_i]$ to the tree.
- If the edges form a spanning tree, we stop. Otherwise, we let j be the smallest index such that x_j is not a vertex in the tree. Start a random walk at x_j and let

it run until it reaches one of the vertices that has already been added. Perform loop-erasure on this path and add the edges in the loop-erasure to the tree.

- Continue until all vertices have been added to the tree.

We claim that for any tree $\mathcal{T}$, the probability that $\mathcal{T}$ is output in this algorithm is

$$q(\mathcal{T}; x_0)\, F(\mathcal{X} \setminus \{x_0\}). \tag{9.13}$$

The result (9.12) follows immediately. To prove (9.13), suppose that a spanning tree $\mathcal{T}$ is given. Then, this gives a collection of self-avoiding paths:

$$\begin{aligned}
\eta_1 &= \left[y_{1,1} = x_1,\ y_{1,2}, \ldots,\ y_{1,k_1},\ z_1 = x_0\right] \\
\eta_2 &= \left[y_{2,1},\ y_{2,2}, \ldots,\ y_{2,k_2},\ z_2\right] \\
&\ \ \vdots \\
\eta_m &= \left[y_{m,1},\ y_{m,2},\ \ldots,\ y_{m,k_m},\ z_m\right].
\end{aligned}$$

Here, η_1 is the unique self-avoiding path in the tree from x_1 to x_0; for $j > 1$, $y_{j,1}$ is the vertex of smallest index (using the ordering $x_0, x_1, \ldots, x_{n-1}$) that has not been listed so far; and η_j is the unique self-avoiding path from $y_{j,1}$ to a vertex z_j in $\eta_1 \cup \cdots \cup \eta_{j-1}$. Then, the probability that $\mathcal{T}$ is chosen is exactly the product of the probabilities that

- if a random walk starting at x_1 is stopped at x_0, the loop-erasure is η_1;
- if a random walk starting at $y_{2,1}$ is stopped at η_1, then the loop-erasure is η_2

$$\vdots$$

- if a random walk starting at $y_{m,1}$ is stopped at $\eta_1 \cup \cdots \cup \eta_{m-1}$, then the loop-erasure is η_m.

With this decomposition, we can now use (9.5) and (9.3) to obtain (9.13). □

Corollary 9.7.2 *If C_n denotes the number of spanning trees of a connected graph with vertices $\{x_0, x_1, \ldots, x_{n-1}\}$, then*

$$\begin{aligned}
\log C_n &= \sum_{j=1}^{n-1} \log d(x_j) - \log F(\mathcal{X} \setminus \{x_0\}) \\
&= \sum_{j=1}^{n-1} \log d(x_j) + \lim_{\lambda \to 1-} [\log g(\lambda; x_0) - \Phi(\lambda)].
\end{aligned}$$

Here, the implicit q is the transition probability for simple random walk on the graph and $d(x_j)$ denotes the degree of x_j. If C_n is a transitive graph of degree d,

$$\log C_n = (n-1)\log d - \log n + \lim_{\lambda \to 1-} [\log g(\lambda) - \Phi(\lambda)]. \tag{9.14}$$

Proof For simple random walk on the graph, for all $\mathcal{T}$,

$$q(\mathcal{T}; x_0) = \left[\prod_{j=1}^{n-1} d(x_j)\right]^{-1}.$$

In particular, it is the same for all trees, and (9.12) implies that the number of spanning trees is

$$[q(\mathcal{T}; x_0)\, F(\mathcal{X} \setminus \{x_0\})]^{-1} = \left[\prod_{j=1}^{n-1} d(x_j)\right] F(\mathcal{X} \setminus \{x_0\})^{-1}.$$

The second equality follows from Proposition 9.3.4 and the relation $\Phi(\lambda) = F(x;\lambda)$. If the graph is transitive, then $g(\lambda) = n\, g(\lambda; x_0)$, from which (9.14) follows. □

♣ If we take a connected graph and add any number of self-loops at vertices, this does not change the number of spanning trees. The last corollary holds, regardless of how many self-loops are added. Note that adding self-loops affects both the value of the degree and the value of $F(\mathcal{X} \setminus \{x_0\})$.

Proposition 9.7.3 *Suppose that $\mathcal{X}$ is a finite, connected graph with n vertices and maximal degree d, and P is the transition matrix for the lazy random walk on $\mathcal{X}$ as in Section 9.2.1. Suppose that the eigenvalues of P are*

$$\alpha_1 = 1,\ \alpha_2,\ \ldots,\ \alpha_n.$$

Then the number of spanning trees of $\mathcal{X}$ is

$$d^{n-1}\, n^{-1} \prod_{j=2}^{n} (1-\alpha_j).$$

Proof Since the invariant probability is $\pi \equiv 1/n$, Proposition 9.3.5 tells us that for each $x \in \mathcal{X}$,

$$\frac{1}{F(\mathcal{X} \setminus \{x\})} = n^{-1} \prod_{j=2}^{n} (1 - \alpha_j).$$

□

♣ The values $1 - \alpha_j$ are the eigenvalues for the (negative of the) Laplacian $I - Q$ for simple random walk on the graph. In graph theory, it is more common to define the Laplacian to be $\pm d(I - Q)$. When looking at formulas, it is important to know which definition of the Laplacian is being used.

9.8 Examples

9.8.1 Complete graph

The complete graph on a collection of vertices is the graph with all (distinct) vertices adjacent.

Proposition 9.8.1 *The number of spanning trees of the complete graph on* $\mathcal{X} = \{x_0, \ldots, x_{n-1}\}$ *is* n^{n-2}.

Proof Consider the Markov chain with transition probabilities $q(x, y) = 1/n$ for all x, y. Let $A_j = \{x_j, \ldots, x_{n-1}\}$. The probability that the chain starting at x_j has its first visit (after time zero) to $\{x_0, \ldots, x_j\}$ at x_j is $1/(j + 1)$ since each vertex is equally likely to be the first one visited. Using the interpretation of $F_{x_j}(A_j)$ as the reciprocal of the probability that the chain starting at x_j visits $\{x_0, \ldots, x_{j-1}\}$ before returning to x_j, we see that

$$F_{x_j}(A_j) = \frac{j+1}{j}, \qquad j = 1, \ldots, n-1$$

and hence (9.3) gives

$$F(\mathcal{X} \setminus \{x_0\}) = n.$$

With the self-loops, each vertex has degree n and hence for each spanning tree

$$q(\mathcal{T}; x_0) = n^{-(n-1)}.$$

Therefore, the number of spanning trees is

$$[q(\mathcal{T}; x_0)\, F(\mathcal{X} \setminus \{x_0\})]^{-1} = n^{n-2}.$$

□

9.8.2 Hypercube

The hypercube $\mathcal{X}_n$ is the graph whose vertices are $\{0, 1\}^n$ with vertices adjacent if they agree in all but one component.

Proposition 9.8.2 *If C_n denotes the number of spanning trees of the hypercube $\mathcal{X}_n := \{0, 1\}^n$, then*

$$\log C_n := (2^n - n - 1) \log 2 + \sum_{k=1}^{n} \binom{n}{k} \log k.$$

By (9.14), Proposition 9.8.2 is equivalent to

$$\lim_{\lambda \to 1-} [\log g(\lambda) - \Phi(\lambda)] = -(2^n - 1) \log n + (2^n - 1) \log 2 + \sum_{k=1}^{n} \binom{n}{k} \log k.$$

where g is the cycle-generating function for simple random walk on $\mathcal{X}_n$. The next proposition computes g.

Proposition 9.8.3 *Let g be the cycle-generating function for simple random walk on the hypercube $\mathcal{X}_n$. Then*

$$g(\lambda) = \sum_{j=0}^{n} \binom{n}{j} \frac{n}{n - \lambda(n - 2j)} = 2^n + \sum_{j=0}^{n} \binom{n}{j} \frac{\lambda(n - 2j)}{n - \lambda(n - 2j)}.$$

Proof of Proposition 9.8.2, given Proposition 9.8.3 Note that

$$\Phi(\lambda) = \int_0^\lambda \frac{g(s) - 2^n}{s} \, ds$$

$$= \int_0^\lambda \left[\sum_{j=0}^{n} \binom{n}{j} \frac{n - 2j}{n - s(n - 2j)} \right] ds$$

$$= (2^n - 1) \log n - \log(1 - \lambda) - \sum_{j=1}^{n} \binom{n}{j} \log[n - \lambda(n - 2j)].$$

Let us write $\lambda = 1 - \epsilon$ so that

$$\log g(\lambda) = \log\left[\sum_{j=0}^{n}\binom{n}{j}\frac{n}{\epsilon\, n + 2j(1-\epsilon)}\right],$$

$$\Phi(\lambda) = (2^n - 1)\,\log n - \log\epsilon - \sum_{j=1}^{n}\binom{n}{j}\log[\epsilon n + (1-\epsilon)\,2j].$$

As $\epsilon \to 0+$,

$$\log g(\lambda) = \log(1/\epsilon) + \log\left[\sum_{j=0}^{n}\binom{n}{j}\frac{n}{n + 2j\frac{1-\epsilon}{\epsilon}}\right] = -\log(\epsilon) + o(1),$$

$$\Phi(\lambda) = (2^n - 1)\,\log n - \log\epsilon - \sum_{j=1}^{n}\binom{n}{j}\log(2j) + o(1)$$

and hence

$$\lim_{\lambda\to 1-}\,[\log g(\lambda) - \Phi(\lambda)] = (1 - 2^n)\,\log n + (2^n - 1)\,\log 2 + \sum_{j=1}^{n}\binom{n}{j}\log j,$$

which is what we needed to show. □

The remainder of this subsection will be devoted to proving Proposition 9.8.3. Let $L(n, 2k)$ denote the number of cycles of length $2k$ in $\mathcal{X}_n$. By definition $L(n, 0) = 2^n$. Let g_n denote the generating function on $\mathcal{X}_n$ using weights 1 (instead of $1/n$) on the edges on the graph and zero otherwise,

$$g_n(\lambda) = \sum_{\omega}\lambda^{-|\omega|} = \sum_{k=0}^{\infty} L(n, 2k)\,\lambda^{2k}.$$

Then, if g is as in Proposition 9.8.3, $g(\lambda) = g_n(\lambda/n)$. Then Proposition 9.8.3 is equivalent to

$$g_n(\lambda) = \sum_{j=0}^{n}\binom{n}{j}\frac{1}{1 - \lambda(n - 2j)}, \tag{9.15}$$

which is what we will prove. By convention we set $L(0,0) = 1$; $L(0,k) = 0$ for $k > 0$, and hence

$$g_0(\lambda) = \sum_{k=0}^{\infty} L(0,2k)\,\lambda^{2k} = 1,$$

which is consistent with (9.15).

Lemma 9.8.4 *If $n, k \geq 0$,*

$$L(n+1,2k) = 2\sum_{j=0}^{k} \binom{2k}{2j} L(n,2j).$$

Proof This is immediate for $n = 0$, since $L(1,2k) = 2$ for every $k \geq 0$. For $n \geq 1$, consider any cycle in $\mathcal{X}_{n+1}$ of length $2k$. Assume that there are $2j$ steps that change one of the first n components and $2(k-j)$ that change the last component. There are $\binom{2k}{2j}$ ways to choose which $2j$ steps make changes in the first n components. Given this choice, there are $L(n,2j)$ ways of moving in the first n components. The movement in the last component is determined once the initial value of the $(n+1)$-component is chosen; the 2 represents the fact that this initial value can equal 0 or 1. □

Lemma 9.8.5 *For all $n \geq 0$,*

$$g_{n+1}(\lambda) = \frac{1}{1-\lambda}\,g_n\left(\frac{\lambda}{1-\lambda}\right) + \frac{1}{1+\lambda}\,g_n\left(\frac{\lambda}{1+\lambda}\right).$$

Proof

$$\begin{aligned}
g_{n+1}(\lambda) &= \sum_{k=0}^{\infty} L(n+1,2k)\,\lambda^{2k} \\
&= 2\sum_{k=0}^{\infty}\sum_{j=0}^{k} \binom{2k}{2j} L(n,2j)\,\lambda^{2k} \\
&= 2\sum_{j=0}^{\infty} L(n,2j) \sum_{k=0}^{\infty} \binom{2j+2k}{2j} \lambda^{2j+2k} \\
&= \frac{2}{1-\lambda}\sum_{j=0}^{\infty} L(n,2j)\left(\frac{\lambda}{1-\lambda}\right)^{2j} \sum_{k=0}^{\infty} \binom{2j+2k}{2j} (1-\lambda)^{2j+1}\,\lambda^{2k}
\end{aligned}$$

Using the identity (see Exercise 9.4).

$$\sum_{k=0}^{\infty} \binom{2j+2k}{2j} p^{2j+1} (1-p)^{2k} = \frac{1}{2} + \frac{1}{2} \left(\frac{p}{2-p} \right)^{2j+1}, \tag{9.16}$$

we see that $g_{n+1}(\lambda)$ equals

$$\frac{1}{1-\lambda} \sum_{j=0}^{\infty} L(n,2j) \left(\frac{\lambda}{1-\lambda} \right)^{2j} + \frac{1}{1+\lambda} \sum_{j=0}^{\infty} L(n,2j) \left(\frac{\lambda}{1+\lambda} \right)^{2j},$$

which gives the result. □

Proof of Proposition 9.8.3. Setting $\lambda = (n+\beta)^{-1}$, we see that it suffices to show that

$$g_n \left(\frac{1}{n+\beta} \right) = \sum_{j=0}^{n} \binom{n}{j} \frac{\beta+n}{\beta+2j}. \tag{9.17}$$

This clearly holds for $n = 0$. Let $H_n(\lambda) = \lambda\, g_n(\lambda)$. Then, the previous lemma gives the recursion relation

$$H_{n+1} \left(\frac{1}{n+1+\beta} \right) = H_n \left(\frac{1}{n+\beta} \right) + H_n \left(\frac{1}{n+2+\beta} \right).$$

Hence, by induction we see that

$$H_n \left(\frac{1}{n+\beta} \right) = \sum_{j=0}^{n} \binom{n}{j} \frac{1}{\beta+2j}.$$

□

9.8.3 Sierpinski graphs

In this subsection we consider the Sierpinski graphs which is a sequence of graphs $V_0, V_1, \ldots$ defined as follows. V_0 is a triangle, i.e. a complete graph on three vertices. For $n > 0$, V_n will be a graph with three vertices of degree 2 (which we call the corner vertices) and $[3^{n+1} - 3]/2$ vertices of degree 4. We define the graph inductively. Suppose that we are given three copies of V_{n-1}, $V_{n-1}^{(1)}, V_{n-1}^{(2)}, V_{n-1}^{(3)}$, with corner vertices $x_1^{(1)}, x_2^{(1)}, x_3^{(1)}, \ldots, x_1^{(3)}, x_2^{(3)}, x_3^{(3)}$. Then V_n is obtained from these three copies by identifying the vertex $x_j^{(k)}$ with the vertex $x_k^{(j)}$. We call the graphs V_n the Sierpinski graphs.

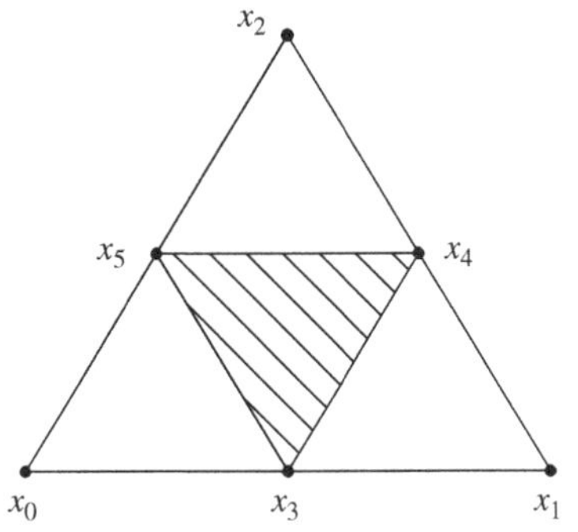

Figure 9.2 The Sierpinski graph $V - 1$.

Proposition 9.8.6 *Let C_n denote the number of spanning trees of the Sierpinski graph V_n. Then C_n satisfies the recursive equation*

$$C_{n+1} = 2\,(5/3)^n\, C_n^3. \tag{9.18}$$

Hence,

$$C_n = (3/20)^{1/4}\,(3/5)^{n/2}\,(540)^{3^n/4}. \tag{9.19}$$

Proof It is clear that $C_0 = 3$, and a simple induction argument shows that the solution to (9.18) with $C_0 = 3$ is given by (9.19). Hence, we need to show the recursive equation $C_{n+1} = 2\,(5/3)^n\, C_n^3$.

For $n \geq 1$, we will write $V_n = \{x_0, x_1, x_2, x_3, x_4, x_5, \ldots, x_{M_n}\}$ where $M_n = [3^n + 3]/2, x_0, x_1, x_2$ are corner vertices of V_n and x_3, x_4, x_5 are the other vertices that are corner vertices for the three copies of V_{n-1}. They are chosen so that x_3 lies between x_0, x_1; x_4 between x_1, x_2; x_5 between x_2, x_0. Using Corollary 9.7.2 and Lemma 9.3.2, we can write that $C_n = \Psi_n\, J_n$ where

$$\Psi_n = \prod_{j=1}^{M_n} d(x_j), \quad J_n = \prod_{j=1}^{M_n} p_{j,n}.$$

Here, $p_{j,n}$ denotes the probability that simple random walk in V_n started at x_j returns to x_j before visiting $\{x_0, \ldots, x_{j-1}\}$. Note that

$$\Psi_n = 2^2\, 4^{(3^{n+1}-3)/2},$$

and hence $\Psi_{n+1} = 4\Psi_n^3$. Therefore, we need to show that

$$J_{n+1} = (1/2)\,(5/3)^n\, J_n^3. \tag{9.20}$$

We can write that

$$J_{n+1} = p_{1,n+1}\, p_{2,n+1}\, J^*_{n+1}$$

where J^*_{n+1} denotes the product over all the other vertices (the noncorner vertices). From this, we see that

$$J_{n+1} = p_{1,n+1}\, p_{2,n+1}, \ldots, p_{5,n+1}\, (J^*_n)^3 = \frac{p_{1,n+1}\, p_{2,n+1}, \ldots, p_{5,n+1}}{p^3_{1,n}\, p^3_{2,n}}\, J^3_n.$$

The computations of $p_{j,n}$ are straightforward computations familiar to those who study random walks on the Sierpinski gasket and are easy exercises in Markov chains. We give the answers here, leaving the details to the reader. By induction on n one can show that $p_{2,n+1} = (3/5)\, p_{2,n}$ and from this one can see that

$$p_{2,n+1} = \left(\frac{3}{5}\right)^{n+1}, \qquad p_{1,n} = \frac{3}{4}\left(\frac{3}{5}\right)^{n+1}.$$

Also,

$$p_{5,n+1} = p_{2,n} = \left(\frac{3}{5}\right)^{n}, \qquad p_{4,n+1} = \frac{15}{16}\left(\frac{3}{5}\right)^{n}, \qquad p_{3,n+1} = \frac{5}{6}\left(\frac{3}{5}\right)^{n}.$$

This gives (9.20). □

9.9 Spanning trees of subsets of $\mathbb{Z}^2$

Suppose that $A \subset \mathbb{Z}^2$ is finite, and let $e(A)$ denote the set of edges with at least one vertex in A. We write that $e(A) = \partial_e A \cup e_o(A)$ where $\partial_e A$ denotes the "boundary edges" with one vertex in ∂A and $e_o(A) = e(A) \setminus \partial_e A$, the "interior edges". There will be two types of spanning tree of A that we will consider.

- **Free** A collection of $\#(A) - 1$ edges from $e_0(A)$ such that the corresponding graph is connected.
- **Wired** The set of vertices is $A \cup \{\Delta\}$ where Δ denotes the boundary. The edges of the graph are the same as $e(A)$ except that each edge in $\partial_e A$ is replaced with an edge connecting the point in A to Δ. (There can be more than one edge connecting a vertex in A to Δ.) A *wired spanning tree* is a collection of edges from $e(A)$ such that the corresponding subgraph of $A \cup \{\Delta\}$ is a spanning tree. Such a tree has $\#(A)$ edges.

In both cases, we will find the number of trees by considering the Markov chain given by simple random walk in $\mathbb{Z}^2$. The different spanning trees correspond to different "boundary conditions" for the random walks.

- **Free** The lazy walker on A as described in Section 9.2.1, i.e.

$$q(x,y) = \frac{1}{4}, \quad x, y \in A, \ |x-y| = 1,$$

  and $q(x,x) = 1 - \sum_y q(x,y)$.
- **Wired** Simple random walk on A killed when it leaves A, i.e.

$$q(x,y) = \frac{1}{4}, \quad x, y \in A, \ |x-y| = 1,$$

  and $q(x,x) = 0$. Equivalently, we can consider this as the Markov chain on $A \cup \{\Delta\}$ where Δ is an absorbing point and

$$q(x, \Delta) = 1 - \sum_{y \in A} q(x,y).$$

♣ In other words, free spanning trees correspond to reflecting or Neumann boundary conditions and wired spanning trees correspond to Dirichlet boundary conditions.

We let $F(A)$ denote the quantity for the wired case. This is the same as $F(A)$ for simple random walk in $\mathbb{Z}^2$. If $x \in A$, we write that $F^*(A \setminus \{x\})$ for the corresponding quantity for the lazy walker. (The lazy walker is a Markov chain on A and hence $F^*(A) = \infty$. In order to get a finite quantity, we need to remove a point x.) The following are immediate corollaries of results in Section 9.7.

Proposition 9.9.1 *If $A \subset \mathbb{Z}^2$ is connected with $\#(A) = n < \infty$, then the number of wired spanning trees of A is*

$$4^n \, F(A)^{-1} = 4^n \prod_{j=1}^{n} (1 - \beta_j),$$

where $\beta_1, \ldots, \beta_n$ denote the eigenvalues of $Q_A = [q(x,y)]_{x,y \in A}$.

Proof This is a particular case of Corollary 9.7.2 using the graph $A \cup \{\Delta\}$ and $x_0 = \Delta$. See also Proposition 9.3.3. □

Proposition 9.9.2 *Suppose that* $\alpha_1 = 1, \ldots, \alpha_n$ *are the eigenvalues of the transition matrix for the lazy walker on a finite, connected* $A \subset \mathbb{Z}^2$ *of cardinality* n. *Then the number of spanning trees of* A *is*

$$4^{n-1} n^{-1} \prod_{j=2}^{n} (1 - \alpha_j).$$

Proof This is a particular case of Proposition 9.7.3. □

Recall that

$$\log F(A) = \sum_{\omega \in \mathcal{L}(A), |\omega| \geq 1} \frac{1}{4^{|\omega|} \, |\omega|}$$

$$= \sum_{x \in A} \sum_{n=1}^{\infty} \frac{1}{2n} \mathbb{P}^x\{S_{2n} = 0; S_j \in A, j = 1, \ldots, 2n\}. \qquad (9.21)$$

The first-order term in an expansion of $\log F(A)$ is

$$\sum_{x \in A} \sum_{n=1}^{\infty} \frac{1}{2n} \mathbb{P}^x\{S_{2n} = 0\},$$

which ignores the restriction that $S_j \in A, j = 1, \ldots, 2n$. The actual value involves a well-known constant C_{cat} called *Catalan's constant*. There are many equivalent definitions of this constant. For our purposes we can use the following

$$C_{cat} = \frac{\pi}{2} \log 2 - \frac{\pi}{4} \sum_{n=1}^{\infty} \frac{1}{2n} 4^{-2n} \binom{2n}{n}^2 = .91596 \cdots .$$

Proposition 9.9.3 *If* $S = (S^1, S^2)$ *is simple random walk in* $\mathbb{Z}^2$, *then*

$$\sum_{n=1}^{\infty} \frac{1}{2n} \mathbb{P}\{S_{2n} = 0\} = \log 4 - \frac{4}{\pi} C_{cat},$$

where C_{cat} *denotes Catalan's constant. In particular, if* $A \subset \mathbb{Z}^2$ *is finite,*

$$\log F(A) = [\log 4 - (4/\pi) \, C_{cat}] \, \#(A) - \sum_{x \in A} \psi(x; A), \qquad (9.22)$$

where

$$\psi(x;A) = \sum_{n=1}^{\infty} \frac{1}{2n} \mathbb{P}^x\{S_{2n} = 0; S_j \notin A \text{ for some } 0 \le j \le 2n\}.$$

Proof Using Exercise 1.7, we get

$$\sum_{n=1}^{\infty} \frac{1}{2n} \mathbb{P}\{S_{2n} = 0\} = \sum_{n=1}^{\infty} \frac{1}{2n} [\mathbb{P}\{S^1_{2n} = 0\}]^2 = \sum_{n=1}^{\infty} \frac{1}{2n} 4^{-2n} \binom{2n}{n}^2.$$

Since $\mathbb{P}\{S(2n) = 0\} \sim c\, n^{-1}$, we can see that the sum is finite. The exact value follows from our (conveniently chosen) definition of C_{cat}. The last assertion then follows from (9.21). □

Lemma 9.9.4 *There exists $c < \infty$ such that if $A \subset \mathbb{Z}^2$, $x \in A$, and $\psi(x;A)$ are defined as in Proposition 9.9.3, then*

$$\psi(x;A) \le \frac{c}{\text{dist}(x, \partial A)^2}.$$

Proof We only sketch the argument, leaving the details for Exercise 9.5. Let $r = \text{dist}(x, \partial A)$. Since it takes about r^2 steps to reach ∂A, the loops with fewer than that many steps rooted at x tend not to leave A. Hence, $\psi(x;A)$ is at most of the order of

$$\sum_{n \ge r^2} \frac{1}{2n} \mathbb{P}\{S_{2n} = 0\} \asymp \sum_{n \ge r^2} n^{-2} \asymp r^{-2}.$$

□

Using this and (9.22) we immediately get the following.

Proposition 9.9.5 *Suppose that A_n is a sequence of finite, connected subsets of $\mathbb{Z}^2$ satisfying the following condition (that roughly means "measure of the boundary goes to zero"). For every $r > 0$,*

$$\lim_{n \to \infty} \frac{\#\{x \in A_n : \text{dist}(x, \partial A_n) \le r\}}{\#(A_n)} = 0.$$

Then,

$$\lim_{n \to \infty} \frac{\log F(A_n)}{\#(A_n)} = \log 4 - \frac{4 C_{\text{cat}}}{\pi}.$$

Suppose that $A_{m,n}$ is the $(m-1)\times(n-1)$ discrete rectangle,

$$A_{m,n} = \{x + iy : 1 \le x \le m-1, 1 \le y \le n-1\}.$$

Note that

$$\#(A_{m,n}) = (m-1)\,(n-1), \qquad \#(\partial A_{m,n}) = 2\,(m-1) + 2\,(n-1).$$

Theorem 9.9.6

$$\frac{4^{(m-1)(n-1)}}{F(A_{m,n})} \asymp e^{4C_{\rm cat} mn/\pi}\,(\sqrt{2}-1)^{m+n}\,n^{-1/2}. \tag{9.23}$$

More precisely, for every $b \in (0,\infty)$ there is a $c_b < \infty$ such that if $b^{-1} \le m/n \le b$, then both sides of (9.23) are bounded above by c_b times the other side. In particular, if $C_{m,n}$ denotes the number of wired spanning trees of $A_{m,n}$,

$$\begin{aligned}\log C_{mn} &= \frac{4C_{\rm cat}}{\pi}\,mn + \log(\sqrt{2}-1)\,(m+n) - \frac{1}{2}\log n + O(1)\\ &= \frac{4C_{\rm cat}}{\pi}\,\#(A_{m,n}) + \left[\frac{2C_{\rm cat}}{\pi} + \frac{1}{2}\log(\sqrt{2}-1)\right]\#(\partial A_{m,n})\\ &\quad - \frac{1}{2}\log n + O(1).\end{aligned}$$

♣ Although our proof will use the exact values of the eigenvalues, it is useful to consider the result in terms of (9.22). The dominant term is already given by (9.22). The correction comes from loops rooted in $A_{m,n}$ that leave A. The biggest contribution to these comes from points near the boundary. It is not surprising, then, that the second term is proportional to the number of points on the boundary. The next correction to this comes from the corners of the rectangle. This turns out to contribute a logarithmic term and after that all other correction terms are $O(1)$. We arbitrarily write $\log n$ rather than $\log m$; note that $\log m = \log n + O(1)$.

Proof The expansion for $\log C_{m,n}$ follows immediately from Proposition 9.9.1 and (9.23), so we only need to establish (9.23). The eigenvalues of $I - Q_A$ can be given explicitly (see Section 8.2),

$$1 - \frac{1}{2}\left[\cos\left(\frac{j\pi}{m}\right) + \cos\left(\frac{k\pi}{n}\right)\right], \quad j = 1,\ldots,m-1; \;\; k = 1,\ldots,n-1,$$

with corresponding eigenfunctions

$$f(x,y) = \sin\left(\frac{j\pi x}{m}\right)\sin\left(\frac{k\pi y}{n}\right),$$

where the eigenfunctions have been chosen so that $f \equiv 0$ on $\partial A_{m,n}$. Therefore,

$$\begin{aligned}-\log F(A_{m,n}) &= \log\det[I - Q_A] \\ &= \sum_{j=1}^{m-1}\sum_{k=1}^{n-1}\log\left(1 - \frac{1}{2}\left[\cos\left(\frac{j\pi}{m}\right) + \cos\left(\frac{k\pi}{n}\right)\right]\right).\end{aligned}$$

Let

$$g(x,y) = \log\left[1 - \frac{\cos(x) + \cos(y)}{2}\right].$$

Then $(mn)^{-1}\log\det[I - Q_A]$ is a Riemann sum approximation of

$$\frac{1}{\pi^2}\int_0^{\pi}\int_0^{\pi} g(x,y)\,dx\,dy.$$

To be more precise, let $V(j,k) = V_{m,n}(j,k)$ denote the rectangle of side lengths π/m and π/n centered at $(j\pi/m) + i(k\pi/n)$. Then we will consider

$$J(j,k) := \frac{1}{mn}\,g\left(\frac{j\pi}{m},\frac{k\pi}{n}\right) = \frac{1}{mn}\left(1 - \frac{1}{2}\left[\cos\left(\frac{j\pi}{m}\right) + \cos\left(\frac{k\pi}{n}\right)\right]\right)$$

as an approximation to

$$\frac{1}{\pi^2}\int_{V(j,k)} g(x,y)\,dx\,dy.$$

Note that

$$\begin{aligned}V &= \bigcup_{j=1}^{m-1}\bigcup_{k=1}^{n-1} V(j,k) \\ &= \left\{x + iy : \frac{\pi}{2m} \le x \le \pi\left(1 - \frac{1}{2m}\right), \frac{\pi}{2n} \le y \le \pi\left(1 - \frac{1}{2n}\right)\right\}.\end{aligned}$$

One can show (using ideas as in Section A.1.1, details omitted), that

$$\log\det[I - Q_A] = mn\int_V g(x,y)\,dx\,dy + O(1).$$

Therefore,

$$\log \det[I - Q_A] = mn \left[\int_{[0,\pi]^2} g(x,y)\,dx\,dy - \int_{[0,\pi]^2 \setminus V} g(x,y)\,dx\,dy \right] + O(1).$$

The result will follow if we show that

$$mn \int_{[0,\pi]^2 \setminus V} g(x,y)\,dx\,dy$$
$$= (m+n)\,\log 4 - (m+n)\,\log(1-\sqrt{2}) + \frac{1}{2}\,\log n + O(1).$$

We now estimate the integral over $[0,\pi]^2 \setminus V$, which we write as the sum of integrals over four thin strips minus the integrals over the "corners" that are doubly counted. One can check (using an integral table, e.g.) that

$$\frac{1}{\pi} \int_0^{\pi} \log \left[1 - \frac{\cos x + \cos y}{2} \right] dy$$
$$= -2 \log 2 + \log[2 - \cos x + \sqrt{2(1-\cos x) + (1-\cos x)^2}].$$

Then,

$$\frac{1}{\pi^2} \int_0^{\epsilon} \int_0^{\pi} \log \left[1 - \frac{\cos x + \cos y}{2} \right] dy\,dx = -\frac{2\epsilon}{\pi}\,\log 2 + \frac{\epsilon^2}{2\pi} + O(\epsilon^3).$$

If we choose $\epsilon = \pi/(2m)$ or $\epsilon = \pi/2n$, this gives

$$\frac{mn}{\pi^2} \int_0^{\pi/(2m)} \int_0^{\pi} \log \left[1 - \frac{\cos x + \cos y}{2} \right] dy\,dx = m\,\log 2 + O(1).$$

$$\frac{mn}{\pi^2} \int_0^{\pi} \int_0^{\pi/(2n)} \log \left[1 - \frac{\cos x + \cos y}{2} \right] dy\,dx = -n\,\log 2 + O(1).$$

Similarly,

$$\frac{1}{\pi^2} \int_{\pi-\epsilon}^{\pi} \int_0^{\pi} \log \left[1 - \frac{\cos x + \cos y}{2} \right] dy\,dx$$
$$= -\frac{2\epsilon}{\pi}\,\log 2 + \frac{\epsilon}{\pi}\,\log[3 + 2\sqrt{2}] + O(\epsilon^3)$$
$$= -\frac{2\epsilon}{\pi}\,\log 2 - \frac{2\epsilon}{\pi}\,\log[\sqrt{2} - 1] + O(\epsilon^3),$$

which gives

$$\frac{mn}{\pi^2}\int_{\pi-\frac{\pi}{2m}}^{\pi}\int_0^{\pi}\log\left[1-\frac{\cos x+\cos y}{2}\right]dy\,dx$$
$$=-\,n\,\log 2-n\log[\sqrt{2}-1]+O(n^{-1}),$$

$$\frac{mn}{\pi^2}\int_0^{\pi}\int_{\pi-\frac{\pi}{2n}}^{\pi}\log\left[1-\frac{\cos x+\cos y}{2}\right]dy\,dx$$
$$=-m\,\log 2-m\log[\sqrt{2}-1]+O(n^{-1}).$$

The only nontrivial "corner" term comes from

$$\int_0^{\epsilon}\int_0^{\delta}\log\left[1-\frac{\cos x+\cos y}{2}\right]dxdy=2\,\epsilon\,\delta\,\log(\epsilon)+O(\epsilon\,\delta).$$

Therefore,

$$\frac{mn}{\pi^2}\int_0^{\frac{\pi}{2m}}\int_0^{\frac{\pi}{2n}}\log\left[1-\frac{\cos x+\cos y}{2}\right]dxdy=-\frac{1}{2}\log n+O(1).$$

All of the other corners give $O(1)$ terms.

Combining it all, we get

$$\sum_{j=1}^{m-1}\sum_{k=1}^{n-1}\log\left(1-\frac{1}{2}\left[\cos\left(\frac{j\pi}{m}\right)+\cos\left(\frac{k\pi}{n}\right)\right]\right)$$

equals

$$Imn+(m+n)\log 4+(m+n)\,\log[1-\sqrt{2}]-\frac{1}{2}\,\log n+O(1),$$

where

$$I=\frac{1}{\pi^2}\int_0^{\pi}\int_0^{\pi}\log\left[1-\frac{\cos x+\cos y}{2}\right]dydx.$$

Proposition 9.9.5 tells us that

$$I=\frac{4\,C_{\rm cat}}{\pi}-\log 4.$$

□

Theorem 9.9.6 allows us to derive some constants for simple random walk that are hard to show directly. Write (9.23) as

$$\log F(A_{m,n}) = B_1\, mn + B_2\,(m+n) + \frac{1}{2}\,\log n + O(1), \tag{9.24}$$

where

$$B_1 = \log 4 - \frac{4C_{\text{cat}}}{\pi}, \qquad B_2 = \log(\sqrt{2}+1) - \log 4.$$

The constant B_1 was obtained by considering the rooted loop measure and B_2 was obtained from the exact value of the eigenvalues. Recall from (9.3) that if we enumerate $A_{m,n}$,

$$A_{m,n} = \{x_1, x_2, \ldots, x_K\}, \quad K = (m-1)\,(n-1),$$

then

$$\log F(A_{m,n}) = \sum_{j=1}^{K} \log F_{x_j}(A_{m,n} \setminus \{x_1, \ldots, x_{j-1}\}),$$

and $F_x(V)$ is the expected number of visits to x for a simple random walk starting at x before leaving V. We will define the *lexicographic order* of $\mathbb{Z}+i\mathbb{Z}$ by $x+iy \prec x_1 + iy_1$ if $x < x_1$ or $x = x_1$ and $y < y_1$.

Proposition 9.9.7 *If*

$$V = \{x+iy : y > 0\} \cup \{0, 1, 2, \ldots\},$$

then

$$F_0(V) = 4\,e^{-4C_{\text{cat}}/\pi}.$$

Proof Choose the lexicographic order for $A_{n,n}$. Then, one can show that

$$F_{x_j}(A_{n,n} \setminus \{x_1, \ldots, x_{j-1}\}) = F_0(V)\;[1 + \text{ error}],$$

where the error term is small for points away from the boundary. Hence

$$\log F(A_{n,n}) = \#(A_{n,n})\; \log F_0(V)\,[1+o(1)].$$

which implies that $\log F_0(V) = B_1$ as in (9.24). □

Proposition 9.9.8 *Let $V \subset \mathbb{Z} \times i\mathbb{Z}$ be the subset*

$$V = (\mathbb{Z} \times i\mathbb{Z}) \setminus \{\ldots, -2, -1\},$$

Then, $F_0(V) = 4\,(\sqrt{2} - 1)$. In other words, the probability that the first return to $\{\ldots, -2, -1, 0\}$ by a simple random walk starting at the origin is at the origin equals

$$1 - \frac{1}{F_0(V)} = \frac{3 - \sqrt{2}}{4}.$$

Proof Consider that

$$A = A_n = \{x + iy : x = 1, \ldots, n-1; -(n-1) \le y \le n-1\}.$$

Then A is a translation of $A_{2n,n}$ and hence (9.24) gives

$$\log F(A) = 2\,B_1\,n^2 + 3\,B_2\,n + \frac{1}{2}\,\log n + O(1).$$

Order A so that the first $n-1$ vertices of A are $1, 2, \ldots, n-1$ in order. Then, we can see that

$$\log F(A) = \left[\sum_{j=1}^{n-1} \log F_j(A \setminus \{1, \ldots, j-1\})\right] + 2\,\log F(A_{n,n}).$$

Using (9.24) again, we see that

$$2\,\log F(A_{n,n}) = 2\,B_1\,n^2 + 4\,B_2\,n + \log n + O(1),$$

and hence

$$\sum_{j=1}^{n-1} \log F_j(A \setminus \{1, \ldots, j-1\}) = -B_2\,n - \frac{1}{2}\,\log n + O(1).$$

Now we use the fact that

$$\log F_j(A \setminus \{1, \ldots, j-1\}) = \log F_0(V)\,\,[1 + \text{ error}]\,,$$

where the error term is small for points away from the boundary to conclude that $F_0(V) = e^{-B_2}$. □

Let $\tilde{A}_{m,n}$ be the $m \times n$ rectangle

$$\tilde{A}_{m,n} = \{x + iy : 0 \le x \le m-1, 0 \le y \le n-1\}.$$

Note that

$$\#(\tilde{A}_{m,n}) = mn, \quad \#(\partial\tilde{A}_{m,n}) = 2(m+n).$$

Let $\tilde{C}_{m,n}$ denote the number of (free) spanning trees of $\tilde{A}_{m,n}$.

Theorem 9.9.9

$$\tilde{C}_{m,n} \asymp e^{4C_{\text{cat}}mn/\pi}\,(\sqrt{2}-1)^{m+n}\,n^{-1/2}.$$

More precisely, for every $b \in (0,\infty)$ there is a $c_b < \infty$ such that if $b^{-1} \le m/n \le b$, then both sides of (9.23) are bounded above by c_b times the other side.

Proof We claim that the eigenvalues for the lazy walker Markov chain on $\tilde{A}_{m,n}$ are:

$$1 - \frac{1}{2}\left[\cos\left(\frac{j\pi}{m}\right) + \cos\left(\frac{k\pi}{n}\right)\right], \quad j = 0, \ldots, m-1; \;\; k = 0, \ldots, n-1,$$

with corresponding eigenfunctions

$$f(x,y) = \cos\left(\frac{j\pi(x+\frac{1}{2})}{m}\right)\cos\left(\frac{k\pi(y+\frac{1}{2})}{n}\right).$$

Indeed, these are eigenvalues and eigenfunctions for the usual discrete Laplacian, but the eigenfunctions have been chosen to have boundary conditions

$$f(0,y) = f(-1,y), f(m-1,y) = f(m,y),$$
$$f(x,0) = f(x,-1), f(x,n-1) = f(x,n).$$

For these reasons we can see that they are also eigenvalues and eigenvalues for the lazy walker.

Using Proposition 9.9.2, we have that

$$\tilde{C}_{mn} = \frac{4^{mn-1}}{mn} \prod_{(j,k)\ne(0,0)} \left(1 - \frac{1}{2}\left[\cos\left(\frac{j\pi}{m}\right) + \cos\left(\frac{k\pi}{n}\right)\right]\right).$$

Recall that if $F(A_{n,m})$ is as in Theorem 9.9.6, then

$$\frac{1}{F(\tilde{A}_{m,n})} = \prod_{1\le j\le m-1, 1\le k\le n-1} \left(1 - \frac{1}{2}\left[\cos\left(\frac{j\pi}{m}\right) + \cos\left(\frac{k\pi}{n}\right)\right]\right).$$

Therefore,

$$\tilde{C}_{mn} = \frac{4^{(m-1)(n-1)}}{F(A_{m,n})} \frac{4^{m+n-1}}{mn} \left(\prod_{j=1}^{n-1}\left[\frac{1}{2} - \frac{1}{2}\cos\left(\frac{j\pi}{n}\right)\right]\right)$$

$$\times \left[\prod_{j=1}^{m-1}\left[\frac{1}{2} - \frac{1}{2}\cos\left(\frac{j\pi}{m}\right)\right]\right).$$

Using (9.23), we see that it suffices to prove that

$$\frac{4^n}{n}\prod_{j=1}^{n-1}\left[\frac{1}{2} - \frac{1}{2}\cos\left(\frac{j\pi}{n}\right)\right] \asymp 1,$$

or equivalently,

$$\sum_{j=1}^{n-1}\log\left[1 - \cos\left(\frac{j\pi}{n}\right)\right] = -n\log 2 + \log n + O(1). \tag{9.25}$$

To establish (9.25), note that

$$\frac{1}{n}\sum_{j=1}^{n-1}\log\left[1 - \cos\left(\frac{j\pi}{n}\right)\right]$$

is a Riemann sum approximation of

$$\frac{1}{\pi}\int_0^\pi f(x)\,dx$$

where $f(x) = \log[1 - \cos x]$. Note that

$$f'(x) = \frac{\sin x}{1 - \cos x}, \qquad f''(x) = -\frac{1}{1 - \cos x}.$$

In particular, $|f''(x)| \le c\,x^{-2}$. Using this we can see that

$$\frac{1}{n}\left[1 - \cos\left(\frac{j\pi}{n}\right)\right] = \frac{1}{n}O(j^{-2}) + \frac{1}{\pi}\int_{\frac{j\pi}{n}-\frac{\pi}{2n}}^{\frac{j\pi}{n}+\frac{\pi}{2n}} f(x)\,dx.$$

Therefore,

$$\frac{1}{n}\sum_{j=1}^{n-1}\log\left[1-\cos\left(\frac{j\pi}{n}\right)\right] = O(n^{-1}) + \frac{1}{\pi}\int_{\frac{\pi}{2n}}^{\pi-\frac{\pi}{2n}} f(x)\,dx$$

$$= O(n^{-1}) + \frac{1}{\pi}\int_0^{\pi} f(x)\,dx - \frac{1}{\pi}\int_0^{\frac{\pi}{2n}} f(x)\,dx$$

$$= O(n^{-1}) - \log 2 + \frac{1}{n}\log n.$$

□

9.10 Gaussian free field

We introduce the Gaussian free field. In this section we assume that q is a symmetric transition probability on the space $\mathcal{X}$. Some of the definitions below are straightforward extensions of definitions for random walk on $\mathbb{Z}^d$.

- We say that $e = \{x, y\}$ is an *edge* if $q(e) := q(x, y) > 0$.
- If $A \subset \mathcal{X}$, let $e(A)$ denote the set of edges with at least one vertex in A. We write $e(A) = \partial_e A \cup e_o(A)$ where $\partial_e A$ are the edges with one vertex in ∂A and $e_o(A)$ are the edges with both vertices in A.
- We let

$$\partial A = \{y \in \mathcal{X} \setminus A : q(x, y) > 0 \text{ for some } x \in A\},$$

$$\overline{A} = A \cup \partial A.$$

- If $f : \overline{A} \to \mathbb{R}$ and $x \in A$, then

$$\Delta f(x) = \sum_y q(x, y)\,[f(y) - f(x)].$$

  We say that f is *harmonic* at x if $\Delta f(x) = 0$, and f is *harmonic on A* if $\Delta f(x) = 0$ for all $x \in A$.
- If $e \in e(A)$, we set $\nabla_e f = f(y) - f(x)$ where $e = \{x, y\}$. This defines $\nabla_e f$ up to a sign. Note that

$$\nabla_e f \; \nabla_e g$$

  is well defined.

Throughout this section we assume that $A \subset \mathcal{X}$ with $\#(\overline{A}) < \infty$.

- If $f, g : \overline{A} \to \mathbb{R}$ are functions, then we define the *energy* or *Dirichlet form* $\mathcal{E}$ to be the quadratic form

$$\mathcal{E}_A(f,g) = \sum_{e \in e(A)} q(e)\, \nabla_e f\, \nabla_e g$$

We let $\mathcal{E}_A(f) = \mathcal{E}_A(f,f)$.

Lemma 9.10.1 (Green's formula) *Suppose that* $f, h : \overline{A} \to \mathbb{R}$. *Then,*

$$\mathcal{E}_A(f,h) = -\sum_{x \in A} f(x)\, \Delta h(x) + \sum_{x \in \partial A} \sum_{y \in A} f(x)\, [h(x) - h(y)]\, q(x,y). \tag{9.26}$$

(a) If h *is harmonic in* A,

$$\mathcal{E}_A(f,h) = \sum_{x \in \partial A} \sum_{y \in A} f(x)\, [h(x) - h(y)]\, q(x,y). \tag{9.27}$$

(b) If $f \equiv 0$ *on* ∂A,

$$\mathcal{E}_A(f,h) = -\sum_{x \in A} f(x)\, \Delta h(x). \tag{9.28}$$

(c) If h *is harmonic in* A *and* $f \equiv 0$ *on* ∂A, *then* $\mathcal{E}_A(f,h) = 0$ *and hence*

$$\mathcal{E}_A(f+h) = \mathcal{E}_A(f) + \mathcal{E}_A(h). \tag{9.29}$$

Proof

$$\begin{aligned}
\mathcal{E}_A(f,h) &= \sum_{e \in e(A)} q(e)\, \nabla_e f\, \nabla_e h \\
&= \frac{1}{2} \sum_{x,y \in A} q(x,y)\, [f(y) - f(x)]\, [h(y) - h(x)] \\
&\quad + \sum_{x \in A} \sum_{y \in \partial A} q(x,y)\, [f(y) - f(x)]\, [h(y) - h(x)] \\
&= -\sum_{x \in A} \sum_{y \in A} q(x,y) f(x)\, [h(y) - h(x)] \\
&\quad - \sum_{x \in A} \sum_{y \in \partial A} q(x,y) f(x)\, [h(y) - h(x)]
\end{aligned}$$

$$+\sum_{x\in A}\sum_{y\in\partial A} q(x,y) f(y)\,[h(y)-h(x)]$$

$$= -\sum_{x\in A} f(x)\,\Delta h(x) + \sum_{y\in\partial A}\sum_{x\in A} q(x,y) f(y)\,[h(y)-h(x)].$$

This gives (9.26) and the final three assertions follow immediately. □

Suppose that $x \in \partial A$ and let h_x denote the function that is harmonic on A with boundary value δ_x on ∂A. Then it follows from (9.27) that

$$\mathcal{E}_A(h_x) = \sum_{y\in A}[1-h_x(y)]\,q(x,y).$$

We extend h_x to $\mathcal{X}$ by setting $h_x \equiv 0$ on $\mathcal{X}\setminus\overline{A}$.

Lemma 9.10.2 *Let Y_j be a Markov chain on $\mathcal{X}$ with transition probability q. Let*

$$T_x = \min\{j \geq 1 : Y_j = x\}, \qquad \tau_A = \min\{j \geq 1 : Y_j \notin A\}.$$

If $A \subset \mathcal{X}, x \in \partial A, A' = A \cup \{x\}$,

$$\mathcal{E}_{A'}(h_x) = \mathbb{P}^x\{T_x \geq \tau_{A'}\} = \frac{1}{F_x(A')}. \tag{9.30}$$

Proof If $y \in A$, then $h_x(y) = \mathbb{P}^y\{T_x = \tau_A\}$. Note that

$$\mathbb{P}^x\{T_x < \tau_{A'}\} = q(x,x) + \sum_{y\in A} q(x,y)\,\mathbb{P}^y\{T_x = \tau_A\} = q(x,x) + \sum_{y\in A} q(x,y)\,h_x(y).$$

Therefore,

$$\begin{aligned}\mathbb{P}^x\{T_x \geq \tau_{A'}\} &= 1 - \mathbb{P}^x\{T_x < \tau_{A'}\}\\ &= \sum_{z\notin A'} q(x,z) + \sum_{y\in A} q(x,y)\,[1-h_x(y)]\\ &= -\Delta h_x(x)\\ &= -\sum_{y\in A'} h_x(y)\,\Delta h_x(y) = \mathcal{E}_{A'}(h_x).\end{aligned}$$

The last equality uses (9.28). The second equality in (9.30) follows from Lemma 9.3.2. □

- If $v : \mathcal{X} \setminus A \to \mathbb{R}, f : A \to \mathbb{R}$, we write that $\mathcal{E}_A(f; v)$ for $\mathcal{E}_A(f_v)$ where $f_v \equiv f$ on A and $f_v \equiv v$ on ∂A. If v is omitted, then $v \equiv 0$ is assumed.
- The *Gaussian free field on A with boundary condition* v is the measure on functions $f : A \to \mathbb{R}$ whose density with respect to the Lebesgue measure on $\mathbb{R}^A$ is

$$(2\pi)^{-\#(A)/2}\, e^{-\mathcal{E}_A(f;v)/2}.$$

- If $v \equiv 0$, we call this the field with *Dirichlet boundary conditions*.
- If $A \subset \mathcal{X}$ is finite and $v : \mathcal{X} \setminus A \to \mathbb{R}$, define the *partition function*

$$\mathcal{C}(A; v) = \int (2\pi)^{-\#(A)/2}\, e^{-\mathcal{E}_A(f;v)/2}\, df,$$

where df indicates that this is an integral with respect to the Lebesgue measure on $\mathbb{R}^A$. If $v \equiv 0$, we write just $\mathcal{C}(A)$. By convention, we set $\mathcal{C}(\emptyset; v) = 1$.

We will give two proofs of the next fact.

Proposition 9.10.3 *For any $A \subset \mathcal{X}$ with $\#(A) < \infty$,*

$$\mathcal{C}(A) = \sqrt{F(A)} = \exp\left\{\frac{1}{2} \sum_{\overline{\omega} \in \overline{\mathcal{L}}(A)} \overline{m}(\overline{\omega})\right\}. \tag{9.31}$$

Proof We prove this inductively on the cardinality of A. If $A = \emptyset$, the result is immediate. From (9.3), we can see that it suffices to show that if $A \subset \mathcal{X}$ is finite, $x \notin A$, and $A' = A \cup \{x\}$,

$$\mathcal{C}(A') = \mathcal{C}(A)\, \sqrt{F_x(A')}.$$

Suppose that $f : A' \to \mathbb{R}$ and extend f to $\mathcal{X}$ by setting $f \equiv 0$ on $\mathcal{X} \setminus A'$. We can write that

$$f = g + t\,h$$

where g vanishes on $\mathcal{X} \setminus A$; $t = f(x)$; and h is the function that is harmonic on A with $h(x) = 1$ and $h \equiv 0$ on $\mathcal{X} \setminus A'$. The edges in $e(A')$ are the edges in $e(A)$ plus those edges of the form $\{x, z\}$ with $z \in \mathcal{X} \setminus A$. Using this, we can see that

$$\mathcal{E}_{A'}(f) = \mathcal{E}_A(f) + \sum_{y \notin A'} q(x, y)\, t^2. \tag{9.32}$$

Also, by (9.29),

$$\mathcal{E}_A(f) = \mathcal{E}_A(g) + \mathcal{E}_A(th) = \mathcal{E}_A(g) + t^2\,\mathcal{E}_A(h),$$

which, combined with (9.32), gives

$$\exp\left\{-\frac{1}{2}\mathcal{E}_{A'}(f)\right\} = \exp\left\{-\frac{1}{2}\mathcal{E}_A(g)\right\}\exp\left\{-\frac{t^2}{2}\,\mathcal{E}_{A'}(h)\right\}.$$

Integrating over A first, we get

$$\begin{aligned} \mathcal{C}(A') &= \mathcal{C}(A)\int_{-\infty}^{\infty}\frac{1}{\sqrt{2\pi}}\,e^{-t^2\mathcal{E}_{A'}(h)/2}\,dt \\ &= \mathcal{C}(A)\int_{-\infty}^{\infty}\frac{1}{\sqrt{2\pi}}\,e^{-t^2/[2F_x(A')]}\,dt \\ &= \mathcal{C}(A)\,\sqrt{F_x(A')}. \end{aligned}$$

The second equality uses (9.30). □

Let $Q = Q_A$ as above and denote the entries of Q^n by $q_n(x, y)$. The Green's function on A is the matrix $G = (I - Q)^{-1}$; in other words, the expected number of visits to y by the chain starting at x equals

$$\sum_{n=0}^{\infty} q_n(x, y)$$

which is the (x, y) entry of $(I - Q)^{-1}$. Since Q is strictly subMarkov, $(I - Q)$ is symmetric, strictly positive definite, and $(I - Q)^{-1}$ is well defined. The next proposition uses the joint normal distribution as discussed in Section A.3.

Proposition 9.10.4 *Suppose that the random variables $\{Z_x : x \in A\}$ have a (mean zero) joint normal distribution with covariance matrix $G = (I - Q)^{-1}$. Then the distribution of the random function $f(x) = Z_x$ is the same as the Gaussian free field on A with Dirichlet boundary conditions.*

Proof Plugging $\Gamma = G = (I - Q)^{-1}$ into (A.14), we see that the joint density of $\{Z_x\}$ is given by

$$(2\pi)^{-\#(A)}\,[\det(I - Q)]^{1/2}\,\exp\left\{-\frac{f\cdot(I - Q)f}{2}\right\}.$$

But (9.28) implies that $f \cdot (I - Q)f = \mathcal{E}_A(f)$. Since this is a probability density, this shows that

$$C(A) = \sqrt{\frac{1}{\det(I - Q)}},$$

and hence (9.31) follows from Proposition 9.3.3. □

♣ The scaling limit of the Gaussian free field for random walk in $\mathbb{Z}^d$ is the Gaussian free field in $\mathbb{R}^d$. There are technical subtleties required in the definition. For example, if $d = 2$ and U is a bounded open set, we would like to define the Gaussian free field $\{Z_z : z \in U\}$ with Dirichlet boundary conditions to be the collection of random variables such that each finite collection $(Z_{z_1}, \ldots, Z_{z_k})$ has a joint normal distribution with covariance matrix $[G_U(z_i, z_j)]$. Here, G_U denotes the Green's function for Brownian motion in the domain. However, the Green's function $G_U(z, w)$ blows up as w approaches z, so this gives an infinite variance for the random variable Z_z. These problems can be overcome, but the collection $\{Z_z\}$ is not a collection of random variables in the usual sense.

♣ The proof of Proposition 9.10.3 is not really needed, given the quick proof in Proposition 9.10.4. However, we choose to include it since it more directly uses the loop measure interpretation of $F(A)$ rather than the interpretation as a determinant. Many computations with the loop measure have interpretations in the scaling limit.

Exercises

Exercise 9.1 Show that for all positive integers k

$$\sum_{j_1+\cdots+j_r=k} \frac{1}{r!\,(j_1, \ldots, j_r)} = 1.$$

Here are two possible approaches.

(i) Show that the number of permutations of k elements with exactly r cycles is

$$\sum_{j_1+\cdots+j_r=k} \frac{k!}{r!\,j_1 j_2, \ldots, j_r}.$$

(ii) Consider the equation

$$\frac{1}{1-t} = \exp\{-\log(1-t)\}.$$

Expand both sides in the power series in t and compare coefficients.

Exercise 9.2 Suppose that X_n is an irreducible Markov chain on a countable state space $\mathcal{X}$ and $A = \{x_1, \ldots, x_k\}$ is a proper subset of $\mathcal{X}$. Let $A_0 = A, A_j = A \setminus \{x_1, \ldots, x_j\}$. If $z \in V \subset \mathcal{X}$, let $g_V(z)$ denote the expected number of visits to z by the chain starting at z before leaving V.

(i) Show that

$$g_A(x_1)\, g_{A\setminus\{x_1\}}(x_2) = g_{A\setminus\{x_2\}}(x_1)\, g_A(x_2). \tag{9.33}$$

(ii) By iterating (9.33), show that the quantity

$$\prod_{j=1}^{k} g_{A_{j-1}}(x_j)$$

is independent of the ordering of $x_1, \ldots, x_k$.

Exercise 9.3 *(Karlin–McGregor)* Suppose that $X_n^1, \ldots, X_n^k$ are independent realizations from a Markov chain with transition probability q on a finite state space $\mathcal{X}$. Assume that $x_1, \ldots, x_k, y_1, \ldots, y_j \in \mathcal{X}$. Consider the event

$$V = V_n(y_1, \ldots, y_k) = \left\{X_m^i \neq X_m^j,\ m = 0, \ldots, n;\ X_n^j = y_j,\ 1 \leq j \leq n\right\}.$$

Show that

$$\mathbb{P}\{V \mid X_0^1 = x_1, \ldots, X_n^1 = x_n\} = \det\left[q_n(x_i, y_j)\right]_{1 \leq i,j \leq k},$$

where

$$q_n(x_i, y_j) = \mathbb{P}\left\{X_n^1 = y_j \mid X_0^1 = x_i\right\}.$$

Exercise 9.4 Suppose that Bernoulli trials are performed with a probability p of success. Let Y_n denote the number of failures before the nth success, and let $r(n)$ be the probability that Y_n is even. By definition, $r(0) = 1$. Give a recursive equation for $r(n)$ and use it to find $r(n)$. Use this to verify (9.16).

Exercise 9.5 Give the details of Lemma 9.9.4.

Exercise 9.6 Suppose that q is the weight arising from simple random walk in $\mathbb{Z}^d$. Suppose that A_1, A_2 are disjoint subsets of $\mathbb{Z}^d$ and $x \in \mathbb{Z}^d$. Let $p(x, A_1, A_2)$ denote the probability that a random walk starting at x enters A_2 and subsequently returns to x all without entering A_1. Let $g(x, A_1)$ denote the expected number of visits to x before entering A_1 for a random walk starting at x. Show that the unrooted loop measure of the set of loops in $\mathbb{Z}^d \setminus A_1$ that intersect both x and A_2 is bounded above by $p(x, A_1, A_2)\, g(x, A_1)$.

(Hint: for each unroooted loop that intersects both x and A_2 choose a (not necessarily unique) representative that is rooted at x and enters A_2 before its first return to x.)

Exercise 9.7 We continue the notation of Exercise 9.6 with $d \geq 3$. Choose an enumeration of $\mathbb{Z}^d = \{x_0, x_1, \ldots\}$ such that $j < k$ implies $|x_j| \leq |x_k|$.

(i) Show that there exists $c < \infty$ such that if $r > 0$, $u \geq 2$, and $|x_j| \leq r$,

$$p(x_j, A_{j-1}, \mathbb{Z}^d \setminus B_{ur}) \leq c_1\, |x_j|^{-2}\, (ur)^{2-d}.$$

(Hint: consider a path that starts at x_j, leaves $\mathcal{B}_{ur}$, and then returns to x_j without visiting A_{j-1}. Split such a curve into three pieces: the "beginning" up to the first visit to $\mathbb{Z}^d \setminus \mathcal{B}_{ur}$; the "end" which (with time reversed) is a walk from x_j to the first (last) visit to $\mathbb{Z}^d \setminus \mathcal{B}_{3|x_j|/2}$; and the "middle" which ties these walks together.)

(ii) Show that there exists $c_1 < \infty$, such that if $r > 0$ and $u \geq 2$, then the (unrooted) loop measure of the set of loops that intersects both $\mathcal{B}_r$ and $\mathbb{Z}^d \setminus \mathcal{B}_{ur}$ is bounded above by $c_1\, u^{2-d}$.

10

Intersection probabilities for random walks

10.1 Long-range estimate

In this section we prove a fundamental inequality concerning the probability of intersection of the paths of two random walks. If S_n is a random walk, we write

$$S[n_1, n_2] = \{S_n : n_1 \leq n \leq n_2\}.$$

Proposition 10.1.1 *If $p \in \mathcal{P}_d$, there exist c_1, c_2 such that for all $n \geq 2$,*

$$\begin{aligned} c_1\, \phi(n) &\leq \mathbb{P}\{S[0,n] \cap S[2n, 3n] \neq \emptyset\} \\ &\leq \mathbb{P}\{S[0,n] \cap S[2n, \infty) \neq \emptyset\} \leq c_2\, \phi(n), \end{aligned}$$

where

$$\phi(n) = \begin{cases} 1, & d < 4, \\ (\log n)^{-1}, & d = 4, \\ n^{(4-d)/2}, & d > 4. \end{cases}$$

♣ As $n \to \infty$, we get a result about Brownian motions. If B is a standard Brownian motion in $\mathbb{R}^d$, then

$$\mathbb{P}\{B[0,1] \cap B[2,3] \neq \emptyset\} \begin{cases} > 0, & d \leq 3 \\ = 0, & d = 4. \end{cases}$$

Four is the critical dimension in which Brownian paths just barely avoid each other.

Proof The upper bound is trivial for $d \leq 3$, and the lower bound for $d \leq 2$ follows from the lower bound for $d = 3$. Hence, we can assume that $d \geq 3$. We will assume that the walk is aperiodic (only a trivial modification is needed for the bipartite case). The basic strategy is to consider the number of intersections of the paths,

$$J_n = \sum_{j=0}^{n} \sum_{k=2n}^{3n} 1\{S_j = S_k\}, \qquad K_n = \sum_{j=0}^{n} \sum_{k=2n}^{\infty} 1\{S_j = S_k\}.$$

Note that

$$\mathbb{P}\{S[0,n] \cap S[2n,3n] \neq \emptyset\} = \mathbb{P}\{J_n \geq 1\},$$
$$\mathbb{P}\{S[0,n] \cap S[2n,\infty) \neq \emptyset\} = \mathbb{P}\{K_n \geq 1\}.$$

We will derive the following inequalities for $d \geq 3$,

$$c_1\, n^{(4-d)/2} \leq \mathbb{E}(J_n) \leq \mathbb{E}(K_n) \leq c_2\, n^{(4-d)/2}, \tag{10.1}$$

$$\mathbb{E}(J_n^2) \leq \begin{cases} c\, n, & d = 3, \\ c\, \log n, & d = 4, \\ c\, n^{(4-d)/2}, & d \geq 5. \end{cases} \tag{10.2}$$

Once these are established, the lower bound follows by the second-moment lemma (Lemma A.6.1),

$$\mathbb{P}\{J_n > 0\} \geq \frac{\mathbb{E}(J_n)^2}{4\,\mathbb{E}(J_n^2)}.$$

Let us write $p(n)$ for $\mathbb{P}\{S_n = 0\}$. Then,

$$\mathbb{E}(J_n) = \sum_{j=0}^{n} \sum_{k=2n}^{3n} p(k-j),$$

and similarly for $\mathbb{E}(K_n)$. Since $p(k-j) \asymp (k-j)^{-d/2}$, we get

$$\begin{aligned} \mathbb{E}(J_n) &\asymp \sum_{j=0}^{n} \sum_{k=2n}^{3n} \frac{1}{(k-j)^{d/2}} \\ &\asymp \sum_{j=0}^{n} \sum_{k=2n}^{3n} \frac{1}{(k-n)^{d/2}} \asymp \sum_{j=0}^{n} n^{1-(d/2)} \asymp n^{2-(d/2)}, \end{aligned}$$

and similarly for $\mathbb{E}(K_n)$. This gives (10.1). To bound the second moments, note that

$$\mathbb{E}(J_n^2) = \sum_{0 \le j,i \le n} \sum_{2n \le k,m \le 3n} \mathbb{P}\{S_j = S_k, S_i = S_m\}$$
$$\le 2 \sum_{0 \le j \le i \le n} \sum_{2n \le k \le m \le 3n} [\mathbb{P}\{S_j = S_k, S_i = S_m\} + \mathbb{P}\{S_j = S_m, S_i = S_k\}].$$

If $0 \le i, j \le n$ and $2n \le k \le m \le 3n$, then

$$\mathbb{P}\{S_j = S_k, S_i = S_m\} \le \left[\max_{l \ge n, x \in \mathbb{Z}^d} \mathbb{P}\{S_l = x\} \right] \left[\max_{x \in \mathbb{Z}^d} \mathbb{P}\{S_{m-k} = x\} \right]$$
$$\le \frac{c}{n^{d/2} (m-k+1)^{d/2}}.$$

The last inequality uses the LCLT. Therefore,

$$\mathbb{E}(J_n^2) \le c\, n^2 \sum_{2n \le k \le m \le 3n} \frac{1}{n^{d/2} (m-k+1)^{d/2}} \le c\, n^{2-(d/2)}$$
$$\times \sum_{0 \le k \le m \le n} \frac{1}{(m-k+1)^{d/2}}.$$

This yields (10.2).

The upper bound is trivial for $d = 3$ and for $d \ge 5$ it follows from (10.1) and the inequality $\mathbb{P}\{K_n \ge 1\} \le \mathbb{E}[K_n]$. Assume that $d = 4$. We will consider $\mathbb{E}[K_n \mid K_n \ge 1]$. On the event $\{K_n \ge 1\}$, let k be the smallest integer $\ge 2n$ such that $S_k \in S[0, n]$. Let j be the smallest index such that $S_k = S_j$. Then by the Markov property, given $[S_0, \ldots, S_k]$ and $S_k = S_j$, the expected value of K_{2n} is

$$\sum_{i=0}^{n} \sum_{l=k}^{\infty} \mathbb{P}\{S_l = S_i \mid S_k = S_j\} = \sum_{i=0}^{n} G(S_i - S_j).$$

Define a random variable, depending on $S_0, \ldots, S_n$,

$$Y_n = \min_{j=0,\ldots,n} \sum_{i=0}^{n} G(S_i - S_j).$$

For any $r > 0$, we have that

$$\mathbb{E}[K_n \mid K_n \ge 1, Y_n \ge r \log n] \ge r \log n.$$

Note that for each r,

$$\mathbb{P}\{Y_n < r \log n\} \leq (n+1)\,\mathbb{P}\left\{\sum_{i \leq n/2} G(S_i) < r\,\log n\right\}.$$

Using Lemma 10.1.2 below, we can find an r such that $\mathbb{P}\{Y_n < r \log n\} = o(1/\log n)$ But,

$$\begin{aligned} c \geq \mathbb{E}[K_n] &\geq \mathbb{P}\{K_n \geq 1; Y_n \geq r \log n\} \mathbb{E}[K_n \mid K_n \geq 1, Y_n \geq r \log n] \\ &\geq \mathbb{P}\{K_n \geq 1; Y_n \geq r \log n\}[r\,\log n]. \end{aligned}$$

Therefore,

$$\mathbb{P}\{K_n \geq 1\} \leq \mathbb{P}\{Y_n < r\,\log n\} + \mathbb{P}\{K_n \geq 1; Y_n \geq r\,\log n\} \leq \frac{c}{\log n}.$$

This finishes the proof except for the one lemma that we will now prove. □

Lemma 10.1.2 *Let $p \in \mathcal{P}_4$.*

(a) For every $\alpha > 0$, there exist c, r such that for all n sufficiently large,

$$\mathbb{P}\left\{\sum_{j=0}^{\xi_n - 1} G(S_j) \leq r\,\log n\right\} \leq c\,n^{-\alpha}.$$

(b) For every $\alpha > 0$, there exist c, r such that for all n sufficiently large,

$$\mathbb{P}\left\{\sum_{j=0}^{n} G(S_j) \leq r\,\log n\right\} \leq c\,n^{-\alpha}.$$

Proof It suffices to prove (a) when $n = 2^l$ for some integer l, and we write $\xi^k = \xi_{2^k}$. Since $G(x) \geq c/(|x|+1)^2$, we have that

$$\sum_{j=0}^{\xi^l - 1} G(S_j) \geq \sum_{k=1}^{l} \sum_{j=\xi^{k-1}}^{\xi^k - 1} G(S_j) \geq c \sum_{k=1}^{l} 2^{-2k}\,[\xi^k - \xi^{k-1}].$$

The reflection principle (Proposition 1.6.2) and the CLT show that for every $\epsilon > 0$, there is a $\delta > 0$ such that if n is sufficiently large, and $x \in \overline{\mathcal{C}}_{n/2}$, then $\mathbb{P}^x\{\xi_n \leq \delta\,n^2\} \leq \epsilon$. Let I_k denote the indicator function of the event

$\{\xi^k - \xi^{k-1} \leq \delta\, 2^{2k}\}$. Then we know that

$$\mathbb{P}(I_k = 1 \mid S_0, \ldots, S_{\xi^{k-1}}) \leq \epsilon.$$

Therefore, $J_l := \sum_{k=1}^{l} I_k$ is stochastically bounded by a binomial random variable with parameters l and ϵ. By exponential estimates for binomial random variables (see Lemma A.2.8), we can find an α such that

$$\mathbb{P}\{J_l \geq l/2\} \leq c\, 2^{-\alpha l}.$$

But on the event $\{J_l < l/2\}$ we know that

$$G(S_j) \geq c(l/2)\,\delta \geq r\, \log n,$$

where the r depends on α.

For part (b) we need only note that $\mathbb{P}\{n < \xi_{n^{1/4}}\}$ decays faster than any power of n and that

$$\mathbb{P}\left\{\sum_{j=0}^{n} G(S_j) \leq \frac{r}{4}\, \log n\right\} \leq \mathbb{P}\left\{\sum_{j=0}^{\xi_{n^{1/4}}} G(S_j) \leq r\, \log n^{1/4}\right\} + \mathbb{P}\{n < \xi_{n^{1/4}}\}.$$

□

♣ The proof of the upper bound for $d = 4$ in Proposition 10.1.1 can be compared to the proof of an easier estimate

$$\mathbb{P}\{0 \in S[n, \infty)\} \leq c\, n^{1-\frac{d}{2}}, \qquad d \geq 3.$$

To prove this, one uses the LCLT to show that the expected number of visits to the origin is $O(n^{1-\frac{d}{2}})$. On the event that $0 \in S[n, \infty)$, we consider the smallest $j \geq n$ such that $S_j = 0$. Then using the strong Markov property, one shows that the expected number of visits given at least one visit is $G(0, 0) < \infty$. In Proposition 10.1.1 we consider the event that $S[0, n] \cap S[2n, \infty) \neq \emptyset$ and try to take the "first" $(j, k) \in [0, n] \times [2n, \infty)$ such that $S_j = S_k$. This is not well defined since, if (i, l) is another pair, it might be the case that $i < j$ and $l > k$. To be specific, we choose the smallest k and then the smallest j with $S_j = S_k$. We then say that the expected number of intersections after this time is the expected number of intersections of $S[k, \infty)$ with $S[0, n]$. Since $S_k = S_j$, this is like the number of intersections of two random walks starting at the origin. In $d = 4$, this is of order $\log n$. However, because S_k, S_j have been chosen specifically, we cannot use a simple strong Markov property argument to assert this. This is why the extra lemma is needed.

10.2 Short-range estimate

We are interested in the probability that the paths of two random walks starting at the origin do not intersect up to some finite time. We discuss only the interesting dimensions $d \leq 4$. Let $S, S^1, S^2, \ldots$ be independent random walks starting at the origin with distribution $p \in \mathcal{P}_d$. If $0 < \lambda < 1$, let $T_\lambda, T_\lambda^1, T_\lambda^2, \ldots$ denote independent geometric random variables with killing rate $1 - \lambda$ and we write $\lambda_n = 1 - 1/n$. We would like to estimate

$$\mathbb{P}\{S(0, n] \cap S^1[0, n] = \emptyset\},$$

or

$$\mathbb{P}\left\{S(0, T_{\lambda_n}] \cap S^1[0, T_{\lambda_n}^2] = \emptyset\right\}.$$

The next proposition uses the long-range estimate to bound a different probability,

$$\mathbb{P}\left\{S(0, T_{\lambda_n}] \cap (S^1[0, T_{\lambda_n}^1] \cup S^2[0, T_{\lambda_n}^2]) = \emptyset\right\}.$$

Let

$$\begin{aligned} Q(\lambda) &= \sum_{y \in \mathbb{Z}^d} \mathbb{P}^{0,y}\{S[0, T_\lambda] \cap S^1[0, T_\lambda^1] \neq \emptyset\} \\ &= (1-\lambda)^2 \sum_{y \in \mathbb{Z}^d} \sum_{j=0}^{\infty} \sum_{k=0}^{\infty} \lambda^{j+k}\, \mathbb{P}^{0,y}\{S[0, j] \cap S^1[0, k] \neq \emptyset\}. \end{aligned}$$

Here, we write $\mathbb{P}^{x,y}$ to denote probabilities assuming that $S_0 = x, S_0^1 = y$. Using Proposition 10.1.1, one can show that as $n \to \infty$ (we omit the details),

$$Q(\lambda_n) \asymp \begin{cases} n^{d/2}, & d < 4 \\ n^2\, [\log n]^{-1}, & d = 4. \end{cases}$$

Proposition 10.2.1 *Suppose that S, S^1, S^2 are independent random walks starting at the origin with increment $p \in \mathcal{P}_d$. Let V_λ be the event that $0 \notin S^1(0, T_\lambda^1]$. Then,*

$$\mathbb{P}\left[V_\lambda \cap \{S(0, T_\lambda] \cap (S^1(0, T_\lambda^1] \cup S^2(0, T_\lambda^2]) = \emptyset\}\right] = (1-\lambda)^2\, Q(\lambda). \quad (10.3)$$

Proof Suppose that $\omega = [\omega_0 = 0, \ldots, \omega_n], \eta = [\eta_0, \ldots, \eta_m]$ are paths in $\mathbb{Z}^d$ with

$$p(\omega) := \prod_{j=1}^{n} p(\omega_{j-1}, \omega_j) > 0 \quad p(\eta) := \prod_{j=1}^{m} p(\eta_{j-1}, \eta_j) > 0.$$

Then, we can write that

$$Q(\lambda) = (1-\lambda)^2 \sum_{n=0}^{\infty} \sum_{m=0}^{\infty} \sum_{\omega,\eta} \lambda^{n+m}\, p(\omega)\, p(\eta),$$

where the last sum is over all paths ω, η with $|\omega| = n, |\eta| = m, \omega_0 = 0$ and $\omega \cap \eta \neq \emptyset$. For each such pair (ω, η) we define a 4-tuple of paths starting at the origin $(\omega^-, \omega^+, \eta^-, \eta^+)$ as follows. Let

$$s = \min\{j : \omega_j \in \eta\}, \qquad t = \min\{k : \eta_k = \omega_s\}.$$

$$\omega^- = [\omega_s - \omega_s, \omega_{s-1} - \omega_s, \ldots, \omega_0 - \omega_s],$$

$$\omega^+ = [\omega_s - \omega_s, \omega_{s+1} - \omega_s, \ldots, \omega_n - \omega_s],$$

$$\eta^- = [\eta_t - \eta_t, \eta_{t-1} - \eta_t, \ldots, \eta_0 - \eta_t],$$

$$\eta^+ = [\eta_t - \eta_t, \eta_{t+1} - \eta_t, \ldots, \eta_m - \eta_t].$$

Note that $p(\omega) = p(\omega^-)\, p(\omega^+), p(\eta) = p(\eta^-)\, p(\eta^+)$. Also,

$$0 \notin [\eta_1^-, \ldots, \eta_t^-], \qquad [\omega_1^-, \ldots, \omega_s^-] \cap [\eta^- \cup \eta^+] = \emptyset. \tag{10.4}$$

Conversely, for each 4-tuple $(\omega^-, \omega^+, \eta^-, \eta^+)$ of paths starting at the origin satisfying (10.4), we can find a corresponding (ω, η) with $\omega_0 = 0$ by inverting this procedure. Therefore,

$$Q(\lambda) = (1-\lambda)^2$$
$$\times \sum_{0 \le n_-, n_+, m_-, m_+} \sum_{\omega, \omega_+, \eta_-, \eta_+} \lambda^{n_- + n_+ + m_- + m_+}\, p(\omega_-) p(\omega_+) p(\eta_-) p(\eta_+),$$

where the last sum is over all $(\omega^-, \omega^+, \eta^-, \eta^+)$ with $|\omega_-| = n_-, |\omega_+| = n_+, |\eta_-| = m_-, |\eta_+| = m_+$ satisfying (10.4). Note that there is no restriction on the path ω^+. Hence, we can sum over n_+ and ω_+ to get

$$Q(\lambda) = (1-\lambda) \sum_{0 \le n, m_-, m_+} \sum_{\omega, \eta_-, \eta_+} \lambda^{n + m_- + m_+}\, p(\omega) p(\eta_-) p(\eta_+),$$

But it is easy to check that the left-hand side of (10.3) equals

$$(1-\lambda)^3 \sum_{0\le n,m_-,m_+} \sum_{\omega,\eta_-,\eta_+} \lambda^{n+m_-+m_+}\, p(\omega)p(\eta_-)p(\eta_+).$$

□

Corollary 10.2.2 *For* $d = 2, 3, 4$,

$$\begin{aligned}
&\mathbb{P}\{S(0,n] \cap (S^1(0,n] \cup S^2[0,n]) = \emptyset\}\\
&\asymp \mathbb{P}\{S(0,T_{\lambda_n}] \cap (S^1(0,T^1_{\lambda_n}] \cup S^2[0,T^2_{\lambda_n}]) = \emptyset\}\\
&\asymp (1-\lambda_n)^2\, Q(\lambda_n)\\
&\asymp \begin{cases} n^{\frac{d-4}{2}}, & d = 2,3\\ (\log n)^{-1}, & d = 4 \end{cases}
\end{aligned}$$

Proof (Sketch) We have already noted the last relation. The previous proposition almost proves the second relation. It gives a lower bound. Since $\mathbb{P}\{T_{\lambda_n} = 0\} = 1/n$, the upper bound will follow if we show that

$$\mathbb{P}[V_{\lambda_n} \mid S(0,T_{\lambda_n}] \cap (S^1(0,T^1_{\lambda_n}] \cup S^2[0,T^2_{\lambda_n}]) = \emptyset, T_{\lambda_n} > 0] \ge c > 0. \quad (10.5)$$

We leave this as an exercise (Exercise 10.1).

One direction of the first relation can be proved by considering the event $\{T_{\lambda_n}, T^1_{\lambda_n}, T^2_{\lambda_n} \le n\}$ which is independent of the random walks and whose probability is bounded below by a $c > 0$ uniformly in n. This shows that

$$\begin{aligned}
&\mathbb{P}\{S(0,T_{\lambda_n}] \cap (S^1(0,T^1_{\lambda_n}] \cup S^2[0,T^2_{\lambda_n}]) = \emptyset\}\\
&\quad \ge c\,\mathbb{P}\{S(0,n] \cap (S^1(0,n] \cup S^2[0,n]) = \emptyset\}.
\end{aligned}$$

For the other direction, it suffices to show that

$$\begin{aligned}
&\mathbb{P}\{S(0,T_{\lambda_n}] \cap (S^1(0,T^1_{\lambda_n}] \cup S^2[0,T^2_{\lambda_n}]) = \emptyset;\\
&\quad T_{\lambda_n}, T^1_{\lambda_n}, T^2_{\lambda_n} \ge n\} \ge c\,(1-\lambda_n)^2\, Q(\lambda_n).
\end{aligned}$$

This can be established by going through the construction in proof of Proposition 10.2.1. We leave this to the interested reader.

□

10.3 One-sided exponent

Let

$$q(n) = \mathbb{P}\{S(0,n] \cap S^1(0,n] = \emptyset\}.$$

This is not an easy quantity to estimate. If we let

$$Y_n = \mathbb{P}\left\{S(0,n] \cap S^1(0,n] = \emptyset \mid S(0,n]\right\},$$

then we can write

$$q(n) = \mathbb{E}[Y_n].$$

Note that if S, S^1, S^2 are independent, then

$$\mathbb{E}[Y_n^2] = \mathbb{P}\left\{S(0,n] \cap (S^1(0,n] \cup S^2(0,n]) = \emptyset\right\}.$$

Hence, we see that

$$\mathbb{E}[Y_n^2] \asymp \begin{cases} (\log n)^{-1}, & d = 4 \\ n^{\frac{d-4}{2}}, & d < 4 \end{cases} \tag{10.6}$$

Since $0 \le Y_n \le 1$, we know that

$$\mathbb{E}[Y_n^2] \le \mathbb{E}[Y_n] \le \sqrt{\mathbb{E}[Y_n^2]}. \tag{10.7}$$

If it were true that $(\mathbb{E}[Y_n])^2 \asymp \mathbb{E}[Y_n^2]$, we would know how $\mathbb{E}[Y_n]$ behaves. Unfortunately, this is not true for small d.

As an example, consider simple random walk on $\mathbb{Z}$. In order for $S(0,n]$ to avoid $S^1[0,n]$, either $S(0,n] \subset \{1,2,\ldots\}$ and $S^1[0,n] \subset \{0,-1,-2,\ldots\}$ or $S(0,n) \subset \{-1,-2,\ldots\}$ and $S^1[0,n] \subset \{0,1,2,\ldots\}$. The gambler's ruin estimate shows that the probability of each of these events is comparable to $n^{-1/2}$ and hence, that

$$\mathbb{E}[Y_n] \asymp n^{-1}, \qquad \mathbb{E}[Y_n^2] \asymp n^{-3/2}.$$

Another way of saying this is that

$$\mathbb{P}\{S(0,n] \cap S^2(0,n] = \emptyset\} \asymp n^{-1},$$

$$\mathbb{P}\{S(0,n] \cap S^2(0,n] = \emptyset \mid S(0,n] \cap S^1(0,n] = \emptyset\} \asymp n^{-1/2}.$$

For $d = 4$, it is true that $(\mathbb{E}[Y_n])^2 \asymp \mathbb{E}[Y_n^2]$. For $d < 4$, the relation $(\mathbb{E}[Y_n])^2 \asymp \mathbb{E}[Y_n^2]$ does not hold. The *intersection exponent* $\zeta = \zeta_d$ is defined by saying $\mathbb{E}[Y_n] \asymp n^{-\zeta}$. One can show the existence of the exponent by first showing the existence of a similar exponent for Brownian motion (this is fairly easy) and then showing that the random walk has the same exponent (this takes more work; see Lawler and Puckette (2000)). This argument does not establish the value of ζ. For $d = 2$, it is known that $\zeta = 5/8$. The techniques of the proof in Lawler *et al.* (2001) use conformal invariance of Brownian motion and a process called the *Schramm–Loewner evolution* (SLE). For $d = 3$, the exact value is not known (and perhaps will never be known). Corollary 10.2.2 and (10.7) imply that $1/4 \le \zeta \le 1/2$, and it has been proved that both inequalities are actually strict. Numerical simulations suggest a value of about 0.29.

♣ The relation $\mathbb{E}[Y_n^2] \approx \mathbb{E}[Y_n]^2$ or equivalently

$$\mathbb{P}\{S(0,n] \cap S^2(0,n] = \emptyset\} \asymp \mathbb{P}\{S(0,n] \cap S^2(0,n] = \emptyset \mid S(0,n] \cap S^1(0,n] = \emptyset\}$$

is sometimes called *mean-field* behavior. Many systems in statistical physics have mean-field behavior above a critical dimension and also exhibit such behavior at the critical dimension with a logarithmic correction. Below the critical dimension they do not have mean-field behavior. The study of the exponents $\mathbb{E}[Y_n^r] \asymp n^{-\zeta(r)}$ sometimes goes under the name of *multifractal analysis*. The function $\zeta_2(r)$ is known for all $r \ge 0$, see Lawler *et al.* (2001).

Exercise 10.1 Prove (10.5).

11

Loop-erased random walk

Loop-erased walks were introduced in Chapter 9 as a measure on self-avoiding paths obtained by starting with a (not necessarily probability) measure on intersecting paths and erasing loops chronologically. In this chapter we will study the corresponding process obtained when one starts with a probability measure on paths; we call this the loop-erased random walk (LERW) or the Laplacian random walk. We will consider only LERW derived from simple random walk in $\mathbb{Z}^d$, but many of the ideas can be extended to LERWs obtained from Markov chains.

♣ The terms loop-erased walk and loop-erased random walk tend to be used synonymously in the literature. We will make a distinction in this book, reserving LERW for a stochastic process associated to a probability measure on paths.

Throughout this section, S_n will denote a simple random walk in $\mathbb{Z}^d$. We write $S_n^1, S_n^2, \ldots$ for independent realizations of the walk. We let $\tau_A, \overline{\tau}_A$ be defined as in (4.27) and we use $\tau_A^j, \overline{\tau}_A^j$ to be the corresponding quantities for S^j. We let

$$p_n^A(x,y) = p_n^A(y,x) = \mathbb{P}^x\{S_n = y : \overline{\tau}_A > n\}.$$

If $x \notin A$, then $p_n^A(x,y) = 0$ for all n, y.

11.1 *h*-processes

We will see that the LERW looks like a random walk conditioned to avoid its past. As the LERW grows, the "past" of the walk also grows; this is an example of what is called a "moving boundary." In this section we consider the process obtained by conditioning random walk to avoid a fixed set. This is a special case of an h-process.

Suppose that $A \subset \mathbb{Z}^d$ and $h : \mathbb{Z}^d \to [0, \infty)$ is a strictly positive and harmonic function on A that vanishes on $\mathbb{Z}^d \setminus \overline{A}$. Let

$$(\partial A)_+ = (\partial A)_{+,h} = \{y \in \partial A : h(y) > 0\} = \{y \in \mathbb{Z}^d \setminus A : h(y) > 0\}.$$

The *(Doob) h-process (with reflecting boundary)* is the Markov chain on $\overline{A}$ with transition probability $\tilde{q} = \tilde{q}^{A,h}$ defined as follows.

- If $x \in A$ and $|x - y| = 1$,

$$\tilde{q}(x, y) = \frac{h(y)}{\sum_{|z-x|=1} h(z)} = \frac{h(y)}{2d\, h(x)}. \tag{11.1}$$

- If $x \in \partial A$ and $|x - y| = 1$,

$$\tilde{q}(x, y) = \frac{h(y)}{\sum_{|z-x|=1} h(z)}.$$

The second equality in (11.1) follows by the fact that $\mathcal{L}h(x) = 0$. The definition of $\tilde{q}(x, y)$ for $x \in \partial A$ is the same as that for $x \in A$, but we write it separately to emphasize that the second equality in (11.1) does not necessarily hold for $x \in \partial A$. The *h-process stopped at* $(\partial A)_+$ is the chain with transition probability $q = q^{A,h}$ which equals $\tilde{q}$ except for

- $q(x, x) = 1, \quad x \in (\partial A)_+$.

Note that if $x \in \partial A \setminus (\partial A)_+$, then $q(y, x) = \tilde{q}(y, x) = 0$ for all $y \in \overline{A}$. In other words, the chain can start in $x \in \partial A \setminus (\partial A)_+$, but it cannot visit there at positive times. Let $\tilde{q}_n = \tilde{q}_n^{A,h}, q_n = q_n^{A,h}$ denote the usual n-step transition probabilities for the Markov chains.

Proposition 11.1.1 *If* $x, y \in A$,

$$q_n^{A,h}(x, y) = p_n^A(x, y)\, \frac{h(y)}{h(x)}.$$

In particular, $q_n^{A,h}(x, x) = p_n^A(x, x)$.

Proof Let

$$\omega = [\omega_0 = x, \omega_1, \ldots, \omega_n = y]$$

be a nearest neighbor path with $\omega_j \in A$ for all j. Then the probability that first n points of the h-process starting at x are $\omega_1, \ldots, \omega_n$ in order is

$$\prod_{j=1}^{n} \frac{h(\omega_j)}{2d\, h(\omega_{j-1})} = (2d)^{-n} \frac{h(y)}{h(x)}.$$

By summing over all paths ω, we get the proposition. □

♣ If we consider $q^{A,h}$ and p^A as measures on finite paths $\omega = [\omega_0, \ldots, \omega_n]$ in A, then we can rephrase the proposition as

$$\frac{dq^{A,h}}{dp^A}(\omega) = \frac{h(\omega_n)}{h(\omega_0)}.$$

Formulations like this in terms of Radon–Nikodym derivatives of measures can be extended to measures on continuous paths such as Brownian motion.

♣ The h-process can be considered as the random walk "weighted by the function h". One can define this for any positive function on A, even if h is not harmonic, using the first equality in (11.1). However, Proposition 11.1.1 will not hold if h is not harmonic.

Examples

- If $A \subset \mathbb{Z}^d$ and $V \subset \partial A$, let

$$h_{V,A}(x) = \mathbb{P}^x\{S_{\overline{\tau}_A} \in V\} = \sum_{y \in V} H_A(x, y).$$

Assume that $h_{V,A}(x) > 0$ for all $x \in A$. By definition, $h_{V,A} \equiv 1$ on V and $h_{V,A} \equiv 0$ on $\mathbb{Z}^d \setminus (A \cup V)$. The $h_{V,A}$-process corresponds to *simple random walk conditioned to leave A at V*. We usually consider the version stopped at $V = (\partial A)_+$.

– Suppose that $x \in \partial A \setminus V$ and

$$H_{\partial A}(x, V) := \mathbb{P}^x\{S_1 \in A; S_{\tau_A} \in V\} = \sum_{y \in A} H_{\partial A}(x, y) > 0.$$

If $|x - y| > 1$ for all $y \in V$, then the excursion measure as defined in Section 9.6 corresponding to paths from x to V in A *normalized to be a probability measure* is the $h_{V,A}$-process. If there is a $y \in V$ with $|x-y| = 1$, the $h_{V,A}$-process allows an immediate transition to y while the normalized excursion measure does not.

- Let $A = \mathcal{H} = \{x + iy \in \mathbb{Z} \times i\mathbb{Z} : y > 0\}$ and $h(z) = \text{Im}(z)$. Then h is a harmonic function on $\mathcal{H}$ that vanishes on ∂A. This h-process corresponds to simple random walk conditioned never to leave $\mathcal{H}$ and is sometimes called an *$\mathcal{H}$-excursion*. With probability one this process never leaves $\mathcal{H}$. Also, if $q = q^{\mathcal{H},h}$ and $x + iy \in \mathcal{H}$,

$$q(x+iy, (x \pm 1) + iy) = \frac{1}{4},$$

$$q(x+iy, x+i(y+1)) = \frac{y+1}{4y}, \qquad q(x+iy, +i(y-1)) = \frac{y-1}{4y}.$$

- Suppose that A is a proper subset of $\mathbb{Z}^d$ and $V = \partial A$. Then, the $h_{V,A}$-process is *simple random walk conditioned to leave A*. If $d = 1, 2$ or $\mathbb{Z}^d \setminus A$ is a recurrent subset of $\mathbb{Z}^d$, then $h_{V,A} \equiv 1$ and the $h_{V,A}$-process is the same as simple random walk.
- Suppose that A is a connected subset of $\mathbb{Z}^d, d \geq 3$ such that

$$h_{\infty,A}(x) := \mathbb{P}^x\{\overline{\tau}_A = \infty\} > 0.$$

  Then the $h_{\infty,A}$-process is *simple random walk conditioned to stay in A*.
- Let A be a connected subset of $\mathbb{Z}^2$ such that $\mathbb{Z}^2 \setminus A$ is finite and nonempty, and let

$$h(x) = a(x) - \mathbb{E}^x\left[a(S_{\overline{\tau}_A})\right],$$

  be the unique function that is harmonic on A, vanishes on ∂A, and satisfies $h(x) \sim (2/\pi) \log |x|$ as $x \to \infty$; see (6.40). Then the h-process is *simple random walk conditioned to stay in A*. Note that this "conditioning" is on an event of probability zero. Using (6.40), we can see that this is the limit as $n \to \infty$ of the h_{V_n,A_n} processes where

$$A_n = A \cap \{|z| < n\}, \qquad V_n = \partial A_n \cap \{|z| \geq n\}.$$

  Note that for large n, $V_n = \partial \mathcal{B}_n$.

11.2 Loop-erased random walk

Suppose that $A \subset \mathbb{Z}^d, V \subset \partial A$, and $x \in A$ with that $h_{V,A}(x) > 0$. The *LERW* from x to V in A is the probability measure on paths obtained by taking the $h_{V,A}$-process stopped at V and erasing loops. We can define the walk equivalently as follows.

- Take a simple random walk S_n started at x and stopped when it reaches ∂A. Condition on the event (of positive probability) $\{S_{\tau_A} \in V\}$. The conditional probability gives a probability measure on (finite) paths

$$\omega = \left[S_0 = x, S_1, \ldots, S_n = S_{\tau_A}\right].$$

- Erase loops from each ω which produces a self-avoiding path

$$\eta = L(\omega) = [\hat{S}_0 = x, \hat{S}_1, \ldots, \hat{S}_m = S_{\tau_A}],$$

  with $\hat{S}_1, \ldots, \hat{S}_{m-1} \in A$. We now have a probability measure on self-avoiding paths from x to V, and this is the LERW.

Similarly, if $x \in \partial A \setminus V$ with $\mathbb{P}^x\{S_{\tau_A} \in V\} > 0$, we define LERW from x to V in A by erasing loops from the $h_{V,A}$-process started at x stopped at V. If $x \in V$, we define LERW from x to V to be the trivial path of length zero.

We write the LERW as

$$\hat{S}_0, \hat{S}_1, \ldots, \hat{S}_\rho.$$

Here, ρ is the length of the loop-erasure of the h-process.

The LERW gives a probability measure on paths which we give explicitly in the next proposition. We will use the results and notations from Chapter 9 where the weight q from that chapter is the weight associated to simple random walk, $q(x, y) = 1/2d$ if $|x - y| = 1$.

Proposition 11.2.1 *Suppose that $V \subset \partial A$, $x \in \overline{A} \setminus V$ and $\hat{S}_0, \hat{S}_1, \ldots, \hat{S}_\rho$ is LERW from x to V in A. Suppose that $\eta = [\eta_0, \ldots, \eta_n]$ is a self-avoiding path with $\eta_0 = x \in A$, $\eta_n \in V$, and $\eta_j \in A$ for $0 < j < n$. Then*

$$\mathbb{P}\{\rho = n; [\hat{S}_0, \ldots, \hat{S}_n] = \eta\} = \frac{1}{(2d)^n \, \mathbb{P}^x\{S_{\tau_A} \in V\}} \, F_\eta(A).$$

Proof This is proved in the same way as Proposition 9.5.1. The extra term $\mathbb{P}^x\{S_{\tau_A} \in V\}$ comes from the normalization to be a probability measure. □

If $\omega = [\omega_0, \omega_1, \ldots, \omega_m]$ and ω^R denotes the reversed path $[\omega_m, \omega_{m-1}, \ldots, \omega_0]$, it is not necessarily true that $L(\omega^R) = [L(\omega)]^R$ (the reader might want to find an example). However, the last proposition shows that for any self-avoiding path η with appropriate endpoints, the probability that LERW produces η depends only on the set $\{\eta_1, \ldots, \eta_{n-1}\}$. For this reason we have the following corollary which shows that the distribution of LERW is reversible.

Corollary 11.2.2 (reversibility of LERW) *Suppose that $x, y \in \partial A$ and $\hat{S}_0, \hat{S}_1, \ldots, \hat{S}_\rho$ is LERW from x to y in A. Then the distribution of $\hat{S}_\rho, \hat{S}_{\rho-1}, \ldots, \hat{S}_0$ is that of LERW from y to x.*

Proposition 11.2.3 *If $x \in A$ with $h_{V,A}(x) > 0$, then the distribution of LERW from x to V in A stopped at V is the same as that of LERW from x to V in $A \setminus \{x\}$ stopped at V.*

Proof Let $X_0, X_1, \ldots$ denote an $h_{V,A}$-process started at x, and let $\tau = \tau_A$ be the first time that the walk leaves A, which with probability one is the first time that the walk visits V. Let

$$\sigma = \max\{m < \tau_A : X_m = x\}.$$

Then, using last-exit decomposition ideas (see Proposition 4.6.5) and Proposition 11.1.1, the distribution of

$$[X_\sigma, X_{\sigma+1}, \ldots, X_{\tau_A}]$$

is the same as that of an $h_{V,A}$-process stopped at V conditioned not to return to x. This is the same as an $h_{V,A\setminus\{x\}}$-process. □

If $x \in \partial A \setminus V$, then the first step $\hat{S}_1$ of the LERW from x to V in A has the same distribution as the first step of the $h_{V,A}$-process from x to V. Hence,

$$\mathbb{P}^x\{\hat{S}_1 = y\} = \frac{h_{V,A}(y)}{\sum_{|z-x|=1} h_{V,A}(z)}.$$

Proposition 11.2.4 *Suppose that $x \in \overline{A} \setminus V$ and $\hat{S}_0, \ldots, \hat{S}_\rho$ denotes LERW from x to V in A. Suppose that $\eta = [\eta_0, \ldots, \eta_m]$ is a self-avoiding path with $\eta_0 = x$ and $\eta_1, \ldots, \eta_m \in A$. Then*

$$\mathbb{P}^x\{\rho > m; [\hat{S}_0, \ldots, \hat{S}_m] = \eta\} = \frac{\mathbb{P}^{\eta_m}\{S_{\tau_{A\setminus\eta}} \in V\}}{(2d)^m\, \mathbb{P}^x\{S_{\tau_A} \in V\}}\, F_\eta(A).$$

Proof Let $\omega = [\omega_0, \ldots, \omega_n]$ be a nearest neighbor path with $\omega_0 = x, \omega_n \in V$ and $\omega_0, \ldots, \omega_{n-1} \in A$ such that the length of $LE(\omega)$ is greater than m and the first m steps of $LE(\omega)$ agree with η. Let

$$s = \max\{j : \omega_j = \eta_m\}$$

and write $\omega = \omega^- \oplus \omega^+$ where

$$\omega^- = [\omega_0, \omega_1, \ldots, \omega_s], \qquad \omega^+ = [\omega_s, \omega_{s+1}, \ldots, \omega_n].$$

Then $L(\omega^-) = \eta$ and ω^+ is a nearest neighbor path from η_m to V with

$$\omega_s = \eta_m, \qquad \{\omega_{s+1}, \ldots, \omega_{n-1}\} \in A \setminus \eta, \qquad \omega_n \in V. \tag{11.2}$$

Every such ω can be obtained by concatenating a ω^- in A with $L(\omega^-) = \eta$ with a ω^+ satisfying (11.2). The total measure of the set of ω^- is given by $(2d)^{-m}\, F_\eta(A)$ and the total measure of the set of ω^+ is given by $\mathbb{P}^{\eta_m}\{S_{\tau_{A\setminus\eta}} \in V\}$. Again, the term $\mathbb{P}^x\{S_{\tau_A} \in V\}$ comes from the normalization to make the LERW a probability measure. $\square$

The LERW is not a Markov process. However, we can consider the LERW from x to V in A as a Markov chain on a different state space. Fix V, and consider the state space $\mathcal{X}$ of ordered pairs (x, A) with $x \in \mathbb{Z}^d, A \subset \mathbb{Z}^d \setminus (V \cup \{x\})$ and either $x \in V$ or $\mathbb{P}^x\{S_{\tau_A} \in V\} > 0$. The states $(x, A), x \in V$ are absorbing states. For other states, the probability of the transition

$$(x, A) \longrightarrow (y, A \setminus \{y\})$$

is the same as the probability that an $h_{V,A}$-process starting at x takes its first step to y. The fact that this is a Markov chain is sometimes called the *domain Markov property* for LERW.

11.3 LERW in $\mathbb{Z}^d$

The loop-erased random walk in $\mathbb{Z}^d$ is the process obtained by erasing the loops from the path of a d-dimensional simple random walk. The $d = 1$ case is trivial, so we will focus on $d \geq 2$. We will use the term self-avoiding path for a nearest neighbor path that is self-avoiding.

11.3.1 $d \geq 3$

The definition of LERW is easier in the transient case $d \geq 3$, for then we can take the infinite path

$$[S_0, S_1, S_2, \ldots]$$

and erase loops chronologically to obtain the path

$$[\hat{S}_0, \hat{S}_1, \hat{S}_2, \ldots].$$

To be precise, we let

$$\sigma_0 = \max\{j \geq 0 : S_j = 0\},$$

and for $k > 0$,

$$\sigma_k = \max\{j > \sigma_{k-1} : S_j = S_{\sigma_{k-1}+1}\},$$

and then

$$[\hat{S}_0, \hat{S}_1, \hat{S}_2, \ldots] = [S_{\sigma_0}, S_{\sigma_1}, S_{\sigma_2}, \ldots].$$

♣ It is convenient to define chronological erasing as above by considering the *last* visit to a point. It is not difficult to see that this gives the same path as obtained by "nonanticipating" loop erasure; i.e. every time one visits a point that is on the path one erases all the points in between.

The following properties follow from the previous sections in this chapter and we omit the proofs.

- Given that $\hat{S}_0, \ldots, \hat{S}_m$, the distribution of $\hat{S}_{m+1}$ is that of the h_{∞,A_m}-process starting at $\hat{S}_m$ where $A_m = \mathbb{Z}^d \setminus \{\hat{S}_0, \ldots, \hat{S}_m\}$. Indeed,

$$\mathbb{P}\{\hat{S}_{m+1} = x \mid [\hat{S}_0, \ldots, \hat{S}_m]\} = \frac{h_{\infty,A_m}(x)}{\sum_{|y-\hat{S}_m|=1} h_{\infty,A_m}(y)}, \qquad |x - \hat{S}_m| = 1.$$

- If $\eta = [\eta_0, \ldots, \eta_m]$ is a self-avoiding path with $\eta_0 = 0$,

$$\mathbb{P}\left\{[\hat{S}_0, \ldots, \hat{S}_m] = \eta\right\} = \frac{\mathrm{Es}_{A_m}(\eta_m)}{(2d)^m} F_\eta(\mathbb{Z}^d) = \frac{\mathrm{Es}_{A_m}(\eta_m)}{(2d)^m} \prod_{j=0}^{m} G_{A_{j-1}}(\eta_j, \eta_j).$$

Here, $A_{-1} = \mathbb{Z}^d$.

- Suppose that $\mathbb{Z}^d \setminus A$ is finite,

$$A^r = A \cap \{|z| < r\}, \qquad V^r = \partial A^r \cap \{|z| \geq r\},$$

and $\hat{S}_0^{(r)}, \ldots, \hat{S}_m^{(r)}$ denotes (the first m steps of) a LERW from 0 to V^r in A^r. Then, for every self-avoiding path η,

$$\mathbb{P}\left\{[\hat{S}_0, \ldots, \hat{S}_m] = \eta\right\} = \lim_{r \to \infty} \mathbb{P}\left\{[\hat{S}_0^{(r)}, \ldots, \hat{S}_m^{(r)}] = \eta\right\}.$$

11.3.2 $d = 2$

There are a number of ways to define LERW in $\mathbb{Z}^2$; all the reasonable ones give the same answer. One possibility (see Exercise 11.2) is to take simple random walk conditioned not to return to the origin and erase loops. We take a different approach in this section and define it as the limit as $N \to \infty$ of the measure obtained by erasing loops from simple random walk stopped when it reaches $\partial \mathcal{B}_N$. This approach has the advantage that we obtain an error estimate on the rate of convergence.

Let S_n denote simple random walk starting at the origin in $\mathbb{Z}^2$. Let $\hat{S}_{0,N}, \ldots, \hat{S}_{\rho_N,N}$ denote LERW from 0 to $\partial \mathcal{B}_N$ in $\mathcal{B}_N$. A This can be obtained by erasing loops from

$$[S_0, S_1, \ldots, S_{\xi_N}].$$

As noted in Section 11.2, if we condition on the event that $\tau_0 > \xi_N$, we get the same distribution on the LERW. Let Ξ_N denote the set of self-avoiding paths $\eta = [0, \eta_1, \ldots, \eta_k]$ with $\eta_1, \ldots, \eta_{k-1} \in \mathcal{B}_N$, and $\eta_N \in \partial \mathcal{B}_N$ and let ν_N denote the corresponding probability measure on Ξ_N,

$$\nu_N(\eta) = \mathbb{P}\{[\hat{S}_{0,N}, \ldots, \hat{S}_{n,N}] = \eta\}.$$

If $n < N$, we can also consider ν_N as a probability measure on Ξ_n, by considering the path η up to the first time it visits $\partial \mathcal{B}_n$ and removing the rest of the path. The goal of this subsection is to prove the following result.

Proposition 11.3.1 *Suppose that* $d = 2$ *and* $n < \infty$. *For each* $N \geq n$, *consider* ν_N *as a probability measure on* Ξ_n. *Then the limit*

$$\nu = \lim_{N \to \infty} \nu_N,$$

exists. Moreover, for every $\eta \in \Xi_n$.

$$\nu_N(\eta) = \nu(\eta)\left[1 + O\left(\frac{1}{\log(N/n)}\right)\right], \qquad N \geq 2n. \tag{11.3}$$

♣ To be more specific, (11.3) means that there is a c such that for all $N \geq 2n$ and all $\eta \in \Xi_n$,

$$\left|\frac{\nu_N(\eta)}{\nu(\eta)} - 1\right| \leq \frac{c}{\log(N/n)}.$$

The proof of this proposition will require an estimate on the loop measure as defined in Chapter 9. We start by stating the following proposition, which is an immediate application of Proposition 11.2.4 to our situation.

Proposition 11.3.2 *If* $n \leq N$ *and* $\eta = [\eta_0, \ldots, \eta_k] \in \Xi_n$,

$$\nu_N(\eta) = \frac{\mathbb{P}^{\eta_k}\left\{\xi_N < \tau_{\mathbb{Z}^d \setminus \eta}\right\}}{(2d)^{|\eta|}} F_\eta(\mathcal{B}_N) = \frac{\mathbb{P}^{\eta_k}\left\{\xi_N < \tau_{\mathbb{Z}^d \setminus \eta}\right\}}{(2d)^{|\eta|}\,\mathbb{P}\{\xi_N < \tau_0\}} F_\eta(\mathcal{B}_N \setminus \{0\}).$$

♣ Since $0 \in \eta$,

$$F_\eta(\mathcal{B}_N) = G_{\mathcal{B}_N}(0,0)\, F_\eta(\mathcal{B}_N \setminus \{0\}) = \mathbb{P}\{\xi_N < \tau_0\}^{-1}\, F_\eta(\mathcal{B}_N \setminus \{0\}),$$

which shows the second equality in the proposition.

We will say that a loop disconnects the origin from a set A if there is no self-avoiding path starting at the origin ending at A that does not intersect the loop; in particular, loops that intersect the origin disconnect the origin from all sets. Let $\overline{m}$ denote the unrooted loop measure for simple random walk as defined in Chapter 9.

Lemma 11.3.3 *There exists* $c < \infty$ *such that the following holds for simple random walk in* $\mathbb{Z}^2$. *For every* $n < N/2 < \infty$ *consider the set* $U = U(n,N)$ *of unrooted loops* $\overline{\omega}$ *satisfying*

$$\overline{\omega} \cap \mathcal{B}_n \neq \emptyset, \qquad \overline{\omega} \cap (\mathbb{Z}^d \setminus \mathcal{B}_N) \neq \emptyset$$

and such that $\overline{\omega}$ *does not disconnect the origin from* $\partial\mathcal{B}_n$. *Then*

$$\overline{m}(U) \leq \frac{c}{\log(N/n)}.$$

Proof Order $\mathbb{Z}^2 = \{x_0 = 0, x_1, x_2, \ldots\}$ so that $j < k$ implies that $|x_j| \le |x_k|$. Let $A_k = \mathbb{Z}^2 \setminus \{x_0, \ldots, x_{k-1}\}$. For each unrooted loop $\overline{\omega}$, let k be the smallest index with $x_k \in \overline{\omega}$ and, as before, let $d_{x_k}(\overline{\omega})$ denote the number of times that $\overline{\omega}$ visits x_k. By choosing the root uniformly among the $d_{x_k}(\overline{\omega})$ visits to x_k, we can see that

$$\overline{m}(U) = \sum_{k=1}^{\infty} \sum_{\omega \in \tilde{U}_k} \frac{1}{(2d)^{|\omega|}\, d_{x_k}(\omega)} \le \sum_{k=1}^{\infty} \sum_{\omega \in \tilde{U}_k} \frac{1}{(2d)^{|\omega|}},$$

where $\tilde{U}_k = \tilde{U}_k(n, N)$ denotes the set of (rooted) loops rooted at x_k satisfying the following three properties:

- $\omega \cap \{x_0, \ldots, x_{k-1}\} = \emptyset$,
- $\omega \cap (\mathbb{Z}^d \setminus \mathcal{B}_N) \neq \emptyset$,
- ω does not disconnect the origin from $\partial \mathcal{B}_n$.

We now give an upper bound for the measure of $\tilde{U}_k$ for $x_k \in \mathcal{B}_n$. Suppose that

$$\omega = [\omega_0, \ldots, \omega_{2l}] \in \tilde{U}_k.$$

Let $s_0 = \omega_0, s_5 = 2l$ and define $s_1, \ldots, s_4$ as follows.

- Let s_1 be the smallest index s such that $|\omega_s| \ge 2|x_k|$.
- Let s_2 be the smallest index $s \ge s_1$ such that $|\omega_s| \ge n$.
- Let s_3 be the smallest index $s \ge s_2$ such that $|\omega_s| \ge N$.
- Let s_4 be the largest index $s \le 2l$ such that $|\omega_s| \ge 2|x_k|$.

Then we can decompose

$$\omega = \omega^1 \oplus \omega^2 \oplus \omega^3 \oplus \omega^4 \oplus \omega^5,$$

where $\omega^j = [\omega_{s_{j-1}}, \ldots, \omega_{s_j}]$. We can use this decomposition to estimate the probability of $\tilde{U}_k$.

- ω^1 is a path from x_k to $\partial \mathcal{B}_{2|x_k|}$ that does not hit $\{x_0, \ldots, x_{k-1}\}$. Using gambler's ruin (or a similar estimate), the probability of such a path is bounded above by $c/|x_k|$.
- ω^2 is a path from $\partial \mathcal{B}_{2|x_k|}$ to $\partial \mathcal{B}_n$ that does not disconnect the origin from $\partial \mathcal{B}_n$. There exists c, α such that the probability of reaching distance n without disconnecting the origin is bounded above by $c\,(|x_k|/n)^{\alpha}$ (see Exercise 3.4).
- ω^3 is a path from $\partial \mathcal{B}_n$ to $\partial \mathcal{B}_N$ that does not disconnect the origin from $\partial \mathcal{B}_n$. The probability of such paths is bounded above by $c/\log(N/n)$ (see Exercise 6.4).

- The reverse of ω^5 is a path like ω^1 and has probability $c/|x_k|$.
- Given that ω^3 and ω^5, ω^4 is a path from $\omega_{s_3} \in \partial\mathcal{B}_N$ to $\omega_{s_4} \in \partial\mathcal{B}_{2|x_k|}$ that does not enter $\{x_0, \ldots, x_{k-1}\}$. The expected number of such paths is $O(1)$.

Combining all of these estimates, we see that the measure of $\tilde{U}_k$ is bounded above by a constant times

$$\frac{1}{|x_k|^{2-\alpha}\, n^{\alpha}} \, \frac{1}{\log(N/n)}.$$

By summing over $x_k \in \mathcal{B}_n$, we get the proposition. □

♣ Being able to verify all the estimates in the last proof is a good test that one has absorbed a lot of material from this book!

Proof of Proposition 11.3.1 Let $\epsilon_r = 1/\log r$. We will show that for $2n \leq N \leq M$ and $\eta = [\eta_0, \ldots, \eta_k] \in \Xi_n$,

$$\nu_M(\eta) = \nu_N(\eta)\,[1 + O(\epsilon_{N/n})]. \tag{11.4}$$

Standard arguments using Cauchy sequences then show the existence of ν satisfying (11.3). Proposition 11.3.2 implies

$$\nu_M(\eta) = \nu_N(\eta)\, \frac{F_\eta(\mathcal{B}_M)}{F_\eta(\mathcal{B}_N)}\, \mathbb{P}^{\eta_k}\left\{\overline{\xi}_M < \tau_{\mathbb{Z}^2\setminus\eta} \mid \overline{\xi}_N < \tau_{\mathbb{Z}^2\setminus\eta}\right\}.$$

The set of loops contributing to the term $F_\eta(\mathcal{B}_M)/F_\eta(\mathcal{B}_N)$ are of two types: those that disconnect the origin from $\partial\mathcal{B}_n$ and those that do not. Loops that disconnect the origin from $\partial\mathcal{B}_n$ intersect every $\tilde{\eta} \in \Xi_n$ and hence contribute a factor $C(n,N,M)$ that is independent of η. Hence, using Lemma 11.3.3, we see that

$$\frac{F_\eta(\mathcal{B}_M)}{F_\eta(\mathcal{B}_N)} = C(n,N,M)\,[1 + O(\epsilon_{N/n})], \tag{11.5}$$

Using Proposition 6.4.1, we can see that for every $x \in \partial\mathcal{B}_N$,

$$\mathbb{P}^x\{\overline{\xi}_M < \tau_{\mathbb{Z}^2\setminus\partial\mathcal{B}_n}\} = \frac{\log(N/n)}{\log(M/n)}\left[1 + O(\epsilon_{N/n})\right]$$

(actually the error is of smaller order than this). Using (6.49), if $x \in \mathcal{B}_n$,

$$\mathbb{P}^x\{\xi_N < \tau_{\mathbb{Z}^2\setminus\eta}\} \leq \frac{c}{\log(N/n)}.$$

We therefore get

$$\mathbb{P}^{\eta_k}\left\{\overline{\xi}_M < \tau_{\mathbb{Z}^2\setminus\eta} \mid \overline{\xi}_N < \tau_{\mathbb{Z}^2\setminus\eta}\right\} = \frac{\log(N/n)}{\log(M/n)}\left[1 + O(\epsilon_{N/n})\right].$$

Combining this with (11.5) we get

$$\nu_M(\eta) = \nu_N(\eta)\, C(n, N, M)\, \frac{\log(N/n)}{\log(M/n)}\left[1 + O(\epsilon_{N/n})\right],$$

where we emphasize that the error term is bounded uniformly in $\eta \in \Xi_n$. However, both ν_N and ν_M are probability measures. By summing over $\eta \in \Xi_n$ on both sides, we get

$$C(n, N, M)\, \frac{\log(N/n)}{\log(M/n)} = 1 + O(\epsilon_{N/n}),$$

which gives (11.4). □

The following is proved similarly (see Exercise 9.7).

Proposition 11.3.4 *Suppose that $d \geq 3$ and $n < \infty$. For each $N \geq n$, consider ν_N as a probability measure on Ξ_n. Then the limit*

$$\nu = \lim_{N\to\infty} \nu_N,$$

exists and is the same as that given by the infinite LERW. Moreover, for every $\eta \in \Xi_n$.

$$\nu_N(\eta) = \nu(\eta)\left[1 + O\left((n/N)^{d-2}\right)\right], \qquad N \geq 2n. \tag{11.6}$$

11.4 Rate of growth

If $\hat{S}_0, \hat{S}_1, \ldots$ denotes LERW in $\mathbb{Z}^d$, $d \geq 2$, we let

$$\hat{\xi}_n = \min\{j : |\hat{S}_j| \geq n\}.$$

Let

$$\hat{F}(n) = \hat{F}_d(n) = \mathbb{E}[\hat{\xi}_n].$$

In other words it takes about $\hat{F}(n)$ steps for the LERW to go distance n. Recall that for simple random walk, $\mathbb{E}[\xi_n] \sim n^2$. Note that

$$\hat{F}(n) = \sum_{x \in \mathcal{B}_n} \mathbb{P}\{\hat{S}_j = x \text{ for some } j < \hat{\xi}_n\}.$$

By Propositions 11.3.1 and 11.3.4, we know that if $x \in \mathcal{B}_n$,

$$\mathbb{P}\{\hat{S}_j = x \text{ for some } j < \hat{\xi}_n\} \asymp \mathbb{P}\{x \in LE(S[0, \xi_{2n}])\}$$
$$= \sum_{j=0}^{\infty} \mathbb{P}\{j < \xi_{2n}; S_j = x; LE(S[0,j]) \cap S[j+1, \xi_{2n}] = \emptyset\}.$$

If S, S^1 are independent random walks, let

$$\hat{Q}(\lambda) = (1-\lambda)^2 \sum_{x \in \mathbb{Z}^d} \sum_{n=0}^{\infty} \sum_{m=0}^{\infty} \lambda^{n+m}\, \mathbb{P}^{0,x}\{LE(S[0,n]) \cap S^1[0,m] = \emptyset\}.$$

In Proposition 10.2.1, a probability of nonintersection of random walks starting at the origin was computed in terms of a "long-range" intersection quantity $Q(\lambda)$. We do something similar for LERW using the quantity $\hat{Q}(\lambda)$. The proof of Proposition 10.2.1 used a path decomposition: given two intersecting paths, the proof focused on the first intersection (using the timescale of one of the paths) and then translating to make that the origin. The proof of the next proposition is similar, given a simple random walk that intersects a loop-erased walk. However, we get two different results depending on whether we focus on the first intersection on the timescale of the simple walk or on the timescale of the loop-erased walk.

Proposition 11.4.1 *Let S, S^1, S^2, S^3 be independent simple random walks starting at the origin in $\mathbb{Z}^d$ with independent geometric killing times $T_\lambda, T_\lambda^1, \ldots, T_\lambda^3$.*

(a) Let $V^1 = V_\lambda^1$ be the event that

$$S^j[1, T_\lambda^1] \cap LE(S[0, T_\lambda]) = \emptyset, \quad j = 1, 2,$$

and

$$S^3[1, T_\lambda^3] \cap [LE(S[0, T_\lambda]) \setminus \{0\}] = \emptyset.$$

Then

$$P(V^1) = (1-\lambda)^2\, \hat{Q}(\lambda). \tag{11.7}$$

(b) Let $V^2 = V^2_\lambda$ be the event that

$$S^1[1, T^1_\lambda] \cap LE(S[0, T_\lambda]) = \emptyset,$$

and

$$S^2[1, T^2_\lambda] \cap \Big[LE(S[0, T_\lambda]) \cup LE(S^1[0, T^1_\lambda]) = \emptyset \Big\}.$$

Then

$$P(V^2) = (1-\lambda)^2\, \hat{Q}(\lambda). \tag{11.8}$$

Proof We use some of the notation from the proof of Proposition 10.2.1. Note that

$$\hat{Q}(\lambda) = (1-\lambda)^2 \sum_{n=0}^{\infty} \sum_{m=0}^{\infty} \sum_{\omega,\eta} \lambda^{n+m}\, p(\omega)\, p(\eta),$$

where the last sum is over all ω, η with $|\omega| = n, |\eta| = m, \omega_0 = 0$ and $L(\omega) \cap \eta \neq \emptyset$. We write

$$\hat{\omega} = L(\omega) = [\hat{\omega}_0, \ldots, \hat{\omega}_l].$$

To prove (11.7), on the event $\hat{\omega} \cap \eta \neq \emptyset$, we let

$$u = \min\{j : \hat{\omega}_j \in \eta\}, \qquad s = \max\{j : \omega_j = \hat{\omega}_u\}, \qquad t = \min\{k : \eta_k = \hat{\omega}_u\}.$$

We define the paths $\omega^-, \omega^+, \eta^-, \eta^+$ as in the proof of Proposition 10.2.1 using these values of s, t. Our definition of s, t implies for $j > 0$, that

$$\omega_j^+ \notin LE^R(\omega^-), \quad \eta_j^- \notin LE^R(\omega^-), \quad \eta_j^+ \notin LE^R(\omega^-) \setminus \{0\}. \tag{11.9}$$

Here, we write LE^R to indicate that one traverses the path in the reverse direction, erases loops, and then reverses the path again — this is not necessarily the same as $LE(\omega_-)$. Conversely, for any 4-tuple $(\omega^-, \omega^+, \eta^-, \eta^+)$ satisfying (11.9), we get a corresponding (ω, η) satisfying $L(\omega) \cap \eta \neq \emptyset$. Therefore,

$$\hat{Q}(\lambda) = (1-\lambda)^2 \times \sum_{0 \le n_-, n_+, m_-, m_+} \sum_{\omega, \omega_+, \eta_-, \eta_+} \lambda^{n_- + n_+ + m_- + m_+}\, p(\omega_-) p(\omega_+) p(\eta_-) p(\eta_+),$$

where the last sum is over all $(\omega^-, \omega^+, \eta^-, \eta^+)$ with $|\omega_-| = n_-, |\omega_+| = n_+, |\eta_-| = m_-, |\eta_+| = m_+$ satisfying (11.9). Using Corollary 11.2.2, we see that the sum is the same if we replace (11.9) with: for $j > 0$,

$$\omega_j^+ \notin LE(\omega_-), \quad \eta_j^- \notin LE(\omega_-), \quad \eta_j^+ \notin LE(\omega_-) \setminus \{0\}.$$

To prove (11.8), on the event $\hat{\omega} \cap \eta \neq \emptyset$, we let

$$t = \min\{k : \eta_k \in \hat{\omega}\}, \quad s = \max\{j : \omega_j = \eta_t\},$$

and define $(\omega^-, \omega^+, \eta^-, \eta^+)$ as before. The conditions now become for $j > 0$,

$$\omega_j^+ \notin LE^R(\omega^-), \quad \eta_j^- \notin [LE^R(\omega^-) \cup LE(\omega^+)],$$

□

It is harder to estimate $\hat{Q}(\lambda)$ than $Q(\lambda)$. We do not give a proof here but we state that if $\lambda_n = 1 - \frac{1}{n}$, then as $n \to \infty$,

$$\hat{Q}(\lambda_n) \asymp \begin{cases} n^{d/2}, & d < 4 \\ n^2 [\log n]^{-1}, & d = 4. \end{cases}$$

This is the same behavior as for $Q(\lambda_n)$. Roughly speaking, if two random walks of length n start distance $\sqrt{n}$ away, then the probability that one walk intersects the loop erasure of the other is of order one for $d \leq 3$ and of order $1/\log n$ for $d = 4$. For $d = 1, 2$, this is almost obvious for topological reasons. The hard cases are $d = 3, 4$. For $d = 3$ the set of "cut points" (i.e. points S_j such that $S[0, j] \cap S[j+1, n] = \emptyset$) has a "fractal dimension" strictly greater than one and hence tends to be hit by a (roughly two-dimensional) simple random walk path. For $d = 4$, one can also show that the probability of hitting the cut points is of order $1/\log n$. Since all cut points are retained in loop erasure, this gives a bound on the probability of hitting the loop erasure. This estimate of $\hat{Q}(\lambda)$ yields

$$\mathbb{P}(V^1_{\lambda_n}) \asymp \mathbb{P}(V^2_{\lambda_n}) \asymp \begin{cases} n^{\frac{d-4}{2}}, & d < 4 \\ \frac{1}{\log n}, & d = 4. \end{cases}$$

To compute the growth rate, we would like to know the asymptotic behavior of

$$\mathbb{P}\{LE(S[0, \lambda_n]) \cap S^1[1, T_{\lambda_n}] = \emptyset\} = \mathbb{E}[Y_n],$$

where

$$Y_n = \mathbb{P}\{LE(S[0,\lambda_n]) \cap S^1[1, T_{\lambda_n}] = \emptyset \mid S[0,\lambda_n]\}.$$

Note that

$$\mathbb{P}(V^1_{\lambda_n}) \asymp \mathbb{E}[Y_n^3].$$

11.5 Short-range intersections

Studying the growth rate for LERW leads one to try to estimate probabilities such as

$$\mathbb{P}\{LE(S[0,n]) \cap S[n+1, 2n] = \emptyset\},$$

which by Corollary 11.2.2 is the same as

$$\hat{q}_n =: \mathbb{P}\{LE(S[0,n]) \cap S^1[1,n] = \emptyset\},$$

where S, S^1 are independent walks starting at the origin. If $d \geq 5$,

$$\hat{q}_n \geq \mathbb{P}\{S[0,n] \cap S^1[1,n] = \emptyset\} \geq c > 0,$$

so we will restrict our discussion to $d \leq 4$. Let

$$\hat{Y}_n = \mathbb{P}\{LE(S[0,n]) \cap S[n+1, 2n] = \emptyset \mid S[0,n]\}.$$

Using ideas similar to those leading up to (11.7), one can show that

$$\mathbb{E}[\hat{Y}_n^3] \asymp \begin{cases} (\log n)^{-1}, & d = 4 \\ n^{\frac{d-4}{2}}, & d = 1,2,3. \end{cases} \tag{11.10}$$

This can be compared to (10.6), where the second moment for an analogous quantity is given. We also know that

$$\mathbb{E}[\hat{Y}_n^3] \leq \mathbb{E}[\hat{Y}_n] \leq \left(\mathbb{E}[\hat{Y}_n]\right)^{1/3}. \tag{11.11}$$

In the "mean-field" case $d = 4$, it can be shown that $\mathbb{E}[\hat{Y}_n^3] \asymp \left(\mathbb{E}[\hat{Y}_n]\right)^3$ and hence, that

$$q_n \asymp (\log n)^{-1/3}.$$

Moreover, if we appropriately scale the process, the LERW converges to a Brownian motion.

For $d = 2, 3$, we do not expect that $\mathbb{E}[\hat{Y}_n^3] \asymp \left(\mathbb{E}[\hat{Y}_n]\right)^3$. Let us define an exponent $\alpha = \alpha_d$ roughly as $q_n \approx n^{-\alpha}$. The relations (11.10) and (11.11) imply that

$$\alpha \geq \frac{4-d}{6}.$$

It has been shown that for $d = 2$ the exponent α exists with $\alpha = 3/8$; in other words, the expected number of points in $LE(S[0,n])$ is of order $n^{5/8}$ (see Kenyon (2000) and also Masson (2009)). This suggests that if we scale appropriately, there should be a limit process whose paths have a fractal dimension of $5/4$. In fact, this has been proved. The limit process is the SLE with parameter $\kappa = 2$ Lawler *et al.* (2004).

For $d = 3$, we get the bound $\alpha \geq 1/6$ which states that the number of points in $LE(S[0,n])$ should be no more than $n^{5/6}$. We also expect that this bound is not sharp. The value of α_3 is any open problem; in fact, the existence of an exponent satisfying $q_n \approx n^{-\alpha}$ has not been established. However, the existence of a scaling limit has been shown in Kozma (2007).

Exercises

Exercise 11.1 Suppose that $d \geq 3$ and X_n is simple random walk in $\mathbb{Z}^d$ conditioned to return to the origin. This is the h-process with

$$h(x) = \mathbb{P}^x\{S_n = 0 \text{ for some } n \geq 0\}.$$

Prove the following

(i) For all $d \geq 3$, X_n is a recurrent Markov chain.
(ii) Assume that $X_0 = 0$ and let $T = \min\{j > 0 : X_j = 0\}$. Show that there is a $c = c_d > 0$ such that

$$\mathbb{P}\{T = 2n\} \asymp n^{-d/2}, \qquad n \to \infty.$$

In particular,

$$\mathbb{E}[T] \begin{cases} = \infty, & d \leq 4, \\ < \infty, & d \geq 5. \end{cases}$$

Exercise 11.2 Suppose that X_n is simple random walk in $\mathbb{Z}^2$ conditioned not to return to the origin. This is the h-process with $h(x) = a(x)$.

(i) Prove that X_n is a transient Markov chain.
(ii) Show that if loops are erased chronologically from this chain, then one gets LERW in $\mathbb{Z}^2$.

A

Appendix

A.1 Some expansions

A.1.1 Riemann sums

In this book we often approximate sums by integrals. Here we give bounds on the size of the error in such approximations.

Lemma A.1.1 *If $f : (0, \infty) \to \mathbb{R}$ is a C^2 function, and b_n is defined by*

$$b_n = f(n) - \int_{n-(1/2)}^{n+(1/2)} f(s)\, ds,$$

then

$$|b_n| \le \frac{1}{24} \sup\left\{ |f''(r)| : |n - r| \le \frac{1}{2} \right\}. \tag{A.1}$$

If $\sum |b_n| < \infty$, let

$$C = \sum_{n=1}^{\infty} b_n, \qquad B_n = \sum_{j=n+1}^{\infty} |b_n|.$$

Then

$$\sum_{j=1}^{n} f(j) = \int_{1/2}^{n+(1/2)} f(s)\, ds + C + O(B_n).$$

Also, for all $m < n$

$$\left| \sum_{j=m}^{n} f(j) - \int_{m-(1/2)}^{n+(1/2)} f(s)\, ds \right| \le B_m.$$

Proof Taylor's theorem shows that for $|s - n| \leq 1/2$,

$$f(s) = f(n) + (s - n)f'(n) + \frac{1}{2}f''(r_s)\,(r_s - n)^2,$$

for some $|n - r_s| < 1/2$. Hence, for such s,

$$|f(s) + f(-s) - 2f(n)| \leq \frac{s^2}{2}\,\sup\{|f''(r)| : |n - r| \leq s\}.$$

Integrating gives (A.1). The rest is straightforward. □

Example. Suppose that $\alpha < 1, \beta \in \mathbb{R}$ and

$$f(n) = n^\alpha \log^\beta n.$$

Note that for $t \geq 2$,

$$|f''(t)| \leq c\, t^{\alpha-2} \log^\beta t.$$

Therefore, there is a $C(\alpha, \beta)$ such that

$$\begin{aligned}\sum_{j=2}^{n} n^\alpha \log^\beta n &= \int_2^{n+(1/2)} t^\alpha \log^\beta t\, dt + C(\alpha,\beta) + O(n^{\alpha-1} \log^\beta n)\\ &= \int_2^{n} t^\alpha \log^\beta t\, dt + \frac{1}{2} n^\alpha \log^\beta n + C(\alpha,\beta) + O(n^{\alpha-1} \log^\beta n)\end{aligned} \tag{A.2}$$

A.1.2 Logarithm

Let log denote the branch of the complex logarithm on $\{z \in \mathbb{C}; \mathrm{Re}(z) > 0\}$ with $\log 1 = 0$. Using the power series

$$\log(1 + z) = \left[\sum_{j=1}^{k} \frac{(-1)^{j+1} z^j}{j}\right] + O_\epsilon(|z|^{k+1}), \qquad |z| \leq 1 - \epsilon.$$

we see that if $r \in (0, 1)$ and $|\xi| \leq rt$,

$$\log\left(1 + \frac{\xi}{t}\right)^t = \xi - \frac{\xi^2}{2t} + \frac{\xi^3}{3t^2} + \cdots + (-1)^{k+1}\frac{\xi^k}{kt^{k-1}} + O_r\left(\frac{|\xi|^{k+1}}{t^k}\right),$$

$$\left(1 + \frac{\xi}{t}\right)^t = e^\xi \exp\left\{-\frac{\xi^2}{2t} + \frac{\xi^3}{3t^2} + \cdots + (-1)^{k+1}\frac{\xi^k}{kt^{k-1}} + O_r\left(\frac{|\xi|^{k+1}}{t^k}\right)\right\}. \tag{A.3}$$

If $|\xi|^2/t$ is not too big, we can expand the exponential in a Taylor series. Recall that for fixed $R < \infty$, we can write

$$e^z = 1 + z + \frac{z^2}{2!} + \cdots + \frac{z^k}{k!} + O_R(|z|^{k+1}), \qquad |z| \le R.$$

Therefore, if $r \in (0,1), R < \infty, |\xi| \le rt, |\xi|^2 \le Rt$, we can write

$$\left(1 + \frac{\xi}{t}\right)^t = e^{\xi}\left[1 - \frac{\xi^2}{2t} + \frac{8\xi^3 + 3\xi^4}{24t^2} + \cdots + \frac{f_k(\xi)}{t^{k-1}} + O\left(\frac{|\xi|^{2k}}{t^k}\right)\right], \tag{A.4}$$

where f_k is a polynomial of degree $2(k-1)$ and the implicit constant in the $O(\cdot)$ term depends only on r, R, and k. In particular,

$$\left(1 + \frac{1}{n}\right)^n = e\left[1 - \frac{1}{2n} + \frac{11}{24\,n^2} + \cdots + \frac{b_k}{n^k} + O\left(\frac{1}{n^{k+1}}\right)\right]. \tag{A.5}$$

Lemma A.1.2 *For every positive integer k, there exist constants $c(k,l), l = k+1, k+2, \ldots$ such that for each $m > k$,*

$$\sum_{j=n}^{\infty} \frac{k}{j^{k+1}} = \frac{1}{n^k} + \sum_{l=k+1}^{m} \frac{c(k,l)}{n^l} + O\left(\frac{1}{n^{m+1}}\right). \tag{A.6}$$

Proof If $n > 1$,

$$n^{-k} = \sum_{j=n}^{\infty}[j^{-k} - (j+1)^{-k}] = \sum_{j=n}^{\infty} j^{-k}\,[1 - (1 + j^{-1})^{-k}]$$

$$= \sum_{l=k}^{\infty} b(k,l) \sum_{j=n}^{\infty} \frac{l}{j^{l+1}},$$

with $b(k,k) = 1$ (the other constants can be given explicitly but we do not need to). In particular,

$$n^{-k} = \sum_{l=k}^{m} b(k,j) \left[\sum_{j=n}^{\infty} \frac{l}{j^{l+1}}\right] + O\left(\frac{1}{n^{m+1}}\right).$$

The expression (A.6) can be obtained by inverting this expression; we omit the details. □

Lemma A.1.3 *There exists a constant γ (called Euler's constant) and $b_2, b_3, \ldots$ such that for every integer $k \geq 2$,*

$$\sum_{j=1}^{n} \frac{1}{j} = \log n + \gamma + \frac{1}{2n} + \sum_{l=2}^{k} \frac{b_l}{n^l} + O\left(\frac{1}{n^{k+1}}\right).$$

In fact,

$$\gamma = \lim_{n\to\infty} \left[\sum_{j=1}^{n} \frac{1}{j}\right] - \log n = \int_0^1 (1 - e^{-t}) \frac{1}{t}\, dt - \int_1^{\infty} e^{-t} \frac{1}{t}\, dt. \tag{A.7}$$

Proof Note that

$$\sum_{j=1}^{n} \frac{1}{j} = \log\left(n + \frac{1}{2}\right) + \log 2 + \sum_{j=1}^{n} \beta_j,$$

where

$$\beta_j = \frac{1}{j} - \log\left(j + \frac{1}{2}\right) + \log\left(j - \frac{1}{2}\right) = -\sum_{k=1}^{\infty} \frac{2}{(2k+1)\,(2j)^{2k+1}}.$$

In particular, $\beta_j = O(j^{-3})$, and hence $\sum \beta_j < \infty$. We can write that

$$\begin{aligned}\sum_{j=1}^{n} \frac{1}{j} &= \log\left(n + \frac{1}{2}\right) + \gamma - \sum_{j=n+1}^{\infty} \beta_j \\ &= \log n + \gamma + \sum_{l=1}^{\infty} \frac{(-1)^{l+1}}{2^l\, n^l} - \sum_{j=n+1}^{\infty} \beta_j,\end{aligned}$$

where γ is the constant

$$\gamma = \log 2 + \sum_{j=1}^{\infty} \beta_j.$$

Using (A.6), we can write

$$\sum_{j=n+1}^{\infty} \beta_j = \sum_{l=3}^{k} \frac{a_l}{n^l} + O\left(\frac{1}{n^{k+1}}\right),$$

for some constants a_l.

We will sketch the proof of (A.7) leaving the details to the reader. By Taylor's series, we know that

$$\log n = -\log\left[1-\left(1-\frac{1}{n}\right)\right] = \sum_{j=1}^{\infty}\left(1-\frac{1}{n}\right)^j \frac{1}{j}.$$

Therefore,

$$\begin{aligned}\gamma &= \lim_{n\to\infty}\left[\sum_{j=1}^{n}\frac{1}{j}\right] - \log n\\ &= \lim_{n\to\infty}\sum_{j=1}^{n}\left[1-\left(1-\frac{1}{n}\right)^j\right]\frac{1}{j} \; - \; \lim_{n\to\infty}\sum_{j=n+1}^{\infty}\left(1-\frac{1}{n}\right)^j\frac{1}{j}.\end{aligned}$$

We now use the approximation $(1-n^{-1})^n \sim e^{-1}$ to get

$$\begin{aligned}\lim_{n\to\infty}\sum_{j=1}^{n}\left[1-\left(1-\frac{1}{n}\right)^j\right]\frac{1}{j} &= \lim_{n\to\infty}\sum_{j=1}^{n}\frac{1}{n}(1-e^{-j/n})\frac{1}{j/n}\\ &= \int_0^1 (1-e^{-t})\,\frac{1}{t}\,dt,\end{aligned}$$

$$\lim_{n\to\infty}\sum_{j=n+1}^{\infty}\left(1-\frac{1}{n}\right)^j\frac{1}{j} = \lim_{n\to\infty}\sum_{j=n+1}^{\infty}\frac{1}{n}e^{-j/n}\frac{1}{j/n} = \int_1^{\infty} e^{-t}\,\frac{1}{t}\,dt.$$

□

Lemma A.1.4 *Suppose that $\alpha \in \mathbb{R}$ and m is a positive integer. There exist constants $r_0, r_1, \ldots$ such that if k is a positive integer and $n \geq m$,*

$$\prod_{j=m}^{n}\left(1-\frac{\alpha}{j}\right) = r_0\, n^{-\alpha}\left[1+\frac{r_1}{n}+\cdots+\frac{r_k}{n^k}+O\left(\frac{1}{n^{k+1}}\right)\right].$$

Proof Without loss of generality we assume that $|\alpha| \leq 2m$; if this does not hold we can factor out the first few terms of the product and then analyze the remaining terms. Note that

$$\begin{aligned}\log\prod_{j=m}^{n}\left(1-\frac{\alpha}{j}\right) &= \sum_{j=m}^{n}\log\left(1-\frac{\alpha}{j}\right)\\ &= -\sum_{j=m}^{n}\sum_{l=1}^{\infty}\frac{\alpha^l}{l\,j^l} = -\sum_{l=1}^{\infty}\sum_{j=m}^{n}\frac{\alpha^l}{l\,j^l}.\end{aligned}$$

For the $l = 1$ term, we have that

$$\sum_{j=m}^{n} \frac{\alpha}{j} = -\sum_{j=1}^{m-1} \frac{\alpha}{j} + \alpha \left[\log n + \gamma + \frac{1}{2n} + \sum_{l=2}^{k} \frac{b_l}{n^l} + O\left(\frac{1}{n^{k+1}}\right) \right].$$

All of the other terms can be written in powers of $(1/n)$. Therefore, we can write that

$$\log \prod_{j=m}^{n} \left(1 - \frac{\alpha}{j}\right) = -\alpha \log n + \sum_{l=0}^{k} \frac{C_l}{n^l} + O\left(\frac{1}{n^{k+1}}\right).$$

The lemma is then obtained by exponentiating both sides. □

A.2 Martingales

A filtration $\mathcal{F}_0 \subset \mathcal{F}_1 \subset \cdots$ is an increasing sequence of σ-algebras.

Definition A sequence of integrable random variables $M_0, M_1, \ldots$ is called a *martingale* with respect to the filtration $\{\mathcal{F}_n\}$ if each M_n is $\mathcal{F}_n$-measurable and for each $m \leq n$,

$$\mathbb{E}[M_n \mid \mathcal{F}_m] = M_m. \tag{A.8}$$

If (A.8) is replaced with $\mathbb{E}[M_n \mid \mathcal{F}_m] \geq M_m$ the sequence is called a *submartingale*. If (A.8) is replaced with $\mathbb{E}[M_n \mid \mathcal{F}_m] \leq M_m$ the sequence is called a *supermartingale*.

Using properties of conditional expectation, it is easy to see that to verify (A.8), it suffices to show for each n that $\mathbb{E}[M_{n+1} \mid \mathcal{F}_n] = M_n$. This equality only needs to hold up to an event of probability zero; in fact, the conditional expectation is only defined up to events of probability zero. If the filtration is not specified, then the assumption is that $\mathcal{F}_n$ is the σ-algebra generated by $M_0, \ldots, M_n$. If $M_0, X_1, X_2, \ldots$ are independent random variables with $\mathbb{E}[|M_0|] < \infty$ and $\mathbb{E}[X_j] = 0$ for $j \geq 1$, and

$$M_n = M_0 + X_1 + \cdots + X_n,$$

then $M_0, M_1, \ldots$ is a martingale. We omit the proof of the next lemma, which is the conditional expectation version of Jensen's inequality.

Lemma A.2.1 (Jensen's inequality) *If X is an integrable random variable; $f : \mathbb{R} \to \mathbb{R}$ is convex with $\mathbb{E}[|f(X)|] < \infty$; and $\mathcal{F}$ is a σ-algebra, then $\mathbb{E}[f(X) \mid \mathcal{F}] \geq f(\mathbb{E}[X \mid \mathcal{F}])$. In particular, if $M_0, M_1, \ldots$ is a martingale, $f : \mathbb{R} \to \mathbb{R}$ is convex with $\mathbb{E}[|f(M_n)|] < \infty$ for all n, and $Y_n = f(M_n)$; then $Y_0, Y_1, \ldots$ is a submartingale.*

In particular, if $M_0, M_1, \ldots$ is a martingale, then

- if $\alpha \geq 1$, $Y_n := |M_n|^\alpha$ is a submartingale;
- if $b \in \mathbb{R}$, $Y_n := e^{bM_n}$ is a submartingale.

In both cases, this is assuming that $\mathbb{E}[Y_n] < \infty$.

A.2.1 Optional sampling theorem

A *stopping time* with respect to a filtration $\{\mathcal{F}_n\}$ is a $\{0, 1, \ldots\} \cup \{\infty\}$-valued random variable T such that for each n, $\{T \leq n\}$ is $\mathcal{F}_n$-measurable. If T is a stopping time and n is a positive integer, then $T_n := T \wedge n$ is a stopping time satisfying $T_n \leq n$.

Proposition A.2.2 *Suppose that $M_0, M_1, \ldots$ is a martingale and T is a stopping time each with respect to the filtration $\mathcal{F}_n$. Then $Y_n := M_{T_n}$ is a martingale with respect to $\mathcal{F}_n$. In particular,*

$$\mathbb{E}[M_0] = \mathbb{E}[M_{T_n}].$$

Proof Note that

$$Y_{n+1} = M_{T_n}\, 1\{T \leq n\} + M_{n+1}\, 1\{T \geq n+1\}.$$

The event $\{T \geq n+1\}$ is the complement of the event $\{T \leq n\}$ and hence is $\mathcal{F}_n$-measurable. Therefore, by properties of conditional expectation,

$$\begin{aligned} \mathbb{E}[M_{n+1}\, 1\{T \geq n+1\} \mid \mathcal{F}_n] &= 1\{T \geq n+1\}\, \mathbb{E}[M_{n+1} \mid \mathcal{F}_n] \\ &= 1\{T \geq n+1\}\, M_n. \end{aligned}$$

Therefore,

$$\mathbb{E}[Y_{n+1} \mid \mathcal{F}_n] = M_{T_n}\, 1\{T \leq n\} + M_n\, 1\{T \geq n+1\} = Y_n.$$

□

The optional sampling theorem states that under certain conditions, if $\mathbb{P}\{T < \infty\} = 1$, then $\mathbb{E}[M_0] = \mathbb{E}[M_T]$. However, this does not hold without some further assumptions. For example, if M_n is one-dimensional simple random walk starting at the origin and T is the first n such that $M_n = 1$, then $\mathbb{P}\{T < \infty\} = 1$, $M_T = 1$, and hence $\mathbb{E}[M_0] \neq \mathbb{E}[M_T]$. In the next theorem we list a number of sufficient conditions under which we can conclude that $\mathbb{E}[M_0] = \mathbb{E}[M_T]$.

Theorem A.2.3 (optional sampling theorem) *Suppose that $M_0, M_1, \ldots$ is a martingale and T is a stopping time with respect to the filtration $\{\mathcal{F}_n\}$. Suppose that $\mathbb{P}\{T < \infty\} = 1$. Suppose also that at least one of the following conditions holds:*

- *There is a $K < \infty$ such that $\mathbb{P}\{T \leq K\} = 1$.*
- *There exists an integrable random variable Y such that for all n, $|M_{T_n}| \leq Y$.*
- $\mathbb{E}[|M_T|] < \infty$ *and* $\lim_{n\to\infty} \mathbb{E}[|M_n|; T > n] = 0$.
- *The random variables $M_0, M_1, \ldots$ are uniformly integrable, i.e. for every $\epsilon > 0$ there is a $K_\epsilon < \infty$ such that for all n,*

$$\mathbb{E}[|M_n|; |M_n| > K_\epsilon] < \epsilon.$$

- *There exist an $\alpha > 1$ and a $K < \infty$ such that for all n, $\mathbb{E}[|M_n|^\alpha] \leq K$.*

Then $\mathbb{E}[M_0] = \mathbb{E}[M_T]$.

Proof We will consider the conditions in order. The sufficiency of the first follows immediately from Proposition A.2.2. We know that $M_{T_n} \to M_T$ with probability one. Proposition A.2.2 gives $\mathbb{E}[M_{T_n}] = \mathbb{E}[M_0]$. Hence, we need to show that

$$\lim_{n\to\infty} \mathbb{E}[M_{T_n}] = \mathbb{E}[M_T]. \tag{A.9}$$

If the second condition holds, then this limit is justified by the dominated convergence theorem. Now, assume the third condition. Note that

$$M_T = M_{T_n} + M_T\, 1\{T > n\} - M_n\, 1\{T > n\}.$$

Since $\mathbb{P}\{T > n\} \to 0$, and $\mathbb{E}[|M_T|] < \infty$, it follows from the dominated convergence theorem that

$$\lim_{n\to\infty} \mathbb{E}[M_T\, 1\{T > n\}] = 0.$$

Hence, if $\mathbb{E}[M_n\, 1\{T > n\}] \to 0$, we have (A.9). Standard exercises show that the fourth condition implies the third and the fifth condition implies the fourth, so either the fourth or fifth condition is sufficient. □

A.2.2 Maximal inequality

Theorem A.2.4 (maximal inequality) *Suppose that $M_0, M_1, \ldots$ is a nonnegative submartingale with respect to $\{\mathcal{F}_n\}$ and $\lambda > 0$. Then*

$$\mathbb{P}\left\{\max_{0\le j\le n} M_j \ge \lambda\right\} \le \frac{\mathbb{E}[M_n]}{\lambda}.$$

Proof Let $T = \min\{j \ge 0 : M_j \ge \lambda\}$. Then,

$$\mathbb{P}\left\{\max_{0\le j\le n} M_j \ge \lambda\right\} = \sum_{j=0}^{n} \mathbb{P}\{T = j\},$$

$$\mathbb{E}[M_n] \ge \mathbb{E}[M_n; T \le n] = \sum_{j=0}^{n} \mathbb{E}[M_n; T = j].$$

Since M_n is a submartingale and $\{T = j\}$ is $\mathcal{F}_j$-measurable,

$$\mathbb{E}[M_n; T = j] = \mathbb{E}[\mathbb{E}[M_n \mid \mathcal{F}_j]; T = j] \ge \mathbb{E}[M_j; T = j] \ge \lambda\, \mathbb{P}\{T = j\}.$$

Combining these estimates gives the theorem. □

Combining Theorem A.2.4 with Lemma A.2.1 gives the following theorem.

Theorem A.2.5 (martingale maximal inequalities) *Suppose that $M_0, M_1, \ldots$ is a martingale with respect to $\{\mathcal{F}_n\}$ and $\lambda > 0$. Then, if $\alpha \ge 1, b \ge 0$,*

$$\mathbb{P}\left\{\max_{0\le j\le n} |M_j| \ge \lambda\right\} \le \frac{\mathbb{E}[|M_n|^\alpha]}{\lambda^\alpha}, \tag{A.10}$$

$$\mathbb{P}\left\{\max_{0\le j\le n} M_j \ge \lambda\right\} \le \frac{\mathbb{E}[e^{bM_n}]}{e^{b\lambda}}.$$

Corollary A.2.6 *Let $X_1, X_2, \ldots$ be independent, identically distributed random variables in $\mathbb{R}$ with mean zero, and let k be a positive integer for which $\mathbb{E}[|X_1|^{2k}] < \infty$. There exists $c < \infty$ such that for all $\lambda > 0$,*

$$\mathbb{P}\left\{\max_{0\le j\le n} |S_j| \ge \lambda\sqrt{n}\right\} \le c\,\lambda^{-2k}. \tag{A.11}$$

Proof Fix k and allow constants to depend on k. Note that

$$\mathbb{E}[S_n^{2k}] = \sum \mathbb{E}[X_{j_1}, \ldots, X_{j_{2k}}],$$

where the sum is over all $(j_1, \ldots, j_{2k}) \in \{1, \ldots, n\}^{2k}$. If there exists an l such that $j_i \neq j_l$ for $i \neq l$, then we can use independence and $\mathbb{E}[X_{j_l}] = 0$ to see that $\mathbb{E}[X_{j_1}, \ldots, X_{j_{2k}}] = 0$. Hence

$$\mathbb{E}[S_n^{2k}] = \sum \mathbb{E}[X_{j_1}, \ldots, X_{j_{2k}}],$$

where the sum is over all $(2k)$-tuples such that if $l \in \{j_1, \ldots, j_{2k}\}$, then l appears at least twice. The number of such $(2k)$-tuples is $O(n^k)$ and hence we can see that

$$\mathbb{E}\left[\left(\frac{S_n}{\sqrt{n}}\right)^{2k}\right] \leq c.$$

Hence, we can apply (A.10) to the martingale $M_j = S_j/\sqrt{n}$. □

Corollary A.2.7 *Let $X_1, X_2, \ldots$ be independent, identically distributed random variables in $\mathbb{R}$ with mean zero, variance σ^2, and such that for some $\delta > 0$, the moment generating function $\psi(t) = \mathbb{E}[e^{tX_j}]$ exists for $|t| < \delta$. Let $S_n = X_1 + \cdots + X_n$. Then for all $0 \leq r \leq \delta\sqrt{n}/2$,*

$$\mathbb{P}\left\{\max_{0 \leq j \leq n} S_j \geq r\sigma\sqrt{n}\right\} \leq e^{-r^2/2} \exp\left\{O\left(\frac{r^3}{\sqrt{n}}\right)\right\}. \tag{A.12}$$

If $\mathbb{P}\{X_1 \geq R\} = 0$ for some R, this holds for all $r > 0$.

Proof Without loss of generality, we may assume that $\sigma^2 = 1$. The moment-generating function of $S_n = X_1 + \cdots + X_n$ is $\psi(t)^n$. Letting $t = r/\sqrt{n}$, we get

$$\mathbb{P}\left\{\max_{0 \leq j \leq n} S_j \geq r\sqrt{n}\right\} \leq e^{-r^2}\, \psi(r/\sqrt{n})^n.$$

Using the expansion for $\psi(t)$ at zero,

$$\psi(t) = 1 + \frac{t^2}{2} + O(t^3), \quad |t| \leq \frac{\delta}{2},$$

we see that for $0 \leq r \leq \delta\sqrt{n}/2$,

$$\psi(r/\sqrt{n})^n = \left[1 + \frac{r^2}{2n} + O\left(\frac{r^3}{n^{3/2}}\right)\right]^n \leq e^{r^2/2} \exp\left\{O\left(\frac{r^3}{\sqrt{n}}\right)\right\}.$$

This gives (A.12). If $\mathbb{P}\{X_1 > R\} = 0$, then (A.12) holds for $r > R\sqrt{n}$ trivially, and we can choose $\delta = 2R$. □

Remark From the last corollary, we also get the following modulus of continuity result for random walk. Let $X_1, X_2, \ldots$ and S_n be as in the previous lemma. There exist c, b such that for every $m \leq n$ and every $0 \leq r \leq \delta\sqrt{m}/2$

$$\mathbb{P}\{\max_{0\leq j\leq n}\max_{1\leq k\leq m}|S_{k+j}-S_j| \geq r\sqrt{m}\} \leq c\, n\, e^{-br^2}. \tag{A.13}$$

This next lemma is not about martingales, but it does concern exponential estimates for probabilities so we will include it here.

Lemma A.2.8 *If $0 < \alpha < \infty$, $0 < r < 1$, and X_n is a binomial random variable with parameters n and $\alpha\, e^{-2\alpha/r}$, then*

$$\mathbb{P}\{X_n \geq rn\} \leq e^{-\alpha n}.$$

Proof

$$\mathbb{P}\{X_n \geq rn\} \leq e^{-2\alpha n}\,\mathbb{E}[e^{(2\alpha/r)X_n}] \leq e^{-2\alpha n}\,[1+\alpha]^n \leq e^{-\alpha n}.$$

□

A.2.3 Continuous martingales

A process M_t adapted to a filtration $\mathcal{F}_t$ is called a *continuous martingale* if for each $s < t$, $\mathbb{E}[M_t \mid M_s] = M_s$ and with probability one the function $t \mapsto M_t$ is continuous. If M_t is a continuous martingale, and $\delta > 0$, then

$$M_n^{(\delta)} := M_{\delta n}$$

is a discrete time martingale. Using this, we can extend results about discrete time martingales to continuous martingales. We state one such result here.

Theorem A.2.9 (optional sampling theorem) *Suppose that M_t is a uniformly integrable continuous martingale and τ is a stopping time with $\mathbb{P}\{\tau < \infty\} = 1$ and $\mathbb{E}[|M_\tau|] < \infty$. Suppose that*

$$\lim_{t\to\infty}\mathbb{E}[|M_t|;\tau > t] = 0.$$

Then

$$\mathbb{E}[M_T] = \mathbb{E}[M_0].$$

A.3 Joint normal distributions

A random vector $Z = (Z_1, \ldots, Z_d) \in \mathbb{R}^d$ is said to have a *(mean zero) joint normal distribution* if there exist independent (one-dimensional) mean zero, variance one normal random variables $N_1, \ldots, N_n$, and scalars a_{jk} such that

$$Z_j = a_{j1} N_1 + \cdots + a_{jn} N_n, \quad j = 1, \ldots, d,$$

or in matrix form

$$Z = AN.$$

Here, $A = (a_{jk})$ is a $d \times n$ matrix and Z, N are column vectors. Note that

$$\mathbb{E}(Z_j Z_k) = \sum_{m=1}^{n} a_{jm} \, a_{km}.$$

In other words, the covariance matrix $\Gamma = [\mathbb{E}(Z_j Z_k)]$ is the $d \times d$ symmetric matrix

$$\Gamma = AA^T.$$

We say that Z has a *nondegenerate* distribution if Γ is invertible.

The characteristic function of Z can be computed using the known formula for the characteristic function of N_k,

$$\mathbb{E}[e^{itN_k}] = e^{-t^2/2},$$

$$\begin{aligned}
\mathbb{E}[\exp\{i\theta \cdot Z\}] &= \mathbb{E}\left[\exp\left\{i \sum_{j=1}^{d} \theta_j \sum_{k=1}^{n} a_{jk} \, N_k\right\}\right] \\
&= \mathbb{E}\left[\exp\left\{i \sum_{k=1}^{n} N_k \sum_{j=1}^{d} \theta_j a_{jk}\right\}\right] \\
&= \prod_{k=1}^{n} \mathbb{E}\left[\exp\left\{iN_k \sum_{j=1}^{d} \theta_j a_{jk}\right\}\right] \\
&= \prod_{k=1}^{n} \exp\left\{-\frac{1}{2}\left(\sum_{j=1}^{d} \theta_j a_{jk}\right)^2\right\}
\end{aligned}$$

$$= \exp\left\{-\frac{1}{2}\sum_{k=1}^{n}\sum_{j=1}^{d}\sum_{l=1}^{d}\theta_j\,\theta_l\,a_{jk}\,a_{lk}\right\}$$

$$= \exp\left\{-\frac{1}{2}\,\theta A A^T \theta^T\right\} = \exp\left\{-\frac{1}{2}\,\theta\Gamma\theta^T\right\}.$$

Since the characteristic function determines the distribution, we see that the distribution of Z depends only on Γ.

The matrix Γ is symmetric and nonnegative definite. Hence, we can find an orthogonal basis $u_1, \ldots, u_d$ of unit vectors in $\mathbb{R}^d$ that are eigenvectors of Γ with nonnegative eigenvalues $\alpha_1, \ldots, \alpha_d$. The random variable

$$Z = \sqrt{\alpha_1}\,N_1\,u_1 + \cdots + \sqrt{\alpha_d}\,N_d\,u_d$$

has a joint normal distribution with covariance matrix Γ. In matrix language, we have written $\Gamma = \Lambda\,\Lambda^T = \Lambda^2$ for a $d \times d$ nonnegative definite symmetric matrix Λ. The distribution is nondegenerate if and only if all of the α_j are strictly positive.

♣ Although we allow the matrix A to have n columns, what we have shown is that there is a symmetric, positive definite $d \times d$ matrix Λ which gives the same distribution. Hence, joint normal distribution in $\mathbb{R}^d$ can be described as linear combinations of d independent one-dimensional normals. Moreover, if we choose the correct orthogonal basis for $\mathbb{R}^d$, the components of Z with respect to that basis are independent normals.

If Γ is invertible, then Z has a density $f(z^1, \ldots, z^d)$ with respect to the Lebesgue measure that can be computed using the inversion formula

$$f(z^1, \ldots, z^d) = \frac{1}{(2\pi)^d}\int e^{-i\theta\cdot z}\,\mathbb{E}[\exp\{i\theta\cdot Z\}]\,d\theta$$

$$= \frac{1}{(2\pi)^d}\int \exp\left\{-i\theta\cdot z - \frac{1}{2}\,\theta\Gamma\theta^T\right\}\,d\theta.$$

(Here and for the remainder of this paragraph the integrals are over $\mathbb{R}^d$ and $d\theta$ represents $d^d\theta$.) To evaluate the integral, we start with the substitution $\theta_1 = \Lambda\theta$, which gives

$$\int \exp\left\{-i\theta\cdot z - \frac{1}{2}\,\theta\Gamma\theta^T\right\}\,d\theta = \frac{1}{\det\Lambda}\int e^{-|\theta_1|^2/2}\,e^{-i(\theta_1\cdot\Lambda^{-1}z)}\,d\theta_1.$$

By completing the square we see that the right-hand side equals

$$\frac{e^{-|\Lambda^{-1}z|^2/2}}{\det \Lambda} \int \exp\left\{\frac{1}{2}\,(i\theta_1 - \Lambda^{-1}z)\cdot(i\theta_1 - \Lambda^{-1}z)\right\}\, d\theta_1 .$$

The substitution $\theta_2 = \theta_1 - i\Lambda^{-1}z$ gives

$$\int \exp\left\{\frac{1}{2}\,(i\theta_1 - \Lambda^{-1}z)\cdot(i\theta_1 - \Lambda^{-1}z)\right\}\, d\theta_1 = \int e^{-|\theta_2|^2/2}\, d\theta_2 = (2\pi)^{d/2}.$$

Hence, the density of Z is

$$f(z) = \frac{1}{(2\pi)^{d/2}\sqrt{\det \Gamma}}\, e^{-|\Lambda^{-1}z|^2/2} = \frac{1}{(2\pi)^{d/2}\sqrt{\det \Gamma}}\, e^{-(z\cdot\Gamma^{-1}z)/2}. \quad \text{(A.14)}$$

Corollary A.3.1 *Suppose that $Z = (Z_1, \ldots, Z_d)$ has a mean zero, joint normal distribution such that $\mathbb{E}[Z_j Z_k] = 0$ for all $j \neq k$. Then $Z_1, \ldots, Z_d$ are independent.*

Proof Suppose that $\mathbb{E}[Z_j Z_k] = 0$ for all $j \neq k$. Then Z has the same distribution as

$$(b_1 N_1, \ldots, b_d N_d),$$

where $b_j = \sqrt{\mathbb{E}[Z_j^2]}$. In this representation, the components are obviously independent. □

♣ If $Z_1, \ldots, Z_d$ are mean zero random variables satisfying $\mathbb{E}[Z_j Z_k] = 0$ for all $j \neq k$, they are called orthogonal. Independence implies orthogonality but the converse is not always true. However, the corollary tells us that the converse is true in the case of joint normal random variables. Orthogonality is often easier to verify than independence.

A.4 Markov chains

A *(time-homogeneous) Markov chain* on a countable state space D is a process X_n taking values in D whose transitions satisfy

$$\mathbb{P}\{X_{n+1} = x_{n+1} \mid X_0 = x_0, \ldots, X_n = x_n\} = p(x_n, x_{n+1})$$

where $p : D \times D \to [0, 1]$ is the *transition function* satisfying $\sum_{y \in D} p(x, y) = 1$ for each x. If A is finite, we call the transition function the transition *matrix* $P = [p(x, y)]_{x,y \in A}$. The n-step transitions are given by the matrix P^n. In other words, if $p_n(x, y)$ is defined as $\mathbb{P}\{X_n = y \mid X_0 = x\}$, then

$$p_n(x, y) = \sum_{z \in D} p(x, z)\, p_{n-1}(z, y) = \sum_{z \in D} p_{n-1}(x, z)\, p(z, y).$$

A Markov chain is called *irreducible* if for each $x, y \in A$, there exists an $n = n(x, y) \geq 0$ with $p_n(x, y) > 0$. The chain is *aperiodic* if for each x there is an N_x such that for $n \geq N_x$, $p_n(x, x) > 0$. If D is finite, then the chain is irreducible and aperiodic if and only if there exists an n such that P^n has strictly positive entries.

Theorem A.4.1 *(Perron–Froebenius theorem) If P is an $m \times m$ matrix such that for some positive integer n, P^n has all entries strictly positive, then there exists $\alpha > 0$ and vectors $\mathbf{v}, \mathbf{w}$, with strictly positive entries such that*

$$\mathbf{v}\, P = \alpha\, \mathbf{v}, \qquad P\, \mathbf{w} = \alpha\, \mathbf{w}.$$

This eigenvalue is simple and all other eigenvalues of P have absolute value strictly less than α. In particular, if P is the transition matrix for an irreducible aperiodic Markov chain, there is a unique invariant probability π satisfying

$$\sum_{x \in D} \pi(x) = 1, \qquad \pi(x) = \sum_{y \in D} \pi(y)\, P(y, x).$$

Proof We first assume that P has all strictly positive entries. It suffices to find a right eigenvector, since the left eigenvector can be handled by considering the transpose of P. We write $\mathbf{w}_1 \geq \mathbf{w}_2$ if every component of $\mathbf{w}_1$ is greater than or equal to the corresponding component of $\mathbf{w}_2$. Similarly, we write $\mathbf{w}_1 > \mathbf{w}_2$ if all the components of $\mathbf{w}_1$ are strictly greater than the corresponding components of $\mathbf{w}_2$. We let $\mathbf{0}$ denote the zero vector and $\mathbf{e}_j$ the vector whose jth component is one and whose other components are zero. If $\mathbf{w} \geq \mathbf{0}$, let

$$\lambda_{\mathbf{w}} = \sup\{\lambda : P\mathbf{w} \geq \lambda \mathbf{w}\}.$$

Clearly, $\lambda_{\mathbf{w}} < \infty$, and since P has strictly positive entries, $\lambda_{\mathbf{w}} > 0$ for all $\mathbf{w} > 0$. Let

$$\alpha = \sup\{\lambda_{\mathbf{w}} : \mathbf{w} \geq \mathbf{0}, \sum_{j=1}^{m} [\mathbf{w}]_j = 1\}.$$

By compactness and continuity arguments, we can see that there exists a $\mathbf{w}$ with $\mathbf{w} \geq \mathbf{0}$, $\sum_j [\mathbf{w}]_j = 1$ and $\lambda_{\mathbf{w}} = \alpha$. We claim that $P\mathbf{w} = \alpha\mathbf{w}$. Indeed, if $[P\mathbf{w}]_j > \alpha[\mathbf{w}]_j$ for some j, one can check that there exist positive ϵ, ρ such that $P[\mathbf{w} + \epsilon\mathbf{e}_j] \geq (\alpha + \rho)\,[\mathbf{w} + \epsilon\mathbf{e}_j]$, which contradicts the maximality of α. If $\mathbf{v}$ is a vector with both positive and negative components, then for each j,

$$|[P\mathbf{v}]_j| < [P|\mathbf{v}|]_j \leq \alpha\,[|\mathbf{v}|]_j.$$

Here, we write $|\mathbf{v}|$ for the vector whose components are the absolute values of the components of $\mathbf{v}$. Hence, any eigenvector with both positive and negative values has an eigenvalue with absolute value strictly less than α. Also, if $\mathbf{w}_1, \mathbf{w}_2$ are positive eigenvectors with eigenvalue α, then $\mathbf{w}_1 - t\mathbf{w}_2$ is an eigenvector for each t. If $\mathbf{w}_1$ is not a multiple of $\mathbf{w}_2$, then there is some value of t such that $\mathbf{w}_1 - t\mathbf{w}_2$ has both positive and negative values. Since this is impossible, we conclude that the eigenvector $\mathbf{w}$ is unique. If $\mathbf{w} \geq 0$ is an eigenvector, then the eigenvalue must be positive. Therefore, α has a unique eigenvector (up to constant), and all other eigenvalues have absolute value strictly less than α. Note that if $\mathbf{v} > 0$, then $P\mathbf{v}$ has all entries strictly positive; hence the eigenvector $\mathbf{w}$ must have all entries strictly positive.

We claim, in fact, that α is a simple eigenvalue. To see this, one can use the argument as in the previous paragraph to all submatrices of the matrix to conclude that all eigenvalues of all submatrices of the matrix are strictly less than α in absolute value. Using this (details omitted), one can see that the derivative of the function $f(\lambda) = \det(\lambda I - P)$ is nonzero at $\lambda = \alpha$, which shows that the eigenvalue is simple.

If P is a matrix such that P^n has all entries strictly positive, and $\mathbf{w}$ is an eigenvector of P with eigenvalue α, then $\mathbf{w}$ is an eigenvector for P^n with eigenvalue α^n. Using this, we can conclude the result for P. The final assertion follows by noting that the vector of all ones is a right eigenvector for a stochastic matrix. □

♣ A different derivation of the Perron–Froebenius theorem which generalizes to some chains on infinite state spaces is done in Exercise A.4.

If P is the transition matrix for an irreducible, aperiodic Markov chain, then $p_n(x,y) \to \pi(y)$ as $n \to \infty$. In fact, this holds for countable state space provided that the chain is positive recurrent, i.e. if there exists an invariant probability measure. The next proposition gives a simple, quantitative version of this fact provided that the chain satisfies a certain condition which always holds for the finite irreducible, aperiodic case.

Proposition A.4.2 *Suppose that $p : D \times D \to [0,1]$ is the transition probability for a positive recurrent, irreducible, aperiodic Markov chain on a countable state space D. Let π denote the invariant probability measure. Suppose that there exist $\epsilon > 0$ and a positive integer k such that for all $x, y \in D$,*

$$\frac{1}{2} \sum_{z \in D} |p_k(x,z) - p_k(y,z)| \leq 1 - \epsilon. \tag{A.15}$$

Then, for all positive integers j and all $x \in A$,

$$\frac{1}{2} \sum_{z \in D} |p_j(x,z) - \pi(z)| \leq c\, e^{-\beta j},$$

where $c = (1-\epsilon)^{-1}$ and $e^{-\beta} = (1-\epsilon)^{1/k}$.

Proof If ν is any probability distribution on D, let

$$\nu_j(x) = \sum_{y \in D} \nu(y) p_j(y,x).$$

Then (A.15) implies that for every ν,

$$\frac{1}{2} \sum_{z \in D} |\nu_k(z) - \pi(z)| \leq 1 - \epsilon.$$

In other words we can write $\nu_k = \epsilon\,\pi + (1-\epsilon)\,\nu^{(1)}$ for some probability measure $\nu^{(1)}$. By iterating (A.15), we can see that for every integer $i \geq 1$ we can write $\nu_{ik} = (1-\epsilon)^i\,\nu^{(i)} + [1-(1-\epsilon)^i]\pi$ for some probability measure $\nu^{(i)}$. This establishes the result for $j = ki$ (with $c = 1$ for these values of j) and for other j we find i with $ik \leq j < (i+1)k$. □

A.4.1 Chains restricted to subsets

We will now consider Markov chains restricted to a subset of the original state space. If X_n is an irreducible, aperiodic Markov chain with state space D and A is a finite proper subset of D, we write $P_A = [p(x,y)]_{x,y \in A}$. Note that $(P_A)^n = [p_n^A(x,y)]_{x,y \in A}$ where

$$p_n^A(x,y) = \mathbb{P}\{X_n = y; X_0, \ldots, X_n \in A \mid X_0 = x\} = \mathbb{P}^x\{X_n = y, \tau_A > n\}, \tag{A.16}$$

where $\tau_A = \inf\{n : X_n \notin A\}$. Note that

$$\mathbb{P}^x\{\tau_A > n\} = \sum_{y \in A} p_n^A(x, y).$$

We call A *connected and aperiodic (with respect to P)* if for each $x, y \in A$, there is an N such that for $n \geq N, p_n^A(x, y) > 0$. If A is finite, then A is connected and aperiodic if and only if there exists an n such that $(P_A)^n$ has all entries strictly positive. In this case, all of the row sums of P_A are less than or equal to one and (since A is a proper subset) there is at least one row whose sum is strictly less than one.

Suppose that X_n is an irreducible, aperiodic Markov chain with state space D and A is a finite, connected, aperiodic proper subset of D. Let α be as in the Perron–Froebenius theorem for the matrix P_A. Then $0 < \alpha < 1$. Let v, w be the corresponding positive eigenvectors which we write as functions,

$$\sum_{x \in A} v(x)\, p(x, y) = \alpha\, v(y), \quad \sum_{y \in A} w(y)\, p(x, y) = \alpha\, w(x).$$

We normalize the functions so that

$$\sum_{x \in A} v(x) = 1, \quad \sum_{x \in A} v(x)\, w(x) = 1,$$

and we let $\pi(x) = v(x)\, w(x)$. Let

$$q^A(x, y) = \alpha^{-1}\, p(x, y)\, \frac{w(y)}{w(x)}.$$

Note that

$$\sum_{y \in A} q^A(x, y) = \frac{\sum_{y \in A} p(x, y)\, w(y)}{\alpha\, w(x)} = 1.$$

In other words, $Q^A := [q^A(x, y)]_{x,y \in A}$ is the transition matrix for a Markov chain which we will denote by Y_n. Note that $(Q^A)^n = [q_n^A(x, y)]_{x,y \in A}$ where

$$q_n^A(x, y) = \alpha^{-n}\, p_n^A(x, y)\, \frac{w(y)}{w(x)}$$

and $p_n^A(x,y)$ is as in (A.16). From this we see that the chain is irreducible and aperiodic. Since

$$\sum_{x \in A} \pi(x)\, q^A(x,y) = \sum_{x \in A} v(x)\, w(x)\, \alpha^{-1}\, p(x,y)\, \frac{w(y)}{w(x)} = \pi(y),$$

we see that π is the invariant probability for this chain.

Proposition A.4.3 *Under the assumptions above, there exist c, β such that for all n,*

$$|\alpha^{-n}\, p_n^A(x,y) - w(x)\, v(y)| \le c\, e^{-\beta n}.$$

In particular,

$$\mathbb{P}\{X_0, \ldots, X_n \in A \mid X_0 = x\} = w(x)\, \alpha^n\, [1 + O(e^{-\beta n})].$$

Proof Consider the Markov chain with transition matrix Q^A. Choose positive integer k and $\epsilon > 0$ such that $q_k^A(x,y) \ge \epsilon\, \pi(y)$ for all $x, y \in A$. Proposition A.4.2 gives

$$|q_n^A(x,y) - \pi(y)| \le c\, e^{-\beta n},$$

for some c, β. Since $\pi(y) = v(y)\, w(y)$ and $q_n^A(x,y) = \alpha^{-n}\, p_n^A(x,y)\, w(y)/w(x)$, we get the first assertion, using the fact that A is finite so that $\inf v > 0$. The second assertion follows from the first using $\sum_y v(y) = 1$ and

$$\mathbb{P}\{X_0, \ldots, X_n \in A \mid X_0 = x\} = \sum_{y \in A} p_n^A(x,y).$$

□

If the Markov chain is symmetric ($p(y,x) = p(x,y)$), then $w(x) = c\, v(x), \pi(x) = c\, v(x)^2$. The function $g(x) = \sqrt{c}\, v(x)$ can be characterized by the fact that g is strictly positive and satisfies

$$P^A g(x) = \alpha\, g(x), \qquad \sum_{x \in A} g(x)^2 = 1.$$

♣ The chain Y_n can be considered the chain derived from X_n by conditioning the chain to "stay in A forever." The probability measures v, π are both "invariant" (sometimes the word quasi-invariant is used) probability measures but with different interpretations. Roughly speaking, the three measures v, w, π can be described as follows.

- Suppose that the chain X_n is observed at a large time n and it is known that the chain has stayed in A for all times up to n. Then, the conditional distribution on X_n given this information approaches v.
- For $x \in A$, the probability that the chain stays in A up to time n is asymptotic to $w(x)\,\alpha^n$.
- Suppose that the chain X_n is observed at a large time n and it is known that the chain has stayed in A and will stay in A for all times up to N where $N \gg n$. Then, the conditional distribution on X_n, given this information, approaches π. We can think of the first term of the product $v(x)\,w(x)$ as the conditional probability of being at x, given that the walk has stayed in A up to time n and the second part of the product as the conditional probability given that the walk stays in A for times between n and N.

The next proposition gives a criterion for determining the rate of convergence to the invariant distribution v. Let us write that

$$\hat{p}_n^A(x,y) = \frac{p_n^A(x,y)}{\sum_{z\in A} p_n^A(x,z)} = \mathbb{P}^x\{X_n = y \mid \tau_A > n\}.$$

Proposition A.4.4 *Suppose that X_n is an irreducible, aperiodic Markov chain on the countable state space D. Suppose that A is a finite, proper subset of D and $A' \subset A$. Suppose that there exist $\epsilon > 0$ and integer $k > 1$ such that the following is true.*

(a) If $x \in A$,

$$\sum_{y\in A'} \hat{p}_k(x,y) \geq \epsilon. \tag{A.17}$$

(b) If $x, x' \in A$,

$$\sum_{y\in A'} [\hat{p}_k^A(x,y) \wedge \hat{p}_k^A(x',y)] \geq \epsilon. \tag{A.18}$$

(c) If $x \in A, y \in A'$ and n is a positive integer

$$\mathbb{P}^y\{\tau_A > n\} \geq \epsilon\, \mathbb{P}^x\{\tau_A > n\}. \tag{A.19}$$

Then there exists $\delta > 0$, depending only on ϵ, such that for all $x, z \in A$ and all integers $m \geq 0$,

$$\frac{1}{2}\sum_{y\in A} \left|\hat{p}_{km}^A(x,y) - \hat{p}_{km}^A(z,y)\right| \leq (1-\delta)^m.$$

Proof We fix ϵ and allow all constants in this proof to depend on ϵ. Let $q_n = \max_{y \in A} \mathbb{P}^y\{\tau_A > n\}$. Then (A.19) implies that for all $y \in A'$ and all n, $\mathbb{P}^y\{\tau_A > n\} \geq \epsilon\, q_n$. Combining this with (A.17) gives for all positive integers k, n,

$$c\, q_n\, \mathbb{P}^x\{\tau_A > k\} \leq \mathbb{P}^x\{\tau_A > k + n\} \leq q_n\, \mathbb{P}^x\{\tau_A > k\}. \tag{A.20}$$

Let m be a positive integer and let

$$Y_0, Y_1, Y_2, \ldots, Y_m$$

be the process corresponding to $X_0, X_k, X_{2k}, \ldots, X_{mk}$ conditioned so that $\tau_A > mk$. This is a time *inhomogeneous* Markov chain with transition probabilities

$$\mathbb{P}\{Y_j = y \mid Y_{j-1} = x\} = \frac{p_k^A(x, y)\, \mathbb{P}^y\{\tau_A > (m - j)k\}}{\mathbb{P}^x\{\tau_A > (m - j + 1)k\}}, \qquad j = 1, 2, \ldots, m.$$

Note that (A.20) implies that for all $y \in A$,

$$\mathbb{P}\{Y_j = y \mid Y_{j-1} = x\} \leq c_2\, \hat{p}_k^A(x, y),$$

and if $y \in A'$,

$$\mathbb{P}\{Y_j = y \mid Y_{j-1} = x\} \geq c_1\, \hat{p}_k^A(x, y).$$

Using this and (A.18), we can see that there is a $\delta > 0$ such that if $x, z \in A$ and $j \leq m$,

$$\frac{1}{2} \sum_{y \in A} \left| \mathbb{P}\{Y_j = y \mid Y_{j-1} = x\} - \mathbb{P}\{Y_j = y \mid Y_{j-1} = z\} \right| \leq 1 - \delta,$$

and using an argument as in the proof of Proposition A.4.2, we can see that

$$\frac{1}{2} \sum_{y \in A} |\mathbb{P}\{Y_m = y \mid Y_0 = x\} - \mathbb{P}\{Y_m = y \mid Y_0 = z\}| \leq (1 - \delta)^m.$$

□

A.4.2 Maximal coupling of Markov chains

Here, we will describe the maximal coupling of a Markov chain. Suppose that $p : D \times D \to [0, 1]$ is the transition probability function for an irreducible, aperiodic

Markov chain with countable state space D. Assume that g_0^1, g_0^2 are two initial probability distributions on D. Let g_n^j denote the corresponding distribution at time n, given recursively by

$$g_n^j(x) = \sum_{z \in D} g_{n-1}^j(z)\, p(z,x).$$

Let $\|\cdot\|$ denote the total variation distance,

$$\|g_n^1 - g_n^2\| = \frac{1}{2} \sum_{x \in D} |g_n^1(x) - g_n^2(x)| = 1 - \sum_{x \in D} [g_n^1(x) \wedge g_n^2(x)].$$

Suppose that

$$X_0^1, X_1^1, X_2^1, \ldots, \qquad X_0^2, X_1^2, X_2^2, \ldots$$

are defined on the same probability space such that for each j, $\{X_n^j : n = 0, 1, \ldots\}$ has the distribution of the Markov chain with initial distribution g_0^j. Then it is clear that

$$\mathbb{P}\{X_n^1 = X_n^2\} \leq 1 - \|g_n^1 - g_n^2\| = \sum_{x \in D} g_n^1(x) \wedge g_n^2(x). \tag{A.21}$$

The following theorem shows that there is a way to define the chains on the same probability space so that equality is obtained in (A.21). This theorem gives one example of the powerful probabilistic technique called *coupling*. Coupling refers to the defining of two or more processes on the same probability space in such a way that each individual process has a certain distribution but the joint distribution has some particularly nice properties. Often, as in this case, the two processes are equal except for an event of small probability.

Theorem A.4.5 *Suppose that p, g_n^1, g_n^2 are as defined in the previous paragraph. We can define $(X_n^1, X_n^2), n = 0, 1, 2, \ldots$ on the same probability space such that:*

- *for each j, $X_0^j, X_1^j, \ldots$ has the distribution of the Markov chain with initial distribution g_0^j;*
- *for each integer $n \geq 0$,*

$$\mathbb{P}\{X_m^1 = X_m^2 \textit{ for all } m \geq n\} = 1 - \|g_n^1 - g_n^2\|.$$

♣ Before doing this proof, let us consider the easier problem of defining (X^1, X^2) on the same probability space so that X^j has distribution g_0^j and

$$\mathbb{P}\{X^1 = X^2\} = 1 - \|g_0^1 - g_0^2\|.$$

Assume that $0 < \|g_0^1 - g_0^2\| < 1$. Let $f^j(x) = g_0^j(x) - [g_0^1(x) \wedge g_0^2(x)]$.

Suppose that J, X, W^1, W^2 are independent random variables with the following distributions.

$$\mathbb{P}\{J = 0\} = 1 - \mathbb{P}\{J = 1\} = \|g_0^1 - g_0^2\|.$$

$$\mathbb{P}\{X = x\} = \frac{g_1(x) \wedge g_2(x)}{1 - \|g_0^1 - g_0^2\|}, \qquad x \in D$$

$$\mathbb{P}\{W^j = x\} = \frac{f^j(x)}{\|g_0^1 - g_0^2\|}, \qquad x \in D.$$

Let $X^j = 1\{J = 1\}\, X + 1\{J = 0\}\, W^j$.
It is easy to check that this construction works.

Proof For ease, we will assume that $\|g_0^1 - g_0^2\| = 1$ and $\|g_n^1 - g_n^2\| \to 0$ as $n \to \infty$; the adjustment needed if this does not hold is left to the reader. Let (Z_n^1, Z_n^2) be independent Markov chains with the appropriate distributions. Let $f_n^j(x) = g_n^j(x) - [g_n^1(x) \wedge g_n^2(x)]$ and define h_n^j by $h_0^j(x) = g_0^j(x) = f_0^j(x)$ and for $n > 1$,

$$h_n^j(x) = \sum_{z \in S} f_{n-1}^j(z)\, p(z, x).$$

Note that $f_{n+1}^j(x) = h_{n+1}^j(x) - [h_n^1(x) \wedge h_n^2(x)]$. Let

$$\rho_n^j(x) = \frac{h_n^1(x) \wedge h_n^2(x)}{h_n^j(x)} \quad \text{if } h_n^j(x) \neq 0.$$

We set $\rho_n^j(x) = 0$ if $h_n^j(x) = 0$. We let $\{Y^j(n, x) : j = 1, 2; n = 1, 2, \ldots \mid x \in D\}$ be independent 0-1 random variables, independent of (Z_n^1, Z_n^2), with $\mathbb{P}\{Y^j(n, x) = 1\} = \rho_n^j(x)$.

We now define 0-1 random variables J_n^j as follows:

- $J_0^j \equiv 0$.
- If $J_n^j = 1$, then $J_m^j = 1$ for all $m \geq n$.
- If $J_n^j = 0$, then $J_{n+1}^j = Y^j(n+1, Z_{n+1}^j)$.

We claim that

$$\mathbb{P}\{J_n^j = 0; Z_n^j = x\} = f_n^j(x).$$

For $n = 0$, this follows immediately from the definition. Also,

$$\begin{aligned}
\mathbb{P}\{J_{n+1}^j = 0; Z_{n+1}^j = x\}& \\
&= \sum_{z \in D} \mathbb{P}\{J_n^j = 0; Z_n^j = z\} \\
&\quad \times \mathbb{P}\{Z_{n+1}^j = x, Y^j(n+1, x) = 0 \mid J_n^j = 0; Z_n^j = z\}.
\end{aligned}$$

The random variable $Y^j(n+1, x)$ is independent of the Markov chain, and the event $\{J_n^j = 0; Z_n^j = z\}$ depends only on the chain up to time n and the values of $\{Y^j(k, y) : k \leq n\}$. Therefore,

$$\mathbb{P}\{Z_{n+1}^j = x, Y^j(n+1, x) = 0 \mid J_n^j = 0; Z_n^j = z\} = p(z, x)\,[1 - \rho_{n+1}^j(x)].$$

Therefore, we have the inductive argument

$$\begin{aligned}
\mathbb{P}\{J_{n+1}^j = 0; Z_{n+1}^j = x\} &= \sum_{z \in D} f_n^j(z)\, p(z, x)\,[1 - \rho_{n+1}^j(x)] \\
&= h_{n+1}^j(x)\,[1 - \rho_{n+1}^j(x)] \\
&= h_{n+1}^j(x) - [h_{n+1}^1(x) \wedge h_{n+1}^2(x)] = f_{n+1}^j(x),
\end{aligned}$$

which establishes the claim.

Let K^j denote the smallest n such that $J_n^j = 1$. The condition $\|g_0^1 - g_0^2\| \to 0$ implies that $K^j < \infty$ with probability one. A key fact is that for each n and each x,

$$\mathbb{P}\{K^1 = n+1; Z_n^1 = x\} = \mathbb{P}\{K^2 = n+1; Z_n^2 = x\} = h_{n+1}^1(x) \wedge h_{n+1}^2(x).$$

This is immediate for $n = 0$ and for $n > 0$,

$$\begin{aligned}
\mathbb{P}\{K^j = n+1; Z_{n+1}^j = x\}& \\
&= \sum_{z \in D} \mathbb{P}\{J_n = 0; Z_n^j = z\}\, \mathbb{P}\{Y^j(n+1, x) = 1; Z_{n+1}^j = x \mid J_n = 0; Z_n^j = z\} \\
&= \sum_{z \in D} f_n^j(z)\, p(z, x)\, \rho_{n+1}^j(x) \\
&= h_{n+1}^j(x)\, \rho_{n+1}^j(x) = h_{n+1}^1(x) \wedge h_{n+1}^2(x).
\end{aligned}$$

The last important observation is that the distribution of $W_m := X^j_{m-n}$ given the event $\{K^j = n; X^j_n = x\}$ is that of a Markov chain with transition probability p starting at x.

The reader may note that for each j, the process (Z^j_n, J^j_n) is a time-inhomogeneous Markov chain with transition probabilities

$$\mathbb{P}\{(Z^j_{n+1}, J^j_{n+1}) = (y, 1) \mid (Z^j_n, J^j_n) = (x, 1)\} = p(x, y),$$
$$\mathbb{P}\{(Z^j_{n+1}, J_{n+1}) = (y, 0) \mid (Z^j_n, J^j_n) = (x, 0)\} = p(x, y)\,[1 - \rho^j_{n+1}(y)],$$
$$\mathbb{P}\{(Z^j_{n+1}, J_{n+1}) = (y, 1) \mid (Z^j_n, J^j_n) = (x, 0)\} = p(x, y)\,\rho^j_{n+1}(y).$$

The chains (Z^1_n, J^1_n) and (Z^2_n, J^2_n) are independent. However, the transition probabilities for these chains depend on both initial distributions and p.

We are now ready to make our construction of (X^1_n, X^2_n).

- Define for each (n, x) a process $\{W^{n,x}_m : m = 0, 1, 2, \ldots\}$ that has the distribution of the Markov chain with initial point x. Assume that all these processes are independent.
- Choose (n, x) according to the probability distribution

$$h^1_{n+1}(x) \wedge h^2_{n+1}(x) = \mathbb{P}\{K^j = n; Z^j_n = x\}.$$

  Set $J^j_m = 1$ for $m \geq n$, $J^j_m = 0$ for $m < n$, and $K^1 = K^2 = n$. Note that K^j is the smallest n such that $J^j_n = 1$.
- Given (n, x), choose $X^1_0, \ldots, X^1_n$ from the conditional distribution of the Markov chain with initial distribution g^1_0 conditioned on the event $\{K^1 = n; Z^1_n = x\}$.
- Given (n, x), choose $X^2_0, \ldots, X^2_n$ (conditionally) independent of $X^1_0, \ldots, X^1_n$ from the conditional distribution of the Markov chain with initial distribution g^2_0 conditioned on the event $\{K^2 = n; Z^2_n = x\}$.
- Let

$$X^j_m = W^{n,x}_{m-n}, \qquad m = n, n+1, \ldots$$

The two conditional distributions above are not easy to express explicitly; fortunately, we do not need to do so.

To finish the proof, we need only check that the above construction satisfies the conditions. For fixed j, the fact that $X^j_0, X^j_1, \ldots$ has the distribution of the chain with initial distribution g^j_0 is immediate from construction and the earlier observation that the distribution of $\{X^j_n, X^j_{n+1}, \ldots\}$, given $\{K^j = n; Z^j_n = x\}$, is that of the Markov chain starting at x. Also, the construction immediately gives $X_m^1 = X_m^2$ if $m \geq K^1 = K^2$. Also,

$$\mathbb{P}\{J_n^j = 0\} = \sum_{x \in D} f_n^j(z) = \|g_1^n - g_2^n\|.$$

□

Remark A review of the proof of Theorem A.4.5 shows that we do not need to assume that the Markov chain is time-homogeneous. However, time-homogeneity makes the notation a little simpler and we use the result only for time-homogeneous chains.

A.5 Some Tauberian theory

Lemma A.5.1 *Suppose that $\alpha > 0$. Then, as $\xi \to 1-$,*

$$\sum_{n=2}^{\infty} \xi^n \, n^{\alpha-1} \sim \frac{\Gamma(\alpha)}{(1-\xi)^\alpha}.$$

Proof Let $\epsilon = 1 - \xi$. First note that

$$\sum_{n \geq \epsilon^{-2}} \xi^n \, n^{\alpha-1} = \sum_{n \geq \epsilon^{-2}} [(1-\epsilon)^{1/\epsilon}]^{n\epsilon} \, n^{\alpha-1}$$
$$\leq \sum_{n \geq \epsilon^{-2}} e^{-n\epsilon} \, n^{\alpha-1},$$

and the right-hand side decays faster than every power of ϵ. For $n \leq \epsilon^{-2}$ we can do the asymptotics

$$\xi^n = \exp\{n \log(1-\epsilon)\} = \exp\{n(-\epsilon - O(\epsilon^2))\}$$
$$= e^{-n\epsilon}[1 + O(n\epsilon^2)].$$

Hence,

$$\sum_{n \leq \epsilon^{-2}} \xi^n \, n^{\alpha-1} = \epsilon^{-\alpha} \sum_{n \leq \epsilon^{-2}} \epsilon \, e^{-n\epsilon} \, (n\epsilon)^{\alpha-1} \, [1 + (n\epsilon) \, O(\epsilon)].$$

Using Riemann sum approximations, we see that

$$\lim_{\epsilon \to 0+} \sum_{n=1}^{\infty} \epsilon \, e^{-n\epsilon} \, (n\epsilon)^{\alpha-1} = \int_0^\infty e^{-t} \, t^{\alpha-1} \, dt = \Gamma(\alpha).$$

□

Proposition A.5.2 *Suppose that u_n is a sequence of nonnegative real numbers. If $\alpha > 0$, the following two statements are equivalent:*

$$\sum_{n=0}^{\infty} \xi^n u_n \sim \frac{\Gamma(\alpha)}{(1-\xi)^\alpha}, \qquad \xi \to 1-, \tag{A.22}$$

$$\sum_{n=1}^{N} u_n \sim \alpha^{-1} N^\alpha, \qquad N \to \infty. \tag{A.23}$$

Moreover, if the sequence is monotone, either of these statements implies that

$$u_n \sim n^{\alpha-1}, \qquad n \to \infty.$$

Proof Let $U_n = \sum_{j \le n} u_j$ where $U_{-1} = 0$. Note that

$$\sum_{n=0}^{\infty} \xi^n u_n = \sum_{n=0}^{\infty} \xi^n [U_n - U_{n-1}] = (1-\xi) \sum_{n=0}^{\infty} \xi^n U_n. \tag{A.24}$$

If (A.23) holds, then by the previous lemma

$$\sum_{n=0}^{\infty} \xi^n u_n \sim (1-\xi) \sum_{n=0}^{\infty} \xi^n \alpha^{-1} n^\alpha \sim \frac{\Gamma(\alpha+1)}{\alpha\,(1-\xi)^\alpha} = \frac{\Gamma(\alpha)}{(1-\xi)^\alpha}.$$

Now, suppose that (A.22) holds. We first give an upper bound on U_n. Using $1 - \xi = 1/n$, we can see as $n \to \infty$,

$$\begin{aligned} U_n &\le n^{-1} \left(1 - \frac{1}{n}\right)^{-2n} \sum_{j=n}^{2n-1} \left(1 - \frac{1}{n}\right)^j U_j \\ &\le n^{-1} \left(1 - \frac{1}{n}\right)^{-2n} \sum_{j=0}^{\infty} \left(1 - \frac{1}{n}\right)^j U_j \sim e^2\,\Gamma(\alpha)\, n^\alpha. \end{aligned}$$

The last relation uses (A.24). Let $\nu^{(j)}$ denote the measure on $[0, \infty)$ that gives measure $j^{-\alpha} u_n$ to the point n/j. Then the last estimate shows that the total mass of $\nu^{(j)}$ is uniformly bounded on each compact interval and hence, there is a subsequence that converges weakly to a measure ν that is finite on each compact interval. Using (A.22) we can see that for each $\lambda > 0$,

$$\int_0^\infty e^{-\lambda x}\, \nu(dx) = \int_0^\infty e^{-\lambda x}\, x^{\alpha-1}\, dx.$$

This implies that ν is $x^{\alpha-1}\,dx$. Since the limit is independent of the subsequence, we can conclude that $\nu^{(j)} \to \nu$ and this implies (A.23).

The fact that (A.23) implies the last assertion if u_n is monotone is straightforward using

$$U_{n(1+\epsilon)} - U_n \sim \alpha^{-1}\left[(n(1+\epsilon))^{\alpha} - n^{\alpha}\right], \quad n \to \infty,$$

□

The following is proved similarly.

Proposition A.5.3 *Suppose that u_n is a sequence of nonnegative real numbers. If $\alpha \in \mathbb{R}$, the following two statements are equivalent:*

$$\sum_{n=0}^{\infty} \xi^n u_n = \left(\frac{1}{1-\xi}\right) \log^{\alpha}\left(\frac{1}{1-\xi}\right), \quad \xi \to 1-, \tag{A.25}$$

$$\sum_{n=1}^{N} u_n \sim N \log^{\alpha} N \quad N \to \infty. \tag{A.26}$$

Moreover, if the sequence is monotone, either of these statements implies that

$$u_n \sim \log^{\alpha} n, \quad n \to \infty.$$

A.6 Second moment method

Lemma A.6.1 *Suppose that X is a nonnegative random variable with $\mathbb{E}[X^2] < \infty$ and $0 < r < 1$. Then*

$$\mathbb{P}\{X \ge r\mathbb{E}(X)\} \ge \frac{(1-r)^2\,\mathbb{E}(X)^2}{\mathbb{E}(X^2)}.$$

Proof Without loss of generality, we may assume that $\mathbb{E}(X) = 1$. Since $\mathbb{E}[X; X < r] \le r$, we know that $\mathbb{E}[X; X \ge r] \ge (1-r)$. Then,

$$\begin{aligned}
\mathbb{E}(X^2) \ge \mathbb{E}[X^2; X \ge r] &= \mathbb{P}\{X \ge r\}\,\mathbb{E}[X^2 \mid X \ge r] \\
&\ge \mathbb{P}\{X \ge r\}\,(\mathbb{E}[X \mid X \ge r])^2 \\
&\ge \frac{\mathbb{E}[X; X \ge r]^2}{\mathbb{P}\{X \ge r\}} \\
&\ge \frac{(1-r)^2}{\mathbb{P}\{X \ge r\}}.
\end{aligned}$$

□

Corollary A.6.2 *Suppose that* $E_1, E_2, \ldots$ *is a collection of events with* $\sum \mathbb{P}(E_n) = \infty$. *Suppose that there is a* $K < \infty$ *such that for all* $j \neq k$, $\mathbb{P}(E_j \cap E_k) \leq K\,\mathbb{P}(E_j)\,\mathbb{P}(E_k)$. *Then*

$$\mathbb{P}\{E_k \; i.o.\} \geq \frac{1}{K}.$$

Proof Let $V_n = \sum_{k=1}^{n} 1_{E_k}$. Then, the assumptions imply that

$$\lim_{n\to\infty} \mathbb{E}(V_n) = \infty,$$

and

$$\mathbb{E}(V_n^2) \leq \sum_{j=1}^{k} \mathbb{P}(E_j) + \sum_{j\neq k} K\,\mathbb{P}(E_j)\,\mathbb{P}(E_k) \leq \mathbb{E}(V_n) + K\,\mathbb{E}(V_n)^2$$
$$= \left[\frac{1}{\mathbb{E}(V_n)} + K\right] \mathbb{E}(V_n)^2.$$

By Lemma A.6.1, for every $r > 0$,

$$\mathbb{P}\{V_n \geq r\mathbb{E}(V_n)\} \geq \frac{(1-r)^2}{K + \mathbb{E}(V_n)^{-1}}.$$

Since $\mathbb{E}(V_n) \to \infty$, this implies that

$$\mathbb{P}\{V_\infty = \infty\} \geq \frac{(1-r)^2}{K}.$$

Since this holds for every $r > 0$, we get the result. □

A.7 Subadditivity

Lemma A.7.1 (subadditivity lemma) *Suppose that* $f : \{1, 2, \ldots\} \to \mathbb{R}$ *is subadditive, i.e. for all* n, m, $f(n+m) \leq f(n) + f(m)$. *Then,*

$$\lim_{n\to\infty} \frac{f(n)}{n} = \inf_{n>0} \frac{f(n)}{n}.$$

Proof Fix integer $N > 0$. We can write any integer n as $jN + k$ where j is a nonnegative integer and $k \in \{1, \ldots, N\}$. Let $b_N = \max\{f(1), \ldots, f(N)\}$. Then, subadditivity implies that

$$\frac{f(n)}{n} \leq \frac{jf(N) + f(k)}{jN} \leq \frac{f(N)}{N} + \frac{b_N}{jN}.$$

Therefore,

$$\limsup_{n\to\infty} \frac{f(n)}{n} \le \frac{f(N)}{N}.$$

Since this is true for every N, we get the lemma. □

Corollary A.7.2 *Suppose that r_n is a sequence of positive numbers and $b_1, b_2 > 0$ such that for every n, m,*

$$b_1\, r_n\, r_m \le r_{n+m} \le b_2\, r_n\, r_m. \tag{A.27}$$

Then there exists $\alpha > 0$ such that for all n,

$$b_2^{-1}\, \alpha^n \le r_n \le b_1^{-1}\, \alpha^n.$$

Proof Let $f(n) = \log r_n + \log b_2$. Then f is subadditive and hence

$$\lim_{n\to\infty} \frac{f(n)}{n} = \frac{f(n)}{n} := \alpha.$$

This shows that $r_n \ge \alpha^n/b_2$. Similarly, by considering the subadditive function $g(n) = -\log r_n - \log b_1$, we get $r_n \le b_1^{-1}\, \alpha^n$. □

Remark Note that if r_n satisfies (A.27), then so does $\beta^n r_n$ for each $\beta > 0$. Therefore, we cannot determine the value of α from (A.27).

Exercises

Exercise A.1 Find $f_3(\xi), f_4(\xi)$ in (A.4).

Exercise A.2 Go through the proof of Lemma A.5.1 carefully and estimate the size of the error term in the asymptotics.

Exercise A.3 Suppose that $E_1 \supset E_2 \supset \cdots$ is a decreasing sequence of events with $\mathbb{P}(E_n) > 0$ for each n. Suppose that there exist $\alpha > 0$ such that

$$\sum_{n=1}^{\infty} |\mathbb{P}(E_n \mid E_{n-1}) - (1 - \alpha n^{-1})| < \infty.$$

Show that there exists c such that

$$\mathbb{P}(E_n) \sim c\, n^{-\alpha}. \tag{A.28}$$

(Hint: use Lemma A.1.4.)

Exercise A.4 In this exercise we will consider an alternative approach to the Perron–Froebenius theorem. Suppose that

$$q : \{1, 2, \ldots\} \times \{1, 2, \ldots\} \to [0, \infty),$$

is a function such that for each $x > 0$,

$$q(x) := \sum_y q(x, y) \leq 1.$$

Define $q_n(x, y)$ by matrix multiplication as usual; that is, $q_1(x, y) = q(x, y)$ and

$$q_n(x, y) = \sum_z q_{n-1}(x, z)\, q(z, y).$$

Assume that for each x, $q_n(x, 1) > 0$ for all n sufficiently large. Define

$$q_n(x) = \sum_y q_n(x, y), \qquad p_n(x, y) = \frac{q_n(x, y)}{q_n(x)},$$

$$\overline{q}_n = \sup_x q_n(x), \qquad \underline{q}(x) = \inf_n \frac{q_n(x)}{\overline{q}_n}.$$

Assume that there is a function $F : \{1, 2, \ldots\} \to [0, 1]$ and a positive integer m such that

$$p_m(x, y) \geq F(y), \qquad 1 \leq x, y < \infty,$$

and

$$\rho := \sum_y F(y)\, \underline{q}(y) > 0.$$

(i) Show that there exists $0 < \alpha \leq 1$ such that

$$\lim_{n \to \infty} \overline{q}_n^{1/n} = \alpha.$$

Moreover, $\overline{q}_n \geq \alpha^n$. (Hint: $\overline{q}_{n+m} \leq \overline{q}_n\, q_m$.)

(ii) Show that

$$p_{n+k}(x, y) = \sum_z \nu_{n,k}(x, z)\, p_k(z, y),$$

where

$$\nu_{n,k}(x,z) = \frac{p_n(x,z)\, q_k(z)}{\sum_w p_n(x,w)\, q_k(w)} = \frac{q_n(x,z)\, q_k(z)}{\sum_w q_n(x,w)\, q_k(w)}.$$

(iii) Show that if k, x are positive integers and $n \geq km$,

$$\frac{1}{2} \sum_y |p_{km}(1,y) - p_n(x,y)| \leq (1-\rho)^k.$$

(iv) Show that the limit

$$v(y) = \lim_{n \to \infty} p_n(1,y)$$

exists and if k, x are positive integers and $n \geq km$,

$$\frac{1}{2} \sum_y |v(y) - p_n(x,y)| \leq (1-\rho)^k.$$

(v) Show that

$$v(y) = \alpha \sum_x v(x)\, q(x,y).$$

(vi) Show that for each x, the limit

$$w(x) = \lim_{n \to \infty} \alpha^{-n}\, q_n(x)$$

exists, is positive, and w satisfies

$$w(x) = \alpha \sum_y q(x,y)\, w(y).$$

(Hint: consider $q_{n+1}(x)/q_n(x)$.)

(vii) Show that there is a $C = C(\rho, \alpha) < \infty$ such that if $\epsilon_n(x)$ is defined by

$$q_n(x) = w(x)\, \alpha^n\, [1 + \epsilon_n(x)],$$

then

$$|\epsilon_n(x)| \leq C\, e^{-\beta n},$$

where $\beta = -\log(1-\rho)/m$.

(viii) Show that there is a $C = C(\rho, \alpha) < \infty$ such that if $\epsilon_n(x, y)$ is defined by

$$q_n(x, y) = w(x)\, \alpha^n \, [v(y) + \epsilon_n(x, y)],$$

then

$$|\epsilon_n(x, y)| \le C\, e^{-\beta n}.$$

(ix) Suppose that Q is an $N \times N$ matrix with nonnegative entries such that Q^m has all positive entries. Suppose that the row sums of Q are bounded by K. For $1 \le j, k \le N$, let $q(j, k) = K^{-1}\, q(j, k)$; set $q(j, k) = 0$ if $k > N$; and $q(k, j) = \delta_{j,1}$ if $k > N$. Show that the conditions are satisfied (and hence we get the Perron–Froebenius theorem).

Exercise A.5 In the previous exercise, let $q(x, 1) = 1/2$ for all k, $q(2, 2) = 1/2$ and $q(x, y) = 0$ for all other x, y. Show that there is no F such that $\rho > 0$.

Exercise A.6 Suppose that $X_1, X_2, \ldots$ are independent identically distributed random variables in $\mathbb{R}$ with mean zero, variance one, and such that for some $t > 0$,

$$\beta := 1 + \mathbb{E}\left[X_1^2 e^{tX_1}; X_1 \ge 0\right] < \infty.$$

Let

$$S_n = X_1 + \cdots + X_n.$$

(i) Show that for all n,

$$\mathbb{E}\left[e^{tS_n}\right] \le e^{\beta n t^2/2}.$$

(Hint: expand the moment-generating function for X_1 about $s = 0$.)

(ii) Show that if $r \le t\beta n$,

$$\mathbb{P}\{S_n \ge r\} \le \exp\left\{-\frac{r^2}{2\beta n}\right\}.$$

Exercise A.7 Suppose that $X_1, X_2, \ldots$ are independent identically distributed random variables in $\mathbb{R}$ with mean zero, variance one, and such that for some $t > 0$ and $0 < \alpha < 1$,

$$\beta := 1 + \mathbb{E}\left[X_1^2 e^{tX_1^\alpha}; X_1 \ge 0\right] < \infty.$$

Let $S_n = X_1 + \cdots + X_n$. Suppose that $r > 0$ and n is a positive integer. Let

$$K = \left(\frac{n\beta t}{r}\right)^{\frac{1}{1-\alpha}}, \qquad \tilde{X}_j = X_j \, 1\{X_j \le K\},$$
$$\tilde{S}_n = \tilde{X}_1 + \cdots + \tilde{X}_n.$$

(i) Show that

$$\mathbb{P}\{X_j \ne \tilde{X}_j\} \le (\beta - 1)\, K^{-2}\, e^{-tK^\alpha}.$$

(ii) Show that

$$\mathbb{E}\left[e^{tK^{\alpha-1}\tilde{S}_n}\right] \le e^{\beta n t^2 K^{2(\alpha-1)}/2}.$$

(iii) Show that

$$\mathbb{P}\{S_n \ge r\} \le \exp\left\{-\frac{r^2}{2\beta n}\right\} + n\,(\beta - 1)\, K^{-2}\, e^{-tK^\alpha}.$$

Bibliography

Bhattacharya, R. and Rao R. (1976). *Normal Approximation and Asymptotic Expansions*, John Wiley & Sons.

Bousquet-Mélou, M. and Schaeffer, G. (2002). Walks on the slit plane, Prob. Theor. Rel. Fields **124**, 305–44.

Duplantier, B. and David, F. (1988). Exact partition functions and correlation functions of multiple Hamilton walks on the Manhattan lattice, J. Stat. Phys. **51**, 327–434.

Fomin, S. (2001). Loop-erased walks and total positivity, Trans. AMS **353**, 3563–83.

Fukai, Y. and Uchiyama, K. (1996). Potential kernel for two-dimensional random walk, Ann. Prob. **24**, 1979–92.

Kenyon, R. (2000). The asymptotic determinant of the discrete Laplacian, Acta Math. **185**, 239–86.

Komlös, J., Major, P., and Tusnády, G. (1975a). An approximation of partial sums of independent rv's and the sample df I, Z. Warsch. verw. Geb. **32**, 111–31.

Komlös, J., Major, P., and Tusnády, G. (1975b). An approximation of partial sums of independent rv's and the sample df II, Z. Warsch. verw. Geb. **34**, 33–58.

Kozma, G. (2007). The scaling limit of loop-erased random walk in three dimensions, Acta Math. **191**, 29–152.

Lawler, G. (1996). *Intersections of Random Walks*, Birkhäuser.

Lawler, G. and Puckette, E. (2000). The intersection exponent for simple random walk, Combin., Prob. Comp. **9**, 441–64.

Lawler, G., Schramm, O., and Werner, W. (2001). Values of Brownian intersection exponents II: plane exponents, Acta Math. **187**, 275–308.

Lawler, G., Schramm, O., and Werner, W. (2004). Conformal invariance of planar loop-erased random walk and uniform spanning trees, Ann. Prob. **32**, 939–95.

Lawler, G. and Trujillo Ferreras, J. A., Random walk loop soup, Trans. Am. Math. Soc. **359**, 767–87.

Masson, R. (2009), The growth exponent for loop-erased walk, Elect. J. Prob. **14**, 1012–73.

Spitzer, F. (1976). *Principles of Random Walk*, Springer-Verlag.

Teufl, E. and Wagner, S. (2006). The number of spanning trees of finite Sierpinski graphs, Fourth Colloquium on Mathematics and Computer Science, DMTCS Proc. AG, 411–14.

Wilson, D. (1996). Generating random spanning trees more quickly than the cover time, Proc. STOC96, 296–303.

Index of Symbols

If an entry is followed by a chapter number, then that notation is used only in that chapter. Otherwise, the notation may appear throughout the book.

Index